METHODS IN MOLECULAR BIOLOGY

For further volumes:
http://www.springer.com/series/7651

For over 35 years, biological scientists have come to rely on the research protocols and methodologies in the critically acclaimed *Methods in Molecular Biology* series. The series was the first to introduce the step-by-step protocols approach that has become the standard in all biomedical protocol publishing. Each protocol is provided in readily-reproducible step-by step fashion, opening with an introductory overview, a list of the materials and reagents needed to complete the experiment, and followed by a detailed procedure that is supported with a helpful notes section offering tips and tricks of the trade as well as troubleshooting advice. These hallmark features were introduced by series editor Dr. John Walker and constitute the key ingredient in each and every volume of the *Methods in Molecular Biology* series. Tested and trusted, comprehensive and reliable, all protocols from the series are indexed in PubMed.

ROS Signaling in Plants

Methods and Protocols

Edited by

Francisco J. Corpas

Spanish National Research Council, Estación Experimental del Zaidín, Granada, Granada, Spain

José M. Palma

Spanish National Research Council, Estación Experimental del Zaidín, Granada, Spain

Humana Press

Editors
Francisco J. Corpas
Spanish National Research Council
Estación Experimental del Zaidín
Granada, Granada, Spain

José M. Palma
Spanish National Research Council
Estación Experimental del Zaidín
Granada, Spain

ISSN 1064-3745 ISSN 1940-6029 (electronic)
Methods in Molecular Biology
ISBN 978-1-0716-3825-5 ISBN 978-1-0716-3826-2 (eBook)
https://doi.org/10.1007/978-1-0716-3826-2

This Humana imprint is published by the registered company Springer Science+Business Media, LLC, part of Springer Nature.
The registered company address is: 1 New York Plaza, New York, NY 10004, U.S.A.

Paper in this product is recyclable.

Preface

Reactive oxygen species (ROS) are produced in all aerobic life forms under both physiological and adverse conditions. In higher plants, the metabolism of ROS is a research area that has attracted the attention of many researchers because these molecules are involved in many facets such as seed germination, formation and development of the root system, leaf growth, stomata movement, senescence, flowering, and fruit ripening. Likewise, ROS participates in the mechanism of response to adverse environmental stressful situations and pathogens. In many cases, the uncontrolled overproduction of ROS can trigger oxidative stress that is characterized by an increase in oxidative damage of macromolecules including lipids, proteins, and nucleic acid, which can finally provoke cell death. As a mechanism for regulating the overproduction of ROS, plant cells have developed a battery of antioxidant mechanisms, both enzymatic and non-enzymatic.

The purpose of this book is to provide a set of methodologies that are used as tools for studying ROS metabolism in plant cells, including techniques for detecting different types of ROS such as superoxide radicals ($O_2^{\bullet -}$), hydrogen peroxide (H_2O_2), or singlet oxygen ($^1O_2$), as well as ROS-generating systems such as xanthine oxidoreductase (XOR). Also, it presents different alternatives to determine malondialdehyde (MDA), which is a highly reactive molecule produced non-enzymatically by ROS or enzymatically by lipoxygenase activity in plant cells. Furthermore, it provides different approaches to determine by electrochemical detection the total antioxidant capacity (TAC), as well as enzymatic antioxidants such as catalase (CAT), superoxide dismutase (SOD), ascorbate peroxidase (APX), or peroxiredoxins (PRX). Also, it supplies different approaches for studying non-enzymatic antioxidants such as ascorbate or dehydroascorbate and some components of the secondary metabolism such as carotenoids, tocopherol, anthocyanins, and molecules with signaling capacity such as the amino acids γ-aminobutyric acid (GABA) or proline. In addition, there are two review chapters focused on the function of reactive carbonyl species (RCS) under abiotic stress and phytomelatonin, a multifunctional molecule with antioxidant capacity as well as signal properties. We believe that this set of chapters written by specialists in each of the described techniques will be of great interest to all those scientists interested in the metabolism of ROS in plant cells.

Francisco J. Corpas and José M. Palma deeply thank all the cooperative authors for contributing their valuable knowledge, time, and enthusiasm to bring this book into its present form. We would like also to express our thankfulness to Prof. John Walker, editor of the series, as well as Anna Rakovsky and David C. Casey from the Springer Nature Group, for giving us the opportunity, professional support, and encouragement, to bring this book to the scientific community.

Granada, Spain

Francisco J. Corpas
José M. Palma

The original version of the book has been revised. A correction to this book can be found at https://doi.org/10.1007/978-1-0716-3826-2_18

Preface

Contents

Contributors

JUAN B. ARELLANO • *Departamento de Estrés Abiótico, Instituto de Recursos Naturales y Agrobiología de Salamanca (IRNASA-CSIC), Salamanca, Spain*

MARINO B. ARNAO • *Phytohormones and Plant Development Lab, Department of Plant Biology (Plant Physiology), Faculty of Biology, University of Murcia, Murcia, Spain*

ANDREA BASSI • *Department of Physics, Politecnico di Milano, Milan, Italy*

OUARDIA BENDOU • *Departamento de Estrés Abiótico, Instituto de Recursos Naturales y Agrobiología de Salamanca (IRNASA-CSIC), Salamanca, Spain*

NARA BUENO-RAMOS • *Departamento de Estrés Abiótico, Instituto de Recursos Naturales y Agrobiología de Salamanca (IRNASA-CSIC), Salamanca, Spain*

STEFANO BURATTI • *Department of Biosciences, Università degli Studi di Milano, Milan, Italy*

MARÍA J. CAMPOS • *Group of Antioxidants, Free Radicals and Nitric Oxide in Biotechnology, Food and Agriculture, Department of Stress, Development and Signaling in Plants, Estación Experimental del Zaidín, Spanish National Research Council (CSIC), Granada, Spain*

ANTONIO CANO • *Phytohormones and Plant Development Lab, Department of Plant Biology (Plant Physiology), Faculty of Biology, University of Murcia, Murcia, Spain*

FRANCISCO J. CORPAS • *Group of Antioxidants, Free Radicals and Nitric Oxide in Biotechnology, Food and Agriculture, Department of Stress, Development and Signaling in Plants, Estación Experimental del Zaidín, Spanish National Research Council (CSIC), Granada, Spain*

SANDRA CORREA • *Abiotic Stress, Production and Quality Laboratory, Department of Stress Biology and Plant Pathology, CEBAS-CSIC, Murcia, Spain*

ALEX COSTA • *Department of Biosciences, Università degli Studi di Milano, Milan, Italy; Institute of Biophysics, National Research Council of Italy (CNR), Milan, Italy*

ALINE BERTINATTO CRUZ • *Departamento de Botânica, Instituto de Biociências, Universidade de São Paulo, São Paulo, SP, Brazil*

ELISA DELL'AGLIO • *Department of Biosciences, Università degli Studi di Milano, Milan, Italy*

CINDY DIAS • *Centro de Biotecnologia e Química Fina (CBQF) – Laboratório Associado, Escola Superior de Biotecnologia, Universidade Católica Portuguesa, Porto, Portugal*

KARL-JOSEF DIETZ • *Biochemistry and Physiology of Plants, Faculty of Biology, Bielefeld University, Bielefeld, Germany*

LUCIANO FRESCHI • *Departamento de Botânica, Instituto de Biociências, Universidade de São Paulo, São Paulo, SP, Brazil*

SALVADOR GONZÁLEZ-GORDO • *Group of Antioxidants, Free Radicals and Nitric Oxide in Biotechnology, Food and Agriculture, Department of Stress, Development and Signaling in Plants, Estación Experimental del Zaidín, Spanish National Research Council (CSIC), Granada, Spain*

ORI GREEN • *Schulich Faculty of Chemistry, Technion – Israel Institute of Technology, Haifa, Israel*

MATTEO GRENZI • *Department of Biosciences, Università degli Studi di Milano, Milan, Italy*

Ismael Gutiérrez-Fernández • *Departamento de Estrés Abiótico, Instituto de Recursos Naturales y Agrobiología de Salamanca (IRNASA-CSIC), Salamanca, Spain*
Tamami Hamada • *Graduate School of Natural Science and Technology, Shimane University, Shimane, Japan*
Nir Hananya • *Schulich Faculty of Chemistry, Technion – Israel Institute of Technology, Haifa, Israel*
Josefa Hernández-Ruiz • *Phytohormones and Plant Development Lab, Department of Plant Biology (Plant Physiology), Faculty of Biology, University of Murcia, Murcia, Spain*
Hayet Houmani • *Laboratory of Extremophile Plants, Center of Biotechnology of Borj Cedria, Hammam-Lif, Tunisia*
Ana Jiménez • *Abiotic Stress, Production and Quality Laboratory, Department of Stress Biology and Plant Pathology, CEBAS-CSIC, Murcia, Spain*
Hitoshi Kusakabe • *Enzyme Sensor Co. Ltd, Tsukuba, Japan*
Bruno Silvestre Lira • *Departamento de Botânica, Instituto de Biociências, Universidade de São Paulo, São Paulo, SP, Brazil*
Emilio L. Marcos-Barbero • *Departamento de Estrés Abiótico, Instituto de Recursos Naturales y Agrobiología de Salamanca (IRNASA-CSIC), Salamanca, Spain*
Takanori Maruta • *Graduate School of Natural Science and Technology, and Institute of Agricultural and Life Sciences, Academic Assembly, Shimane University, Shimane, Japan*
Jorge Monza • *Food and Plant Biology Group, Departamento de Biología Vegetal, Facultad de Agronomía, Universidad de la República, Montevideo, Uruguay*
Melanie Morales • *Departament de Biologia Evolutiva, Ecologia i Ciències Ambientals, Universitat de Barcelona, Barcelona, Spain*
Rosa Morcuende • *Departamento de Estrés Abiótico, Instituto de Recursos Naturales y Agrobiología de Salamanca (IRNASA-CSIC), Salamanca, Spain*
Sergi Munné-Bosch • *Departament de Biologia Evolutiva, Ecologia i Ciències Ambientals, Universitat de Barcelona, Barcelona, Spain*
María A. Muñoz-Varga • *Group of Antioxidants, Free Radicals and Nitric Oxide in Biotechnology, Food and Agriculture, Department of Stress, Development and Signaling in Plants, Estación Experimental del Zaidín, Spanish National Research Council (CSIC), Granada, Spain*
Sara Paola Nastasi • *Department of Biosciences, Università degli Studi di Milano, Milan, Italy*
Tatsuya Nishiyama • *Life Science Research Centre, College of Bioresource Sciences, Nihon University, Fujisawa, Japan*
Rengin Ozgur • *Faculty of Science, Department of Biology, Ege University, Bornova, Izmir, Turkey*
José M. Palma • *Spanish National Research Council, Estación Experimental del Zaidín, Granada, Spain*
Manuela Pintado • *Centro de Biotecnologia e Química Fina (CBQF) – Laboratório Associado, Escola Superior de Biotecnologia, Universidade Católica Portuguesa, Porto, Portugal*
Marta Rodríguez-Ruiz • *Group of Antioxidants, Free Radicals and Nitric Oxide in Biotechnology, Food and Agriculture, Department of Stress, Development and Signaling in Plants, Estación Experimental del Zaidín, Spanish National Research Council (CSIC), Granada, Spain*
Magdalena Rossi • *Departamento de Botânica, Instituto de Biociências, Universidade de São Paulo, São Paulo, SP, Brazil*

Moshe Sagi • *The Albert Katz Department of Dryland Biotechnologies, French Associates Institute for Agriculture and Biotechnology of Dryland, The Jacob Blaustein Institutes for Dessert Research, Ben-Gurion University of the Negev, Beersheba, Israel; Katif Research Center for Development of Coastal Deserts, Netivot, Israel*

Florencia Sena • *Food and Plant Biology Group, Departamento de Biología Vegetal, Facultad de Agronomía, Universidad de la República, Montevideo, Uruguay; Laboratory of Apicomplexan Biology, Institut Pasteur Montevideo, Montevideo, Uruguay*

Francisca Sevilla • *Abiotic Stress, Production and Quality Laboratory, Department of Stress Biology and Plant Pathology, CEBAS-CSIC, Murcia, Spain*

Doron Shabat • *School of Chemistry, Faculty of Exact Sciences, Tel Aviv University, Tel Aviv, Israel*

Santiago Signorelli • *Food and Plant Biology Group, Departamento de Biología Vegetal, Facultad de Agronomía, Universidad de la República, Montevideo, Uruguay; ARC Centre of Excellence in Plant Energy Biology, School of Molecular Sciences, The University of Western Australia, Crawley, WA, Australia*

Aigerim Soltabayeva • *Biology Department, School of Science and Humanities, Nazarbayev University, Astana, Kazakhstan*

Mustafa Cemre Sonmez • *Faculty of Science, Department of Biology, Ege University, Bornova, Izmir, Turkey*

Clara Sousa • *Centro de Biotecnologia e Química Fina (CBQF) – Laboratório Associado, Escola Superior de Biotecnologia, Universidade Católica Portuguesa, Porto, Portugal*

Jorge Taboada • *Group of Antioxidants, Free Radicals and Nitric Oxide in Biotechnology, Food and Agriculture, Department of Stress, Development and Signaling in Plants, Estación Experimental del Zaidín, Spanish National Research Council (CSIC), Granada, Spain*

Giorgia Tortora • *Department of Physics, Politecnico di Milano, Milan, Italy*

Ismail Turkan • *Faculty of Science, Department of Biology, Ege University, Bornova, Izmir, Turkey*

Kenji Ueda • *Life Science Research Centre, College of Bioresource Sciences, Nihon University, Fujisawa, Japan*

Baris Uzilday • *Faculty of Science, Department of Biology, Ege University, Bornova, Izmir, Turkey*

Lara Vogelsang • *Biochemistry and Physiology of Plants, Faculty of Biology, Bielefeld University, Bielefeld, Germany*

Noriko Wada • *Life Science Research Centre, College of Bioresource Sciences, Nihon University, Fujisawa, Japan*

Side Selin Su Yirmibesoglu • *Faculty of Science, Department of Biology, Ege University, Bornova, Izmir, Turkey*

Chapter 1

Electrochemical Detection of Total Antioxidant Capacity (TAC) in Plant Tissues from Different Origins

Francisco J. Corpas, Marta Rodríguez-Ruiz, María J. Campos, Jorge Taboada, and José M. Palma

Abstract

Total antioxidant capacity (TAC) is a reliable indicator of antioxidant content in animal and plant samples. The different experimental approaches available allow the determination of TAC using, as a reference, diverse compounds with recognized antioxidant capacities such as Trolox, ascorbic acid, gallic acid, or melatonin. A new portable device, named BRS (BQC redox system), is now commercially available that, through an electrochemical approach, allows the determination of TAC in a simple, fast, reproducible, and robust way. In this chapter, using this portable device, a comparative analysis of the TAC is assayed in different red, citrus, and Solanaceae fruits, several *Allium* species, and organs of different plant species, including *Arabidopsis thaliana*. The obtained results demonstrate the versatility of the BRS portable device.

Key words Arabidopsis, Garlic, Melatonin, Pepper, Total antioxidant capacity, Electrochemistry, Tomato, Ascorbate, Trolox, Gallic acid, Fruits, Pepper

1 Introduction

Reactive oxygen species (ROS) are a group of molecules generated during the partial reduction of oxygen (O_2) that can be grouped into radical and nonradical species. The main radical molecules are superoxide ($O_2^{\bullet -}$), hydroxyl ($^{\bullet}OH$), alkoxyl ($RO^{\bullet}$), and peroxyl ($ROO^{\bullet}$), whereas the nonradicals include hydrogen peroxide (H_2O_2), singlet oxygen ($^1O_2$) and ozone (O_3) [1]. Although ROS have been considered an undesirable byproduct of the cellular metabolism of aerobic organisms, it is known that some of these molecules have signaling and regulatory functions [2–4]. However, an uncontrolled overproduction of ROS can lead to oxidative damage of the main

The original version of the chapter has been revised. A correction to this chapter can be found at https://doi.org/10.1007/978-1-0716-3826-2_18

Francisco J. Corpas and José M. Palma (eds.), *ROS Signaling in Plants: Methods and Protocols*,
Methods in Molecular Biology, vol. 2798, https://doi.org/10.1007/978-1-0716-3826-2_1,

biomolecules (proteins, lipids, and nucleic acids), causing their malfunction, triggering oxidative stress, and, in extreme circumstances, causing cell death. Consequently, cells have developed enzymatic and nonenzymatic systems with antioxidant capacities to keep the physiological production of ROS under control.

From the human point of view, it is well known that the intake of fruits and vegetables provides beneficial effects to maintain a healthy status. Among the components that contribute to such wellness are vitamins, minerals, fibers, and a wide range of antioxidant compounds including phenolics, terpenes, and sulfur derivatives, with vitamins C and E playing a noteworthy role. The nonenzymatic antioxidants that overcome oxidative stress are made up of different families of both hydrophilic and lipophilic components that exert their regulatory functions in plants [5, 6], and both types of antioxidants and concentrations vary widely among the different plant species [7, 8]. Moreover, many of these compounds with antioxidant capacity have associated beneficial pharmacological properties such as anticancer, antidiabetic, and anti-inflammatory potentialities. Therefore, the regular consumption of antioxidant-rich fleshy fruits has significant benefits for human health [9–13].

Total antioxidant capacity (TAC) is a global measure of the nonenzymatic antioxidant activity that integrates the individual effect of all antioxidants in a given matrix, and their additive, synergistic, or antagonistic interactions. TAC methods allow, therefore, to measure the ability of antioxidants to control the oxidation of molecules present in a specific biological sample [14]. TAC methods may be broadly classified as electron transfer (ET)– and hydrogen atom transfer (HAT)–based methods [15]. They can be grouped into three categories, namely chromatographic, spectrophotometric, and electrochemical assays [16–20]. Most popular TAC methods are based on spectrophotometric techniques, either fluorescence, chemiluminescence, or colorimetry. A wide range of spectrophotometric TAC methods such as ORAC (oxygen radical absorption capacity), HORAC (hydroxyl radical antioxidant capacity), TRAP (total peroxyl radical trapping antioxidant parameter), CUPRAC (cupric reducing antioxidant power), FRAP (ferric reducing antioxidant power), PFRAP (potassium ferricyanide reducing power), TEAC (trolox equivalent antioxidant capacity, named also ABTS assay because uses [2,2′-azinobis(3-ethylbenzothiazoline-6-sulfonic acid)], and DPPH [2,2-di(4-tert-octylphenyl)-1-picrylhydrazyl] have been described in the literature. Although all the mentioned methods have their advantages and disadvantages, one common characteristic is that they are laborious and time consuming. The chromatographic methods comprising gas and high-performance liquid chromatography are also complex and require the use of sophisticated instrumentation. Electrochemical methods based on voltammetry, amperometry, and coulometry allow direct, simple, and rapid TAC determination even in turbid or colored samples [21, 22].

Consequently, the portable BRC device described in this work is a new alternative to measure the TAC of a specific sample in less than a minute with a single drop. Here, we evaluate this small portable equipment (6.5 × 11.5 cm) in the measurement of the TAC for different fruits and vegetables available in the market and how this TAC can be referred to different reference compounds with recognized antioxidant capacity that are used as standards including ascorbic acid, gallic acid, melatonin, or Trolox.

2 Materials

2.1 Plant Materials

1. All plant materials (fruits and vegetables) were obtained from the local market. *Allium* species include leek (*Allium ampeloprasum* var. porrum), Welsh onion (*A. fistulosum*), purple onion (*A. cepa*), and garlic (*A. sativum L.)* cloves. Red fruits include blackberry (*Rubus ulmifoli*), raspberry (*Rubus idaeus*), blueberries (*Vaccinium corymbosum*), strawberries (*Fragaria* × *ananassa*), black plum (*Prunus domestica*), and red grape (*Vitis vinifera* cv Ralli seedless). Citrus fruits include orange (*Citrus sinensis* 'Valencia'), tangerine (*Citrus reticulate*), lemon (*Citrus limon*), lime (*Citrus* × *aurantiifolia*), and grapefruit (*Citrus* × *paradisi*). Solanaceae fruits include cherry- and pear-type tomato (*Solanum lycopersicum*) and California-type sweet pepper (*Capsicum annuum* L.) fruits at distinct ripening stages (green, yellow, and red). Leaf samples from spearmint (*Mentha spicata*) were obtained from a local market; goji (*Lycium barbarum*) berries were harvested from a shrub in the gardens of our research center (Estación Experimental del Zaidín, Granada, Spain); and 30-day-old *Arabidopsis thaliana* plants were grown in a growth chamber under optimal light and temperature conditions.

2.2 BRS Device, Test Strips, and Other Solutions

1. BRS (BQC redox system) device (BQC Redox Technologies, Oviedo, Spain).
2. BRS electrolyte©. It is a commercially available solution (BQC Redox Technologies) whose composition is optimized for the best sample conductivity when using the BRS device. Alternatively, Milli-Q water or a standard buffer could be used, although the conductivity of the measurement is not guaranteed (*see* **Note 1**).
3. Standard antioxidant solution stock.
4. TAC test strips for BRS device.

Table 1
Preparation of the calibration curve from a stock solution that is diluted with the BRS electrolyte

Antioxidant standard (mM)	Stock solution (μL)	BRS electrolyte (μL)
0	0	1000
0.05	5	995
0.10	10	990
0.25	25	975
0.50	50	950
1.00	100	900
5.00	500	500
10.0	1000	0

3 Methods

All the procedures are done at room temperature.

3.1 Calibration Curve

1. Select a standard compound with recognized antioxidant capacity such as ascorbic acid, gallic acid melatonin, and Trolox, among others, either using a commercial standard solution or being prepared in the laboratory.
2. Prepare a calibration curve, for example, in the range of 0–10 mM (*see* **Note 2,** Table 1).

3.2 Plant Samples

1. Ground the plant samples using a garlic press (fruits) or a mortar with a pestle (leaf of vegetables) (*see* **Note 3,** Fig. 1).
2. Filter the obtained juice through two layers of nylon.
3. Mix the resulting juice (approximately 100 μL) in the proportion 1:1 with BRS electrolyte.

3.3 TAC Measurement of Antioxidant Standards and Plant Samples with the BRS Device

1. Turn on the BRS device by pressing the central button.
2. Connect through Bluetooth your electronic device (either Android mobile phone or tablet where the application has to be installed) to the BRS and open the BRS *app*.
3. Select to measure TAC. The device has different options, for example, *Agrifood* or *Extracts.* In our case, due to the sample nature, the option *Extracts* was selected.
4. Insert the test strip in the BRS device and deposit a drop (around 40 μL) for each point of the calibration curve (*see* **Note 3**, Fig. 2) and the plant samples (*see* **Note 3**, Fig. 3).

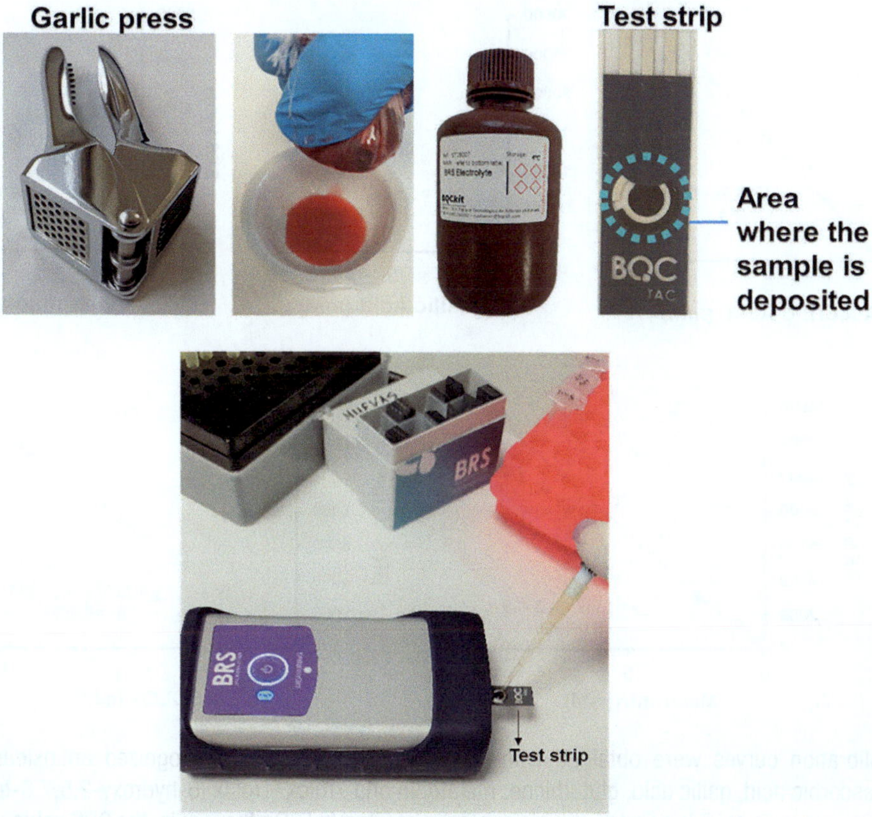

Fig. 1 Pictures of the different elements necessary to prepare the samples and to perform the TAC measurements: garlic press, mortar, BRS electrolyte, test strip, and BRS device

5. Press *Start measurement* and get the BRS value in less than 10 s. This value can be saved in an Excel table. It is recommended to measure each sample at least in triplicate.
6. The TAC results obtained with the BRS device are expressed in terms of BRS value. TAC in BRS value can be converted into antioxidant equivalents by extrapolating the BRS value of the samples over the selected standard calibration curve (*see* **Note 4**, Fig. 4).

4 Notes

1. BRS electrolyte, which is commercially available, is a solution that optimizes sample conductivity when using the BRS device. Some compounds used as standards are commercially available, but they can also be prepared in the laboratory just like the samples, using pure compounds.

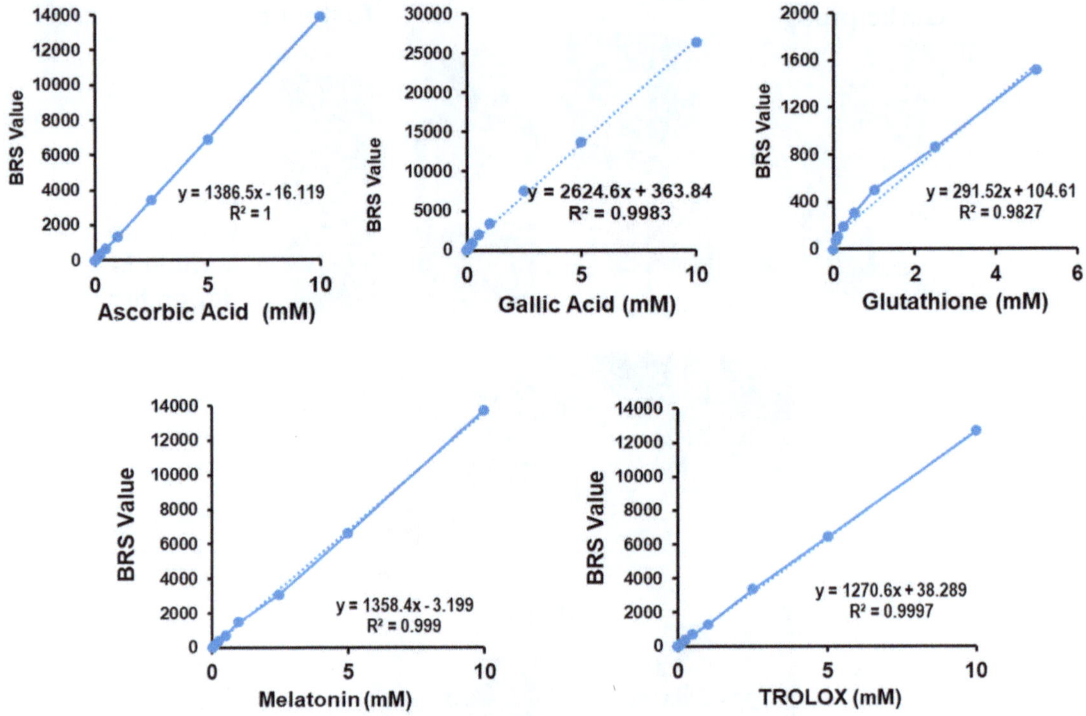

Fig. 2 Calibration curves were obtained with different compounds with recognized antioxidant capacity including ascorbic acid, gallic acid, glutathione, melatonin and Trolox. Trolox (6-hydroxy-2,5,7,8-tetramethyl-chroman-2-carboxylic acid) is a water-soluble analogue of vitamin E. On the *y*-axis, the BRS value obtained for each compound is represented against the corresponding concentration (0–10 mM) (*see* **Note 5**)

2. Select the standard antioxidant compound and prepare the calibration curve according to Table 1.
3. Figure 1 shows some representative pictures of the different utensils used for the preparation of the extracts as well as the portable device, named BRS (BQC redox system), and a test trip where samples are deposited. The BRX device uses a voltammetric technique to measure TAC. The sample is electrochemically oxidized by applying a potential scanning. The BRS value is an arbitrary unit provided by the equipment which is directly related to the electrical charge transferred during the electrochemical measurement.
4. TAC measurements obtained with the BRS device are given in BRS value. This value can be easily converted into antioxidant equivalents (e.g., TEAC, Trolox equivalents antioxidant capacity; CEAC, vitamin C equivalents antioxidant capacity; GAE, gallic acid equivalents; or melatonin equivalents) by simply performing a standard curve with the desired antioxidant compound (e.g., Trolox, ascorbic acid, gallic acid, or melatonin).

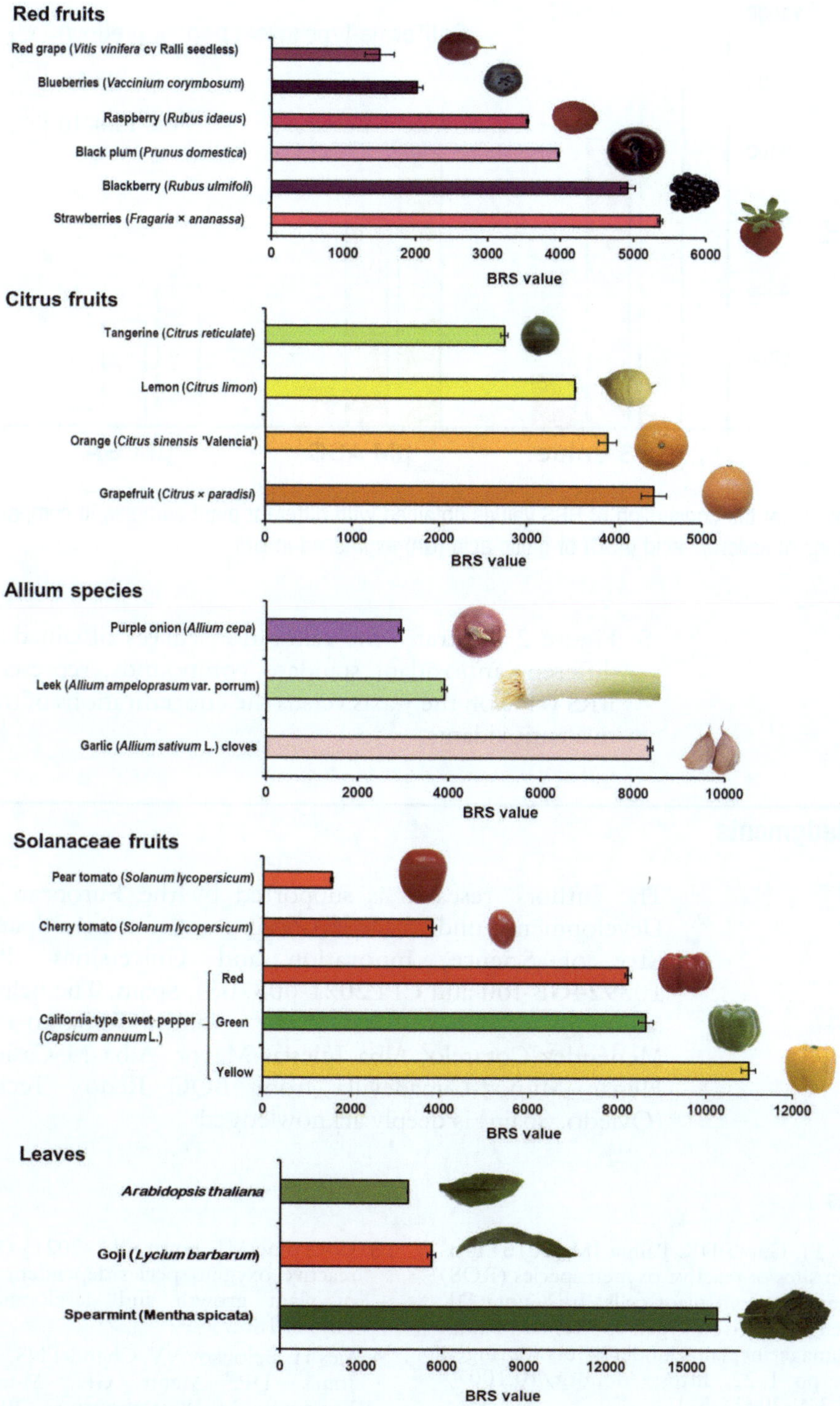

Fig. 3 BRS values obtained in different plant samples

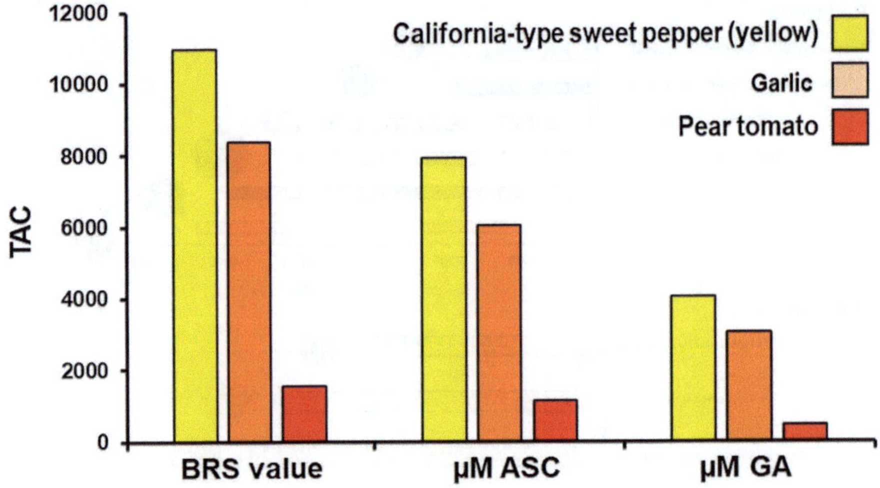

Fig. 4 Examples of the conversion of BRS values obtained with different plant samples in comparison to the concentrations of ascorbic acid (ASC) or gallic acid (GA) expressed in µM

5. Figure 2 illustrates the calibration curves obtained from the different antioxidant standard compounds, representing the BRS value on the *y*-axis versus the concentrations of the respective antioxidants.

Acknowledgments

The authors' research is supported by the European Regional Development Fund with cofinanced grants from the Spanish Ministry of Science, Innovation and Universities (PID2019-103924GB-I00 and CPP2021-008703), Spain. The help and collaboration of María Díaz-González, Sandra Tamargo-Díaz, Sara Menéndez-Cotarelo, Alba Iglesias-Mayor, Alba M Casielles, and Henar Muñoz-Cimadevilla from BQC Redox Technologies (Oviedo, Spain) is deeply acknowledged.

References

1. Corpas FJ, Gupta DK, Palma JM (2015) Production sites of reactive oxygen species (ROS) in organelles from plant cells. In: Gupta DK et al (eds) Reactive oxygen species and oxidative damage in plants under stress. Springer, Cham, pp 1–22. https://doi.org/10.1007/978-3-319-20421-5_1
2. Sies H, Jones DP (2020) Reactive oxygen species (ROS) as pleiotropic physiological signalling agents. Nat Rev Mol Cell Biol 21(7): 363–383
3. Considine MJ, Foyer CH (2021) Oxygen and reactive oxygen species-dependent regulation of plant growth and development. Plant Physiol 186:79–92
4. Sies H, Belousov VV, Chandel NS, Davies MJ, Jones DP, Mann GE, Murphy MP, Yamamoto M, Winterbourn C (2022) Defining roles of specific reactive oxygen species (ROS) in cell biology and physiology. Nat Rev Mol Cell Biol 23:499–515

5. Zhou X, Hao T, Zhou Y, Tang W, Xiao Y, Meng X, Fang X (2015) Relationships between antioxidant compounds and antioxidant activities of tartary buckwheat during germination. J Food Sci Technol 52:2458–2463
6. Correia S, Schouten R, Silva AP, Gonçalves B (2017) Factors affecting quality and health promoting compounds during growth and postharvest life of sweet cherry (*Prunus avium* L.). Front Plant Sci 8:2166
7. Agati G, Azzarello E, Pollastri S, Tattini M (2012) Flavonoids as antioxidants in plants: location and functional significance. Plant Sci 196:67–76
8. Corpas FJ, Freschi L, Palma FJ (2023) ROS metabolism and ripening of fleshy fruits. Adv Bot Res 105:205–238
9. Wahyuni Y, Ballester AR, Sudarmonowati E, Bino RJ, Bovy AG (2013) Secondary metabolites of capsicum species and their importance in the human diet. J Nat Prod 76:783–793
10. Raiola A, Tenore GC, Barone A, Frusciante L, Rigano MM (2015) Vitamin E content and composition in tomato fruits: beneficial roles and bio-fortification. Int J Mol Sci 16:29250–29264
11. Zhao X, Yuan Z (2021) Anthocyanins from pomegranate (*Punica granatum* L.) and their role in antioxidant capacities in vitro. Chem Biodivers 18(10):e2100399
12. Merecz-Sadowska A, Sitarek P, Kowalczyk T, Zajdel K, Jęcek M, Nowak P, Zajdel R (2023) Food anthocyanins: Malvidin and its glycosides as promising antioxidant and anti-inflammatory agents with potential health benefits. Nutrients 15(13):3016
13. Rodríguez-Ruiz M, Ramos MC, Campos MJ, Díaz-Sánchez I, Cautain B, Mackenzie TA, Vicente F, Corpas FJ, Palma JM (2023) Pepper fruit extracts show anti-proliferative activity against tumor cells altering their NADPH-generating dehydrogenase and catalase profiles. Antioxidants 12:1461
14. Gökmen V, Serpen A, Fogliano V (2009) Direct measurement of the total antioxidant capacity of foods: the "QUENCHER" approach. Trends Food Sci 20:278–288
15. Mendonça JDS, Guimarães RCA, Zorgetto-Pinheiro VA, Fernandes CDP, Marcelino G, Bogo D, Freitas KC, Hiane PA, de Pádua Melo ES, Vilela MLB, Nascimento VAD (2022) Natural antioxidant evaluation: a review of detection methods. Molecules 27:3563
16. Antolovich M, Prenzler PD, Patsalides E, McDonald S, Robards K (2002) Methods for testing antioxidant activity. Analyst 127:183–198
17. Niki E (2010) Assessment of antioxidant capacity in vitro and in vivo. Free Radic Biol Med 49(4):503–515
18. Munteanu IG, Apetrei C (2021) Analytical methods used in determining antioxidant activity: a review. Int J Mol Sci 22:3380
19. Chiorcea-Paquim AM, Enache TA, De Souza GE, Oliveira-Brett AM (2020) Natural phenolic antioxidants electrochemistry: towards a new food science methodology. Compr Rev Food Sci Food Saf 19:1680–1726
20. Silvestrini A, Meucci E, Ricerca BM, Mancini A (2023) Total antioxidant capacity: biochemical aspects and clinical significance. Int J Mol Sci 24:10978
21. Brainina KZ, Shpigun LK (2022) State-of-the-art electrochemistry for the assessment of oxidative stress and integral antioxidant activity of biological environments. Electrochem Sci Adv 1
22. Munteanu IG, Apetrei C (2022) A review on electrochemical sensors and biosensors used in assessing antioxidant activity. Antioxidants 11:584

Chapter 2

Singlet Oxygen and Superoxide Anion Radical Detection by EPR Spin Trapping in Thylakoid Preparations

Ouardia Bendou, Nara Bueno-Ramos, Emilio L. Marcos-Barbero, Rosa Morcuende, and Juan B. Arellano

Abstract

Reactive oxygen species (ROS) are produced by energy transfer and electron transport in plant chloroplast thylakoids at non-toxic levels under normal growth conditions, but at threatening levels under adverse or fluctuating environmental conditions. Among chloroplast ROS, singlet oxygen and superoxide anion radical, respectively, produced by photosystem II (PSII) and PSI, are known to be the major ROS under several stress conditions. Both are very unlikely to diffuse out of chloroplasts, but they are instead capable of triggering ROS-mediated chloroplast operational retrograde signalling to activate defence gene expression in concert with hormones and other molecular compounds. Therefore, their detection, identification and localization in vivo or in biological preparations is a priority for a deeper understanding of their role in (concurrent) regulation of plant growth and defence responses. Here, we present two EPR spin traps, abbreviated as TEMPD-HCl and DEPMPO, to detect and identify ROS in complex systems, such as isolated thylakoids, together with some hints and cautions to perform reliable spin trapping experiments.

Key words Chloroplasts, DEPMPO, EPR, Operational retrograde signalling, Photosynthesis, ROS, Singlet oxygen, Superoxide anion radical, TEMPD-HCl, Thylakoid

1 Introduction

Reactive oxygen species (ROS) are essential by-products of aerobic metabolism in living organisms, including plants, where they act as both cytotoxic and signalling molecules [1, 2]. When ROS are overproduced in plant cells in an uncontrolled manner, they disrupt cellular redox homeostasis and cause indiscriminate oxidative damage to key biomolecules (DNA, microRNA, proteins and lipids), leading to biological and molecular dysfunctions and eventual necrotic cell death [3]. Despite this apparently life-threatening reactivity, ROS play important biological roles. For instance, plants constitutively produce controlled, non-cytotoxic levels of ROS under normal growth conditions to regulate basic biological

Francisco J. Corpas and José M. Palma (eds.), *ROS Signaling in Plants: Methods and Protocols*,
Methods in Molecular Biology, vol. 2798, https://doi.org/10.1007/978-1-0716-3826-2_2,

processes throughout their entirely seasonal plant life cycle from germination to senescence [4]. Even more fascinating, plant cells are able to produce controlled waves of NADPH oxidase-mediated ROS bursts under fluctuating/adverse environmental conditions [5]. When this happens, ROS activate spatiotemporal defence responses in coordination with hormones and other compounds to arrest growth and possibly induce programmed cell death, deterring and preventing further local/systemic damage.

Among the distinct living cell-produced ROS, singlet oxygen ($^1O_2$) is molecular oxygen in its lowest singlet excited state; superoxide anion radical ($O_2^{\bullet-}$) and hydrogen peroxide (H_2O_2), respectively, are the one- and two-electron reduction products of molecular oxygen; and $^{\bullet}OH$ (together with and OH^-) is the product of the one-electron reduction of H_2O_2. $^1O_2$ is mainly photosensitized by photosystem II (PSII) after the molecular collision of O_2 with triplet excited state chlorophyll (Chl) [6], although it can also be photosensitized by the red Chl catabolite in mitochondria after migrating from chloroplasts or chemically produced by the Haber-Weiss and Russell mechanisms in other plant tissues [7, 8]. $^1O_2$ is a short-lived ROS in neat water (~ μs), but it becomes extremely short-lived (< μs) in crowding systems and is therefore rather unlikely to escape from chloroplasts when it is photosensitized in the PSII reaction centre. $O_2^{\bullet-}$ is produced by several complexes and components of the photosynthetic electron transport chain, but mainly by PSI (Mehler reaction) [9], the complexes I-III of the mitochondria respiratory chain [10, 11] and plasma membrane-located NADPH oxidases [12]. $O_2^{\bullet-}$ is a short-lived ROS (~ μs) showing low reactivity, but it becomes highly reactive when protonated ($HO_2^{\bullet}$). The protonated form can cross the chloroplast envelope; however, its high avidity for hydrogen atom abstraction makes it quite unlikely to reach the cytoplasm [13]. H_2O_2 is produced by spontaneous or superoxide dismutase (SOD)-catalyzed disproportionation of $O_2^{\bullet-}$ formed by O_2 reduction in chloroplasts and mitochondria [14]. H_2O_2 is also produced during photorespiration in peroxisomes and other cellular compartments in which $O_2^{\bullet-}/H_2O_2$-forming oxidases or peroxidases are also present [14, 15]. Likewise, H_2O_2 can be generated in thylakoids by incomplete oxidation of water in the PSII Mn-cluster and be converted into $^{\bullet}OH$ by PSII Mn or non-heme Fe [13]; however, because $^{\bullet}OH$, like $^1O_2$, is very reactive and extremely short-lived (~ ns), it is improbable indeed that $^{\bullet}OH$ leaves the chloroplast. In comparison with other ROS, H_2O_2 is a long-lived ROS (~ ms) with lower reactivity and high ability to flow through aquaporins embedded in chloroplast and nuclear envelopes [16].

Among chloroplast ROS, H_2O_2 can directly contribute to operational retrograde signalling, in which thiol peroxidases are recognized to play an essential role in sensing and transducing the

H_2O_2 signal [14]. Outside chloroplasts, H_2O_2 can feed into the cytosolic mitogen-activated protein kinase pathway, which controls gene expression for defence and hypersensitive responses [14, 17]. H_2O_2 can also reach the nucleus, modify gene expression through its interaction with redox-sensitive transcription factors and provide organelle specificity to the operational signalling when chloroplasts are particularly closely associated with the nucleus [16, 18]. Moreover, H_2O_2 can be converted into $^{\bullet}OH$ in the nucleus [19], adding further complexity to the direct H_2O_2 signalling and sensing pathways. Inside chloroplasts, H_2O_2 acts as a signalling molecule required for ATP/NADPH balance in photosynthesis. H_2O_2 accumulation activates the thylakoid cyclic electron flow, but it inactivates the Calvin-Benson cycle by thiol oxidation [20]. Besides, oxidative stress enhancement in chloroplasts by $O_2^{\bullet -}$ and H_2O_2 inactivates, on the one hand, adenosine bisphosphate phosphatase 1 (SAL1), a negative regulator of abscisic acid (ABA) that converts 3′(2′)-phosphoadenosine 5′- phosphate (PAP) into AMP [21], and, on the other hand, ferredoxin-dependent hydroxymethyl-butenyl diphosphate synthase (HDS), an enzyme that belongs to the MEP pathway and converts methylerythritol cyclodiphosphate (MEcPP) into 1-hydroxy-2-methyl-2-(*E*)-butenyl 4-diphosphate. PAP and MEcPP are involved in an overlapping chloroplast operational retrograde signalling cascade that controls the expression of stress-related, redox-, salicylic acid- and jasmonic acid–responsive genes, while blocking those related to ethylene and auxin signalling [22]. In addition to all this, operational retrograde signalling from mitochondria and chloroplasts can be coordinated through the nuclear protein radical-induced cell death 1 (RCD1). RCD1 itself suppresses the activity of endoplasmic reticulum-bound transcription factors of the NAC family, which mediate ROS-dependent mitochondrial retrograde signalling, while, at the same time, RCD1 regulation is sensitive to chloroplast H_2O_2-dependent PAP accumulation [23]. Inactivation of RCD1 results in chloroplast-to-mitochondria electron flow and a suppression of chloroplast ROS production. A specific signalling pathway mediated by $O_2^{\bullet -}$ from chloroplasts was also proposed in which several genes encoding small heat shock proteins and WRKY transcription factors were upregulated [24]. With regard to $^{\bullet}OH$, this type of ROS is not a signalling molecule that can escape from chloroplasts and its involvement in operational retrograde signalling needs to be experimentally proven [19]. Finally, the operational retrograde signalling pathways with origin in chloroplasts and activated by $^1O_2$-oxidized products are presented in Chap. 3, and the readers are thus referred to it for the purpose of avoiding repetition between chapters.

Among ROS produced in chloroplast thylakoids under stress conditions, $^1O_2$ and $O_2^{\bullet -}$ are the ones produced at higher levels and predominantly in PSII and PSI, respectively. In this chapter, after briefly introducing basic concepts of EPR spectroscopy, the protocols to detect $^1O_2$ and $O_2^{\bullet -}$ produced by thylakoid membranes under illumination using two commercial spin traps TEMPD-HCl and DEPMPO are described together with cautions and hints to perform reliable experiments.

Electron paramagnetic resonance (EPR) or electron spin resonance (ESR) spectroscopy [25] is a very valuable technique to detect and identify free radicals and molecular species with one or more unpaired electrons in complex systems. In Materials, Life and Health Sciences, the EPR technique is used to investigate, for instance, photopolymerization processes [26], photocatalysis in TiO_2 nanomaterials [27], oxidative stress in living cells [28], metal toxicity in plants [29], Alzheimer's disease progression [30] and metalloproteins and metallodrugs [31]. In brief, when an unpaired free-spinning electron with a magnetic moment μ_s is placed in a magnetic field of flux density B_0, the field lifts the degeneracy of the electron spin, yielding two spin states (Zeeman effect). The magnetic moment μ_s of the electron can be aligned either parallel (spin up, $m_s = +1/2$) or antiparallel (spin down, $m_s = -1/2$) to B_0, where m_s is the component of the spin angular momentum (s) of the electron parallel to the z-axis by convention. The solution of the electron Zeeman Hamiltonian gives the fundamental equation of EPR spectroscopy, which establishes that the resonance energy gap between the two electron spin states is $\Delta E = h\nu = g_e\mu_B B_0$, if B_0 is also parallel to the z-axis, where h is the Planck constant, ν is the frequency of the magnetic field of the microwave electromagnetic radiation absorbed (or emitted) by the electron, g_e is the Lande's g-factor for the free electron—whose value is altered when the unpaired electron is confined to a molecule in which extensive spin delocalization occurs—and μ_B is the Bohr magneton. The degeneracy of the unpaired electron confined to a molecule may also be lifted if the molecule contains nuclei with spin magnetic moment (μ_I). Each of the non-zero μ_I creates a magnetic field that also interacts with μ_s. As a consequence, the interaction Hamiltonian now includes the interactions of the electron and nuclei with B_0 (electron and nuclear Zeeman effects) and the hyperfine interaction(s) of μ_s with μ_I. In liquid media, in which the molecules rotate freely, the solution of the hyperfine Hamiltonian is used to determine the isotropic hyperfine splitting constant (HFSC, denoted here with the letter a) and so the extent of the delocalization of the unpaired electron over the radical. The selection rules for electron spin state transitions establish that the z-component of the total spin angular momentum (I) of the nuclei remains unchanged (i.e. $\Delta M_I = 0$), while the electron spin flips (i.e. $\Delta M_S = \pm 1$). The number of lines in the EPR spectrum resulting from the interaction of μ_s with n equivalent nuclei with μ_I is $2nI + 1$. These lines are separated by the value of a and their intensities follow the Pascal's triangle rule. Some nuclei with non-zero μ_I of interest for this chapter are ^1H ($I = 1/2$), ^{14}N ($I = 1$) and ^{31}P ($I = 1/2$).

Although $O_2^{\bullet-}$ and $^{\bullet}OH$ with $S = 1/2$ are oxygen-centred paramagnetic ROS, and $^1O_2$ with $S = 0$ exhibits paramagnetic properties because of its net electron orbital angular momentum [32], the direct detection of $O_2^{\bullet-}$, $^{\bullet}OH$ and $^1O_2$ by EPR is very unusual because the experimental conditions and technical requirements are very demanding [27]. Instead, spin traps are used. Spin traps are diamagnetic compounds that covalently bind free radicals to form stable paramagnetic adducts that can accumulate and be detected by EPR spectroscopy. Finally, it is customary that the EPR spectrum is recorded as the first derivative while the strength of B_0 sweeps and the microwave frequency of the electromagnetic radiation remains fixed.

2 Materials

Prepare all the buffers and stock solutions with ultrapure water and analytical-grade reagents. Adhere to all waste and biohazard disposal regulations while handling the chemical material for proper and safe disposal.

2.1 Buffer and Stock Solutions for ROS Detection by EPR

1. Thylakoid suspension buffer: 25 mM HEPES (mw 238.30 g mol^{-1}), pH 7.6, 0.3 M sorbitol (mw 182.17 g mol^{-1}), 50 mM KCl (mw 74.55 g mol^{-1}) and $MgCl_2$ (mw 95.21 g mol^{-1}) (*see* **Note 1**).
2. Stock solution of 2,2,6,6-tetramethyl-4-piperidone hydrochloride (TEMPD-HCl): 1 M TEMPD-HCl (191.70 g mol^{-1}). This nitroxide spin trap is easily dissolved in aqueous buffers (*see* **Note 2**).
3. Stock solution of 2-(diethoxyphosphoryl)-2-methyl-3,4-dihydro-2*H*-pyrrole 1-oxide (DEPMPO): 1 M DEPMPO (mw 235.22 g mol^{-1}). This nitrone spin trap has a very small partition coefficient in a 1-octanol/water system with no EPR signal in the dark or under illumination.
4. Stock solution of diethylenetriaminepentaacetic acid (DTPA): 10 mM DTPA (mw 393.35 g mol^{-1}).
5. Stock solution of NH_4Cl: 0.5 M NH_4Cl (mw 53.49 g mol^{-1}).
6. Other chemicals such as deuterium oxide (D_2O, mw 20.03 g mol^{-1}), NaN_3 (mw 65.01 g mol^{-1}) and SOD (lyophilized powder, ≥2500 units/mg protein, purity ≥90%) were directly mixed with solutions containing either free Chl or thylakoid membranes. Measure the right volume, weight or number of units of D_2O, NaN_3 or SOD to make the final solutions 50%

D_2O, 1 mM NaN_3 or 50 U SOD when the procedure requires them.

Keep standard chemicals at room temperature, buffer solutions at 4 °C and lyophilized SOD at −20 °C. Acquire EPR spectra in an ESP-300 X-band spectrometer (Bruker, Rheinstetten, Germany) at room temperature and utilize a standard quartz flat cell to place the thylakoid membrane suspensions or free Chl with the right mixture of chemicals. Employ a molecule editor for molecular drawing and an open-source toolbox for EPR spectrum simulation.

3 Methods

3.1 Detection of Singlet Oxygen Photosensitized by Thylakoid Membranes

1. Add the right volume of 1 M TEMPD-HCl to thylakoid membranes dispersed in the thylakoid suspension buffer pH 7.6 to reach a final concentration of 100 mM TEMPD-HCl and 10 μg of Chl per mL (*see* **Note 3**). Shake the solution gently to ensure full sample homogenization.
2. Load the standard quartz flat cell with the thylakoid membrane suspension containing 100 mM TEMPD-HCl (*see* **Note 4**).
3. Place the flat cell in the EPR spectrometer cavity and tune the EPR spectrometer.
4. Select the EPR settings. For the signal size shown in Figs. 1 and 3, the following EPR settings were selected: microwave frequency, 9.73 GHz; modulation frequency, 100 kHz; modulation amplitude, 1 G; microwave power, 63 mW; receiver gain, 2×10^4; time constant, 40.96 ms and number of scans, 4.
5. Acquire the EPR spectrum of TEMPD-HCl in the dark to check whether there is any background EPR signal before illumination.
6. Acquire the EPR spectrum while illuminating the thylakoid membrane suspension with red light (using a long-pass optical filter RG 630 or similar) for 2 min at 1000 μE m^{-2} s^{-1} (Fig. 1a) (*see* **Note 5**). The reaction of TEMPD-HCl with $^1O_2$ yields 2,2,6,6-tetramethyl-4-piperidone-*N*-oxyl (4-oxo-TEMPO) (*see* **Note 6**).
7. Analyze the HFSC from N (a_N) of the photoproduced radical to confirm that the EPR spectrum corresponds to the stable, kinetically persistent nitroxide radical 4-oxo-TEMPO. The EPR spectrum of 4-oxo-TEMPO is characterized by having a 1:1:1 triplet signal of lines with equal intensity and a value of 16.1 G for a_N in water/aqueous buffers (*see* **Note** 7).
8. Check whether the experimental EPR spectrum of 4-oxo--TEMPO shows a satisfactory match with its computer EPR simulation using the value of the a_N parameter (Fig. 2).

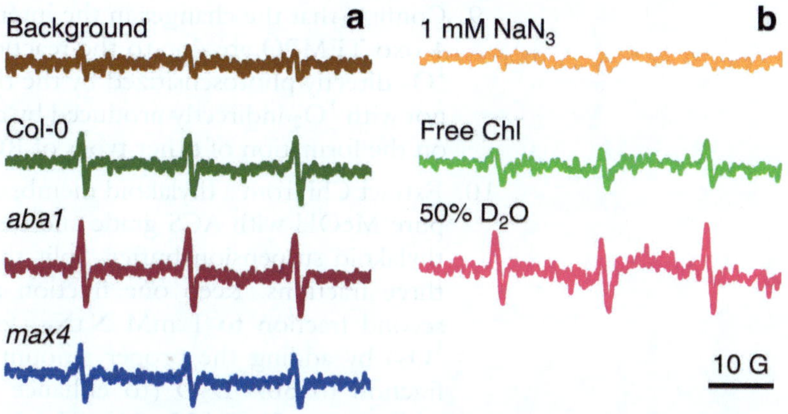

Fig. 1 Singlet oxygen detection by EPR spectroscopy using 100 mM TEMPD-HCl as spin trap in thylakoid membrane suspensions of wild-type *Arabidopsis thaliana* ecotype Col-0 and the *max4* and *aba1* mutants illuminated with red light at 1000 μE m^{-2} s^{-1} for 2 min (**a**) and MeOH-extracted Chl from thylakoids with 1 mM NaN_3, no additions and 50% D_2O at 500 μE m^{-2} s^{-1} for 2 min (**b**). The background EPR signal of 100 mM TEMP-HCl in the thylakoid suspension buffer is also shown under illumination, but in the absence of thylakoid membranes. Chl concentration in thylakoid membrane suspensions and the solutions containing free Chl was 10 mg mL^{-1}. (Adapted from [35] with permission from Elsevier GmbH 2017)

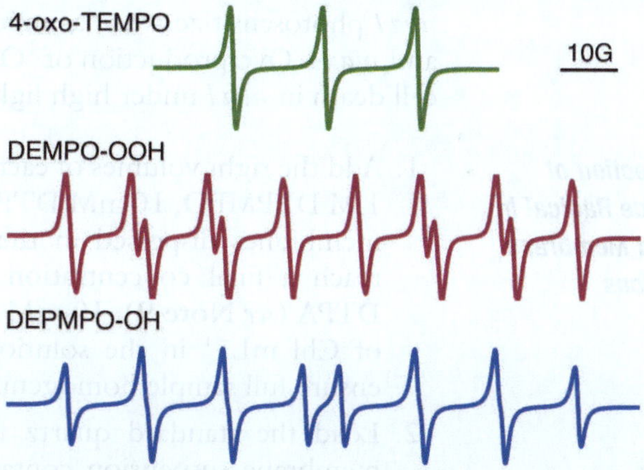

Fig. 2 Simulated EPR spectra of the spin trapping adducts 4-oxo-TEMPO, DEPMPO-OOH and DEPMPO-OH after the reactions of TEMP with $^1O_2$ and DEPMPO with $O_2^{\bullet -}$ or $^{\bullet}OH$. The HFSC parameters were taken from references [43, 51, 53]. A fast exchange between two equally populated radical species of DEPMPO-OOH diastereomers with slightly different values for the HFSC parameters was assumed [51]; therefore, the computer simulation was made using the average of the corresponding two sets of HFSC parameters. An open toolbox for EPR spectrum simulation can be found at https:/www.eprsimulator.org/

9. Confirm that the changes in the intensity of the EPR signal of 4-oxo-TEMPO are due to the reaction of TEMPD-HCl with $^1O_2$ directly photosensitized by the thylakoid preparation, but not with $^1O_2$ indirectly produced by chemical reactions relying on the formation of other types of ROS (*see* **Note 8**).
10. Extract Chl from a thylakoid membrane suspension with 100% pure MeOH with ACS grade and disperse the free Chl in the thylakoid suspension buffer. Split the free Chl suspension in three fractions. Keep one fraction as a control. Adjust the second fraction to 1 mM NaN_3 (for physical quenching of $^1O_2$) by adding the proper amount of NaN_3 and the third fraction to 50% D_2O (to enhance the lifetime of $^1O_2$) by adding pure D_2O [33, 34]. The final concentration of free Chl in the three fractions must be equal (~10 mg mL^{-1}) to compare the EPR spectra.
11. Acquire the EPR spectra of the free Chl solutions as described in **steps 1–6** and investigate the changes in the intensity of the EPR signals in the absence or presence of 1 mM NaN_3 and 50% D_2O under illumination (Fig. 1b).

EPR spectra of 4-oxo-TEMPO in three different preparations of thylakoid membranes isolated from wild-type *Arabidopsis thaliana* ecotype Col-0 and the *aba1* and *max4* mutants under illumination are displayed in Fig. 1a. The results show that thylakoids of *aba1* photosensitize twice as much $^1O_2$ as thylakoids of wild type and *max4*. Overproduction of $^1O_2$ leads to chloroplast rupture and cell death in *aba1* under high light stress [35].

3.2 Detection of Superoxide Radical in Thylakoid Membrane Suspensions

1. Add the right volumes of each of the following stock solutions: 1 M DEPMPO, 10 mM DTPA and 0.5 M NH_4Cl to thylakoid membranes dispersed in the thylakoid suspension buffer to reach a final concentration of 50 mM DEPMPO, 1 mM DTPA (*see* **Note 9**), 10 mM NH_4Cl (*see* **Note 10**) and 10 μg of Chl mL^{-1} in the solution. Shake gently the solution to ensure full sample homogenization.
2. Load the standard quartz flat cell with the final thylakoid membrane suspension containing all the chemicals indicated in the previous step to proceed with the EPR analysis.
3. Reproduce **steps 3** and **4** described in the method for $^1O_2$ detection to tune the EPR spectrometer and select the EPR settings.
4. Acquire the EPR spectrum in the dark to confirm there is no background signal by DEPMPO or any impurity with EPR signal in the reaction mixture (Fig. 3).
5. Acquire the EPR spectrum while illuminating the thylakoid membrane suspension with red light (using a long-pass optical

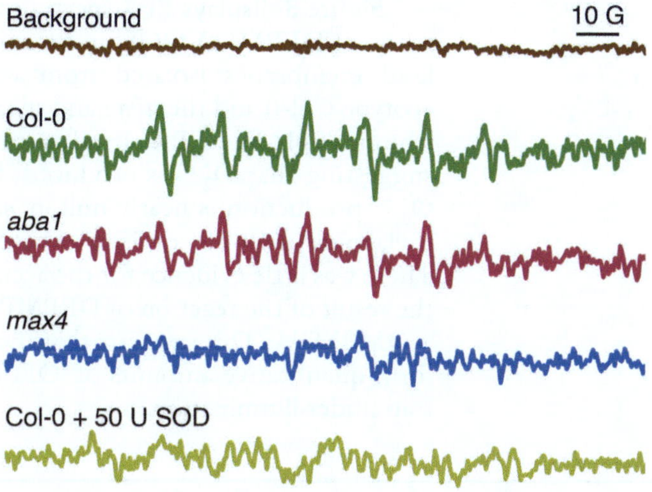

Fig. 3 ROS detection by spin trapping EPR using 50 mM DEPMPO in thylakoid membrane suspensions of wild-type *Arabidopsis thaliana* ecotype Col-0 and the *max4* and *aba1* mutants. Thylakoids were illuminated with high light (1000 μE m^{-2} s^{-1}) for 2 min. Approximately 50 U of SOD was added to thylakoids of Arabidopsis Col-0 as a control experiment and the background signal of 50 mM DEPMPO illuminated in the suspension buffer, but in the absence of thylakoid, was shown for comparison. Chl concentration in thylakoid membrane suspensions was 10 mg mL^{-1}. (Adapted from [35] with permission from Elsevier GmbH 2017)

filter RG 630 or similar) for 2 min at 1000 μE m^{-2} s^{-1} (Fig. 3) (*see* **Note 11**).

6. Analyze the HFSC from N (a_N), H ($a_{H\beta}$) and P (a_P) of the photoproduced radical to confirm that the EPR spectrum corresponds with the 2-diethoxyphosphoryl-2-methyl-5-hydroperoxy-l-pyrrolidinyloxyl (DEPMPO-OOH). Keep in mind that diastereomers of DEPMPO-OOH corresponding with two different radical species with very similar HFSC parameters are expected to be formed (*see* **Note 12**). A computer-simulated EPR spectrum for DEPMPO-OOH is shown in Fig. 2.
7. Confirm that the changes in the intensity of the EPR signal of the photoproduced radical is the product of the reaction of DEPMPO with $O_2^{\bullet -}$ by adding ~50 U of lyophilized SOD to the reaction mixture. A decrease in the intensity of the EPR radical signifies a competitive reaction between SOD and DEPMPO for $O_2^{\bullet -}$ (Fig. 3) (*see* **Note 13**).
8. Confirm whether DEPMPO might have reacted with $^{\bullet}$OH to form 2-diethoxyphosphoryl-2-methyl-5-hydroxy-l-pyrrolidinyloxyl (DEPMPO-OH) (*see* **Note 14**). A simulation of the EPR spectrum of DEPMPO-OH is also shown in Fig. 2.

Figure 3 displays EPR spectra of (the mixture of the diastereomers) of DEPMPO-OOH in three different preparations of thylakoid membranes isolated from wild-type *Arabidopsis thaliana* ecotype Col-0 and the *aba1* and *max4* mutants under illumination. The multiple-line EPR signal is rather weak in Col-0 and *aba1*, suggesting that $O_2^{\bullet -}$ is produced, but at low levels. Intriguingly, $O_2^{\bullet -}$ production is nearly null in *max4*. The addition of SOD to thylakoids of Col-0 confirms $O_2^{\bullet -}$ production under illumination. There was little evidence for the accumulation of DEPMPO-OH as the result of the reaction of DEPMPO with $^{\bullet}OH$ in the presence of 1 mM DTPA. This suggests that the thylakoid membranes did not form quantitative amounts of $^{\bullet}OH$ that could be detected by spin trap under illumination.

4 Notes

1. Thylakoid membranes from leaves of *Arabidopsis thaliana* were prepared following standard procedures [36]. In this study, thylakoid membranes were prepared from the leaves of wild-type Arabidopsis ecotype Col-0 and the *aba1* and *max4* mutants with defects in the carotenoid-dependent biosynthesis of the hormones, ABA and strigolactones, respectively. These two mutants were hypothesized to be prone to overproduce ROS [37]. The isolated thylakoid membranes were dispersed in the suspension buffer, frozen in liquid nitrogen and stored at −80 °C until use. Chl content in thylakoid membranes was determined in 80% (v/v) acetone with ACS grade as described in [38].
2. The background EPR signal of TEMPD-HCl in the dark should be checked to confirm whether there are impurities in the commercial batches under use. TEMPD-HCl should not give any EPR signal in the dark [39]. If TEMP is used instead of TEMPD-HCl, it is highly recommended to further purify it by vacuum distillation [39]. TEMPD-HCl itself can give a background EPR signal under illumination. This background signal is small when red light is applied for illumination, but it increases when using white light.
3. The reaction of $^1O_2$ with TEMPD-HCl is pH-dependent and occurs when the sterically hindered cyclic amine of TEMPD-HCl is deprotonated [33]. The amine of TEMPD-HCl has a p*K*a of 7.6. Therefore, basic buffers with pH values beyond the former p*K*a should be avoided.
4. TEMPD-HCl at a concentration of 100 mM does not inhibit oxygen evolution in thylakoid membranes after a 30-min incubation, but the ratio between the variable fluorescence and maximal fluorescence decreases slightly [39]. It is thus

recommended to acquire the EPR spectra in a time frame of 30 min after the addition of TEMPD-HCl.

5. TEMPD-HCl requires prolonged and strong illumination in comparison with other EPR probes used for $^1O_2$ detection [40].
6. TEMPD-HCl reacts with $^1O_2$ to form 4-oxo-TEMPO after the initial formation of a charge transfer complex followed by the hydroperoxidation of the nitrogen [41] as shown in Fig. 4.

   The stability of the EPR signal of nitroxide radicals can be affected by the presence of redox compounds, with which the aminoxyl group of the nitroxide radical can undergo one-electron reduction or one-electron oxidation to form hydroxylamine or an oxoammonium cation, respectively [42, 43] as shown in Fig. 5. Accumulation of reducing agents in the reaction mixture can underestimate $^1O_2$ photosensitization. Therefore, the application of TEMP-HCl has some limitations in vivo [44].
7. The EPR spectral features of 4-oxo-TEMPO have been described in reference [33, 40]. The magnitude of a_N for nitroxide radicals depends on the molecular geometry and the solvent polarity. In particular, the value of a_N of 4-oxo--TEMPO increases through the relative stabilization of its zwitterionic resonance structure in polar solvents [43] as shown in Fig. 6.
8. ROS such as $O_2^{\bullet-}$, $^{\bullet}OH$ and H_2O_2 do not give nitroxides by reaction with sterically hindered cyclic amines [45]. However, ROS-dependent formation of $^1O_2$ is possible and was described in several biological systems [46, 47]. If needed, control experiments with SOD or catalase can be performed to eliminate or prevent the formation of $O_2^{\bullet-}$, $^{\bullet}OH$ and H_2O_2.
9. DTPA is a metal ion chelator added to the solution to inhibit the Fe-catalyzed Haber-Weiss reaction [48], in which $^{\bullet}OH$ is formed after the reaction of $O_2^{\bullet-}$ with H_2O_2. In the event that DEPMPO-OH was observed by EPR in the presence of DTPA, this would mean that $^{\bullet}OH$ was directly formed by the biological system under study.
10. NH_4Cl decouples the proton gradient across the thylakoid membrane from the thylakoid electron transport [49].
11. DEPMPO reacts with $O_2^{\bullet-}$ to yield the DEPMPO-superoxide spin adduct (DEPMPO-OOH) as shown in Fig. 7.
12. The EPR spectrum of DEPMPO-OOH should have given rise to 12 lines of equal intensity [50, 51]. Indeed, computation simulations exhibit 12 lines (Fig. 2), but the shape and intensity of some of them are unusual due to overlap of two different radical species of DEPMPO-OOH diastereomers **1** and **2** (Fig. 8).

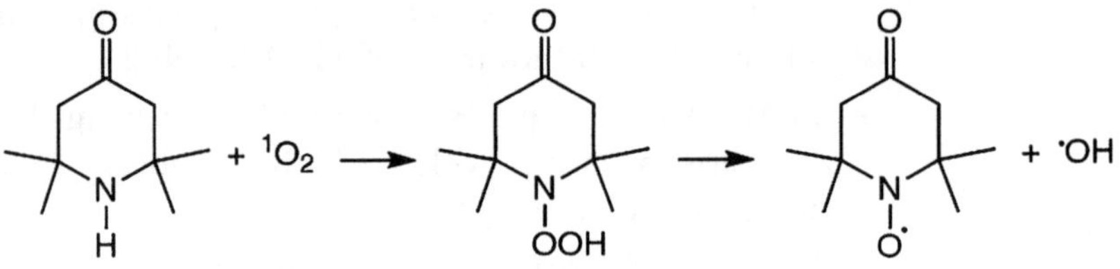

Fig. 4 Chemical reaction between TEMPD and $^{1}O_2$

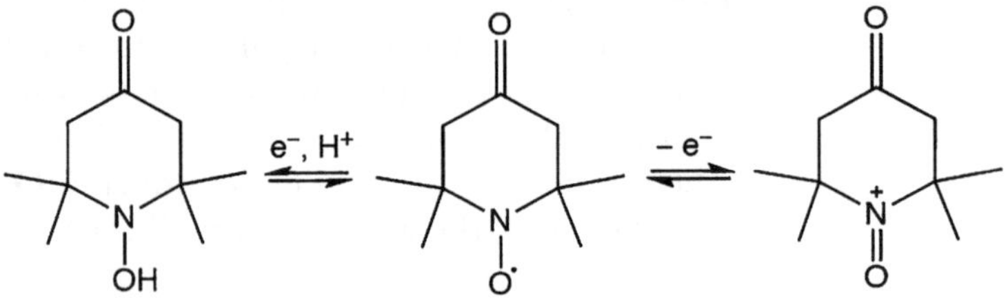

Fig. 5 One-electron redox reactions of 4-oxo-TEMPO

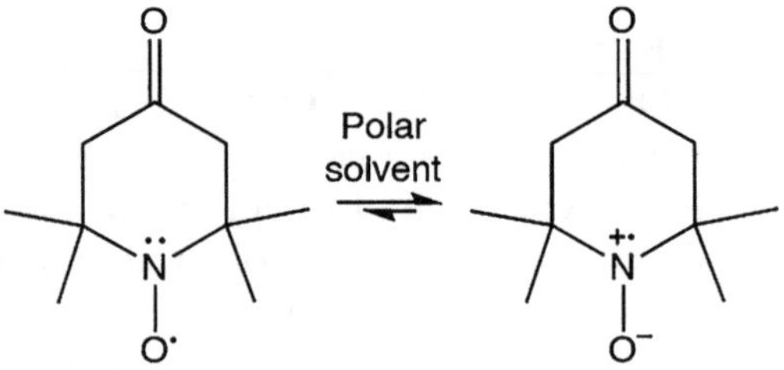

Fig. 6 Resonance structures of the aminoxyl group of 4-oxo-TEMPO

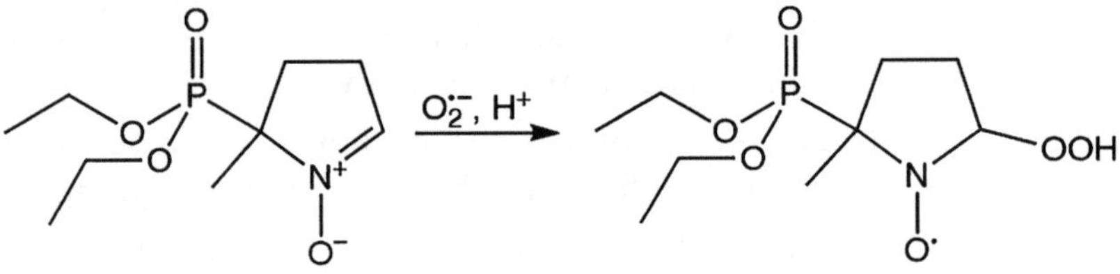

Fig. 7 Chemical reaction between DEPMPO and $O_2^{\bullet -}$

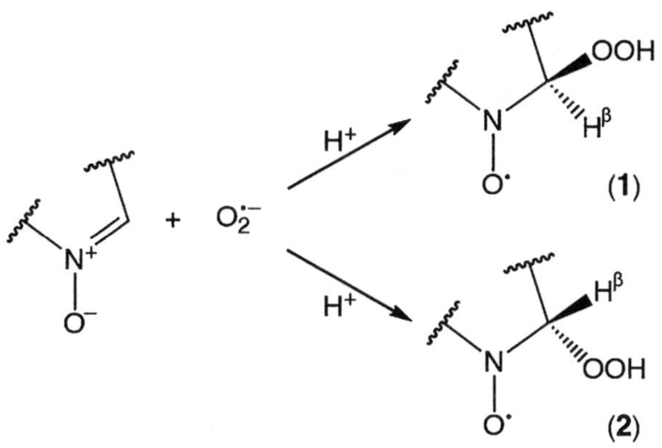

Fig. 8 Formation of DEPMPO-OOH diastereomers

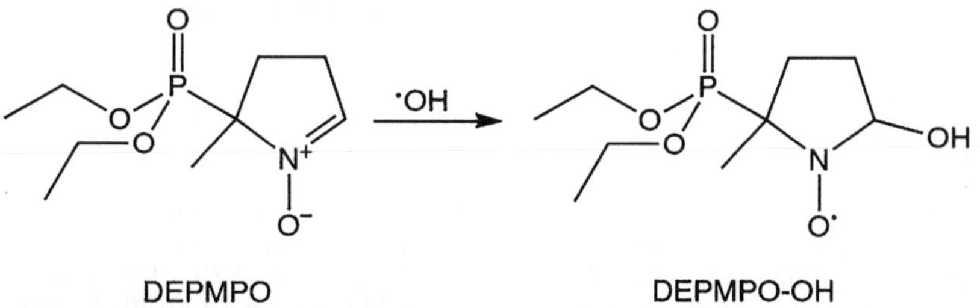

Fig. 9 Chemical reaction between DEPMPO and •OH

The two sets of HFSC parameters (a_N, $a_{H\beta}$ and a_P) for the two radical species of the DEPMPO-OOH diastereomers are slightly different and each set has quite similar values for a_N and $a_{H\beta}$ [51–53]. Moreover, the decomposition of DEPMPO-OOH into DEPMPO-OH and other spin adducts has also been described, adding more complexity to the interpretation of the EPR spectrum [54, 55]. DEPMPO-OOH is only stable in a range of few minutes and exhibits both first- and second-order decay kinetic components [50, 53]. The intensity of the EPR signal of DEPMPO-OOH notably decreases in the presence of reductants such as ascorbic acid and glutathione [28].

13. SOD catalyzes the disproportionation of $O_2^{\bullet -}$ formed by the reduction of molecular oxygen by thylakoid membranes during strong illumination.
14. The reaction product of DEPMPO with •OH (Fig. 9) yields diastereomers of DEPMPO-OH as observed for DEPMPO-OOH. However, this spin radical is persistent and has EPR spectral features different from those of DEPMPO-OOH (Fig. 2) [50, 52, 53]. The two sets of the HFSC parameters (a_N, $a_{H\beta}$ and a_P) for the two radical species are identical and a_N and $a_{H\beta}$ are even more similar to each other than

observed for DEPMPO-OH. This gives rise to a quartet form that is split by the large P coupling, resulting in eight lines with intensity ratios of 1:2:2:1:1.2.2.1.

Acknowledgements

We thank Dr. Anja Krieger-Liszkay for her careful and critical reading. The research was funded by MCIN/AEI/10.13039/501100011033 (Project n° PID2019-107154RB-100) and the regional government of Castilla y León (Project n° CSI260P20). The Project "CLU-2019–05—IRNASA/CSIC Unit of Excellence" funded by the Junta de Castilla y León and co-financed by the European Union (ERDF, "Europe Drives Our Growth") and the CSIC Interdisciplinary Thematic Platform (PTI) Optimization of Agricultural and Forestry Systems (PTI-AGROFOR) are also acknowledged.

References

1. Das K, Roychoudhury A (2014) Reactive oxygen species (ROS) and response of antioxidants as ROS-scavengers during environmental stress in plants. Front Environ Sci 2:53. https://doi.org/10.3389/fenvs.2014.00053
2. Foyer CH, Hanke G (2022) ROS production and signalling in chloroplasts: cornerstones and evolving concepts. Plant J 111:642–661. https://doi.org/10.1111/tpj.15856
3. Van Breusegem F, Dat JF (2006) Reactive oxygen species in plant cell death. Plant Physiol 141:384–390. https://doi.org/10.1104/pp.106.078295
4. Huang H, Ullah F, Zhou DX et al (2019) Mechanisms of ROS regulation of plant development and stress responses. Front Plant Sci 10:800. https://doi.org/10.3389/fpls.2019.00800
5. Baxter A, Mittler R, Suzuki N (2014) ROS as key players in plant stress signalling. J Exp Bot 65:1229–1240. https://doi.org/10.1093/jxb/ert375
6. Arellano JB, Naqvi KR (2016) Endogenous singlet oxygen photosensitizers in plants. In: Nonell S, Flors C (eds) Singlet oxygen: applications in biosciences and nanosciences. Royal Society of Chemistry, London, pp 239–269
7. Khorobrykh S, Havurinne V, Mattila H, Tyystjärvi E (2020) Oxygen and ROS in photosynthesis. Plan Theory 9:91. https://doi.org/10.3390/plants9010091
8. Pattanayak GK, Venkataramani S, Hortensteiner S et al (2012) ACCELERATED CELL DEATH 2 suppresses mitochondrial oxidative bursts and modulates cell death in Arabidopsis. Plant J 69:589–600. https://doi.org/10.1111/j.1365-313X.2011.04814.x
9. Kozuleva M (2022) Recent advances in the understanding of superoxide anion radical formation in the photosynthetic electron transport chain. Acta Physiol Plant 44:1–17. https://doi.org/10.1007/s11738-022-03428-0
10. Vanlerberghe GC (2013) The signaling role of a mitochondrial superoxide burst during stress. Plant Signal Behav 8:e22749. https://doi.org/10.4161/psb.22749
11. Turrens JF (2003) Mitochondrial formation of reactive oxygen species. J Physiol 552:335–344. https://doi.org/10.1113/jphysiol.2003.049478
12. Torres MA, Dangl JL (2005) Functions of the respiratory burst oxidase in biotic interactions, abiotic stress and development. Curr Opin Plant Biol 8:397–403. https://doi.org/10.1016/j.pbi.2005.05.014
13. Pospíšil P, Kumar A, Prasad A (2022) Reactive oxygen species in photosystem II: relevance for oxidative signaling. Photosynth Res 152:245–260. https://doi.org/10.1007/s11120-022-00922-x
14. Smirnoff N, Arnaud D (2019) Hydrogen peroxide metabolism and functions in plants. New Phytol 221:1197–1214. https://doi.org/10.1111/nph.15488

15. Corpas FJ, del Río LA, Palma JM (2019) Plant peroxisomes at the crossroad of NO and H_2O_2 metabolism. J Integr Plant Biol 61:803–816. https://doi.org/10.1111/jipb.12772
16. Breeze E, Mullineaux PM (2022) The passage of H_2O_2 from chloroplasts to their associated nucleus during retrograde signalling: reflections on the role of the nuclear envelope. Plan Theory 11:552. https://doi.org/10.3390/plants11040552
17. Dietz K-J, Turkan I, Krieger-Liszkay A (2016) Redox- and reactive oxygen species-dependent signaling into and out of the photosynthesizing chloroplast. Plant Physiol 171:1541–1550. https://doi.org/10.1104/pp.16.00375
18. Exposito-Rodriguez M, Laissue PP, Yvon-Durocher G et al (2017) Photosynthesis-dependent H_2O_2 transfer from chloroplasts to nuclei provides a high-light signalling mechanism. Nat Commun 8:1–11. https://doi.org/10.1038/s41467-017-00074-w
19. Leister D (2019) Piecing the puzzle together: the central role of reactive oxygen species and redox hubs in chloroplast retrograde signaling. Antioxid Redox Signal 30:1206–1219. https://doi.org/10.1089/ars.2017.7392
20. Strand DD, Livingston AK, Satoh-Cruz M et al (2015) Activation of cyclic electron flow by hydrogen peroxide in vivo. Proc Natl Acad Sci USA 112:5539–5544. https://doi.org/10.1073/pnas.1418223112
21. Wilson PB, Estavillo GM, Field KJ et al (2009) The nucleotidase/phosphatase SAL1 is a negative regulator of drought tolerance in Arabidopsis. Plant J 58:299–317. https://doi.org/10.1111/J.1365-313X.2008.03780.x
22. Li M, Kim C (2022) Chloroplast ROS and stress signaling. Plant Commun 3:100264. https://doi.org/10.1016/j.xplc.2021.100264
23. Shapiguzov A, Vainonen JP, Hunter K et al (2019) Arabidopsis RCD1 coordinates chloroplast and mitochondrial functions through interaction with ANAC transcription factors. elife 8:e43284. https://doi.org/10.7554/ELIFE.43284
24. Scarpeci TE, Zanor MI, Carrillo N et al (2008) Generation of superoxide anion in chloroplasts of *Arabidopsis thaliana* during active photosynthesis: a focus on rapidly induced genes. Plant Mol Biol 66:361–378. https://doi.org/10.1007/S11103-007-9274-4
25. Banwell CN, McCash EM (1994) Fundamentals of molecular spectroscopy. McGraw-Hill, New York
26. Peyrot F, Lajnef S, Versace DL (2022) Electron paramagnetic resonance spin trapping (EPR-ST) technique in photopolymerization processes. Catalysts 12:772. https://doi.org/10.3390/catal12070772
27. Nosaka Y, Nosaka AY (2017) Generation and detection of reactive oxygen species in photocatalysis. Chem Rev 117:11302–11336. https://doi.org/10.1021/acs.chemrev.7b00161
28. Saito K, Takahashi M, Kamibayashi M et al (2009) Comparison of superoxide detection abilities of newly developed spin traps in the living cells. Free Radic Res 43:668–676. https://doi.org/10.1080/10715760902988850
29. Jegerschöld C, Arellano JB, Schröder WP et al (1995) Copper(II) inhibition of electron transfer through photosystem II studied by EPR spectroscopy. Biochemistry 34:12747–12754. https://doi.org/10.1021/bi00039a034
30. Fang D, Zhang Z, Li H et al (2016) Increased electron paramagnetic resonance signal correlates with mitochondrial dysfunction and oxidative stress in an Alzheimer's disease mouse brain. J Alzheimers Dis 51:571–580. https://doi.org/10.3233/JAD-150917
31. Eisermann J, Seif-Eddine M, Roessler MM (2021) Insights into metalloproteins and metallodrugs from electron paramagnetic resonance spectroscopy. Curr Opin Chem Biol 61:114–122. https://doi.org/10.1016/J.CBPA.2020.11.005
32. Ruzzi M, Sartori E, Moscatelli A et al (2013) Time-resolved EPR study of singlet oxygen in the gas phase. J Phys Chem A 117:5232–5240. https://doi.org/10.1021/jp403648d
33. Moan J, Wold E (1979) Detection of singlet oxygen production by ESR. Nature 279:450–451. https://doi.org/10.1038/279450a0
34. Harbour JR, Issler SL (1982) Involvement of the azide radical in the quenching of singlet oxygen by azide anion in water. J Am Chem Soc 104:903–905. https://doi.org/10.1021/ja00367a066
35. Sánchez-Corrionero Á, Sánchez-Vicente I, González-Pérez S et al (2017) Singlet oxygen triggers chloroplast rupture and cell death in the zeaxanthin epoxidase defective mutant *aba1* of *Arabidopsis thaliana* under high light stress. J Plant Physiol 216:188–196. https://doi.org/10.1016/j.jplph.2017.06.009
36. Arellano JB, Schröder WP, Sandmann G et al (1994) Removal of nuclear contaminants and of non-specifically photosystem II-bound copper from photosystem II preparations. Physiol Plant 91:369–374. https://doi.org/10.1111/j.1399-3054.1994.tb02962.x
37. Gutiérrez J, González-Pérez S, García-García F et al (2011) Does singlet oxygen activate cell death in Arabidopsis cell suspension cultures?

Analysis of the early transcriptional defense responses to high light stress. Plant Signal Behav 6:1937–1942. https://doi.org/10.4161/psb.6.12.18264

38. Porra RJ, Thompson WA, Kriedemann PE (1989) Determination of accurate extinction coefficients and simultaneous equations for assaying chlorophylls *a* and *b* extracted with four different solvents: verification of the concentration of chlorophyll standards by atomic absorption spectroscopy. Biochim Biophys Acta 975:384–394
39. Hideg É, Deák Z, Hakala-Yatkin M et al (2011) Pure forms of the singlet oxygen sensors TEMP and TEMPD do not inhibit photosystem II. Biochim Biophys Acta 1807:1658–1661. https://doi.org/10.1016/j.bbabio.2011.09.009
40. Igarashi T, Sakurai K, Oi T et al (1999) New sensitive agents for detecting singlet oxygen by electron spin resonance spectroscopy. Free Radic Biol Med 26:1339–1345. https://doi.org/10.1016/S0891-5849(98)00291-3
41. Lion Y, Gandin E, Van de Vorst A (1980) On the production of nitroxide radicals by singlet oxygen reaction: an EPR study. Photochem Photobiol 31:305–309. https://doi.org/10.1111/j.1751-1097.1980.tb02545.x
42. Prescott C, Bottle SE (2017) Biological relevance of free radicals and nitroxides. Cell Biochem Biophys 75:227–240. https://doi.org/10.1007/s12013-016-0759-0
43. Karoui H, Le Moigne F, Ouari O, Tordo P (2010) Nitroxide radicals: properties, synthesis and applications. In: Hicks RG (ed) Stable radicals: fundamentals and applied aspects of odd-electron compounds. Wiley, Hoboken, pp 173–229
44. Fischer BB, Hideg E, Krieger-Liszkay A (2013) Production, detection, and signaling of singlet oxygen in photosynthetic organisms. Antioxid Redox Signal 18:2145–2162. https://doi.org/10.1089/ars.2012.5124
45. Rigo A, Argese E, Stevanato R et al (1977) A new method of detecting O_2^- production. Inorganica Chim Acta 24:L71–L73. https://doi.org/10.1016/S0020-1693(00)93831-3
46. Inoue S, Kawanishi S (1989) ESR evidence for superoxide, hydroxyl radicals and singlet oxygen produced from hydrogen peroxide and nickel(II) complex of glycylglycyl-L-histidine. Biochem Biophys Res Commun 159:445–451. https://doi.org/10.1016/0006-291X(89)90012-0
47. Miyaji A, Gabe Y, Kohno M, Baba T (2017) Generation of hydroxyl radicals and singlet oxygen during oxidation of rhododendrol and rhododendrol-catechol. J Clin Biochem Nutr 60:86–92. https://doi.org/10.3164/jcbn.16-38
48. Halliwell B (1978) Superoxide-dependent formation of hydroxyl radicals in the presence of iron chelates. Is it a mechanism for hydroxyl radical production in biochemical systems? FEBS Lett 92:321–326. https://doi.org/10.1016/0014-5793(78)80779-0
49. Dean RL, Miskiewicz E (2003) Rates of electron transport in the thylakoid membranes of isolated, illuminated chloroplasts are enhanced in the presence of ammonium chloride. Biochem Mol Biol Educ 31:410–417. https://doi.org/10.1002/bmb.2003.494031060265
50. Frejaville C, Karoui H, Tuccio B et al (1995) 5-(Diethoxyphosphoryl)-5-methyl-1-pyrroline *N*-oxide: a new efficient phosphorylated nitrone for the *in vitro* and *in vivo* spin trapping of oxygen-centered radicals. J Med Chem 38:258–265. https://doi.org/10.1021/jm00002a007
51. Frejaville C, Karoui H, Tuccio B et al (1994) 5-Diethoxyphosphoryl-5-methyl-1-pyrroline *N*-oxide (DEPMPO): a new phosphorylated nitrone for the efficient *in vitro* and *in vivo* spin trapping of oxygen-centred radicals. J Chem Soc Chem Commun 1793–1794. https://doi.org/10.1039/C39940001793
52. Heyno E, Gross CM, Laureau C et al (2009) Plastid alternative oxidase (PTOX) promotes oxidative stress when overexpressed in tobacco. J Biol Chem 284:31174–31180. https://doi.org/10.1074/jbc.M109.021667
53. Kamibayashi M, Oowada S, Kameda H et al (2006) Synthesis and characterization of a practically better DEPMPO-type spin trap, 5-(2,2-dimethyl-1,3-propoxy cyclophosphoryl)-5-methyl-1-pyrroline *N*-oxide (CYPMPO). Free Radic Res 40:1166–1172. https://doi.org/10.1080/10715760600883254
54. Dikalov S, Jiang J, Mason RP (2005) Characterization of the high-resolution ESR spectra of superoxide radical adducts of 5-(diethoxyphosphoryl)-5-methyl-1-pyrroline *N*-oxide (DEPMPO) and 5,5-dimethyl-1-pyrroline *N*-oxide (DMPO). Analysis of conformational exchange. Free Radic Res 39:825–836. https://doi.org/10.1080/10715760500155688
55. Barbati S, Clément JL, Olive G et al (1997) ^{31}P labeled cyclic nitrones: a new class of spin traps for free radicals in biological milieu. In: Minisci F (ed) Free radicals in biology and environment. NATO ASI series, vol 27. Springer, Dordrecht, pp 39–47

Chapter 3

Singlet Oxygen Detection by Chemiluminescence Probes in Living Cells

Nir Hananya, Ori Green, Ismael Gutiérrez-Fernández, Doron Shabat, and Juan B. Arellano

Abstract

Singlet oxygen is a reactive oxygen species that causes oxidative damage to plant cells, but intriguingly it can also act as a signalling molecule to reprogram gene expression required to induce plant physiological/cellular responses. Singlet oxygen photosensitization in plants mainly occurs in chloroplasts after the molecular collision of ground-state molecular oxygen with triplet-excited-state chlorophyll. Singlet oxygen direct detection through phosphorescence emission in chloroplasts is a herculean task due to its extremely low luminescence quantum yield. Because of this, indirect alternative methods have been developed for its detection in biological systems, for example, by measuring the changes in the EPR signal or fluorescence intensity of singlet oxygen reaction–based probes. The singlet oxygen chemiluminescence (SOCL) is a chemiluminescence probe with high sensitivity and selectivity towards singlet oxygen and promising use to detect it in living cells without the inconvenience of low stability of the EPR signal of spin probes in the presence of redox compounds, spurious light scattering coming from the light source required for the excitation of fluorescence probes or the light emission of endogenous fluorescent molecules like chlorophyll in chloroplasts. The protocol presented in this chapter describes the first steps to characterizing singlet oxygen production within the biological system under study; this is accomplished through monitoring molecular oxygen consumption by SOCL using a Clark-type oxygen electrode and measuring the chemiluminescence generated by SOCL 1,2-dioxetane using a spectrofluorometer. For singlet oxygen detection within living cells, a version of SOCL with increased membrane permeability (SOCL-CPP) is described.

Key words Chemiluminescence, Chloroplast, Photosynthesis, Schaap's dioxetanes, Singlet oxygen, SOCL, Thylakoid

1 Introduction

Singlet oxygen ($^1O_2$) photosensitization in plants mainly occurs in chloroplasts after the molecular collision of the ground state O_2 with chlorophyll (Chl) or certain Chl precursors and catabolic products while they remain in their triplet-excited state. The triplet-excited species are formed either by intersystem crossing, when they are free or bound to light-harvesting antenna complexes,

Francisco J. Corpas and José M. Palma (eds.), *ROS Signaling in Plants: Methods and Protocols*,
Methods in Molecular Biology, vol. 2798, https://doi.org/10.1007/978-1-0716-3826-2_3,

or by the radical pair mechanism, when Chl forms the special pigment P_{680} in the reaction centre (RC) of photosystem II (PSII) [1]. Chemical production of $^1O_2$ in the chloroplast envelope or non-photosynthetic organelles following the Russell or Haber-Weiss mechanisms, in which peroxyl radicals or reduced forms of ROS act as reactants, was also reported [2–5]. The non-radical $^1O_2$ is constitutively photosensitized in PSII and is considered the major ROS involved in photooxidative damage to plants [6, 7], particularly when it is overproduced under high light (HL). Despite its high cytotoxicity, $^1O_2$ is involved in operational chloroplast-to-nucleus retrograde signalling under stress conditions to reprogram gene expression required for activating physiological/cellular responses including stress acclimation, chloroplast dismantling, and programmed cell death (PCD) [8–12]. Interestingly, indiscriminate cellular $^1O_2$ photosensitization by Rose Bengal in *Arabidopsis* suspension cell cultures showed the need for functional chloroplasts for PCD activation [13]. Overall, chloroplasts are environmental sensor organelles that help maintain cell homeostasis [12].

$^1O_2$ is very unlikely to play a direct role in chloroplast signal transduction cascade as its diffusion distance is very short; this is due to its extremely short lifetime in macromolecular crowding systems and its high reactivity [1, 14–16]. $^1O_2$ specially reacts with electron-enriched compounds containing double bonds, including aromatic residues of proteins, polyunsaturated fatty acids, β-carotene molecules, and certain hydrophilic and hydrophobic antioxidants [1, 12, 17]. In the conditional *fluorescent* (*flu*) mutant of *Arabidopsis thaliana*, protochlorophyllide (PChlide), a potent $^1O_2$ photosensitizer, accumulates in chloroplast thylakoids in the dark. After a dark to (moderate) light shift, a surge of $^1O_2$ is photosensitized by PChlide. The nuclear-encoded EXECUTER 1 (EX1) protein, located in thylakoid grana margins, is susceptible to oxidation at a tryptophan residue by $^1O_2$ [18]. When EX1 oxidation occurs, proteolytic degradation of EX1 by the FtsH metalloprotease happens. This event promotes the initiation of the operational $^1O_2$-mediated retrograde signal cascade in *flu* responsible for chloroplast degradation and eventual PCD after the concurrent activation of several hormones embracing jasmonic acid (JA), ethylene, and salicylic acid (SA) [18–20]. The EX1-paralog protein EX2 similarly undergoes $^1O_2$-mediated tryptophan oxidation, but intriguingly EX2 oxidation negatively regulates EX1 signalling [18]. $^1O_2$-mediated lipid peroxidation also occurs in the *flu* mutant, and oxylipins such as 12-oxo-phytodienoic acid (OPDA)—a type of reactive carbonyl species (RCS) [21]—are formed soon after shifting to light [19]. OPDA is a JA precursor that induces defense genes that antagonizes the PCD-inducing activity of JA [19]. An EX1/EX2-independent $^1O_2$-mediated retrograde signalling pathway was described in *Arabidopsis thaliana* wild-type plants and *chlorina1* (*ch1*), a mutant with defects in the biosynthesis of Chl *b*, in which $^1O_2$ photosensitization occurs at the core of PSII [8, 22, 23]. Under HL, the overproduction of $^1O_2$ in PSII RC of wild-type plants causes β-carotene oxidation and the formation of several volatile derivatives such as β-cyclocitral, β-ionone and dihydroactinidiolide [22], all of them RCS. β-cyclocitral induces cellular detoxification-responsive genes and SA-mediated acclimation to HL in wild plants [22, 24]. PCD was induced in *ch1* under HL, but it was markedly reduced in the double mutant *ch1/oxi1* in comparison with *ch1*, indicating that the response to $^1O_2$ overproduction is mediated by the OXIDATIVE SIGNAL-INDUCIBLE 1 (OXI1) Ser/Thr kinase and that JA acts downstream of OXI1 [23]. Ferrochelatase 2 (FC2) catalyses the insertion of Fe^{2+} into protoporphyrin (Proto) IX to produce protoheme and is located in both the chloroplast envelope and thylakoids. *fc2* is a conditionally lethal mutant of *Arabidopsis thaliana* that accumulates excess Proto IX in chloroplasts immediately after dawn [25]. $^1O_2$ photosensitization by Proto IX in the chloroplast envelope of the mutant *fc2* was proposed to induce chloroplast degradation and eventual cell death. Cell death in *fc2* adult plants was not attributed to direct damage by $^1O_2$, but rather to a $^1O_2$-mediated signalling involving the cytoplasmic plant U-box 4 E3 ubiquitin ligase (PUB4). This pathway is independent of the EX signalling (when chloroplast are fully developed), but dependent of the OXI1 signalling [25, 26]. Recently, accumulation of the Chl precursor uroporphyrinogen III in the *PCD8* mutant of *Arabidopsis thaliana* was also described to cause EX1-independent PCD after a burst of $^1O_2$ upon illumination [27].

Direct detection of the phosphorescence emission at 1270 nm of $^1O_2$ photosensitized in chloroplasts is very challenging due to the extremely low phosphorescence quantum yield of $^1O_2$ in a macromolecular crowding system like its interior, which contains multiple physical and chemical quenchers of $^1O_2$ [1]. While time-resolved measurements of $^1O_2$ phosphorescence have been achieved in an aqueous environment using PSII RC isolated from thylakoid membranes, quantifying the faint luminescence signal is severely prone to errors [15]. For this reason, scientists developed alternative methods in which fluorescent and EPR probes react with $^1O_2$ to indirectly detect and identify its production site in living cells [28–31]. In this chapter, we present the Singlet Oxygen ChemiLuminescence (SOCL) probe and its cell-permeable version SOCL-CPP (where CPP stands for cell-penetrating peptide), for real-time detection of $^1O_2$ in living cells. SOCL is an enol-ether that reacts with $^1O_2$ with high selectivity, forming a phenoxy-dioxetane. The latter decomposes to yield an electronically excited, highly emissive benzoate ester that produces chemiluminescence (Fig. 1) [32, 33]. Since chemiluminescence measurements do not require excitation light, artefacts due to spurious light scattering or

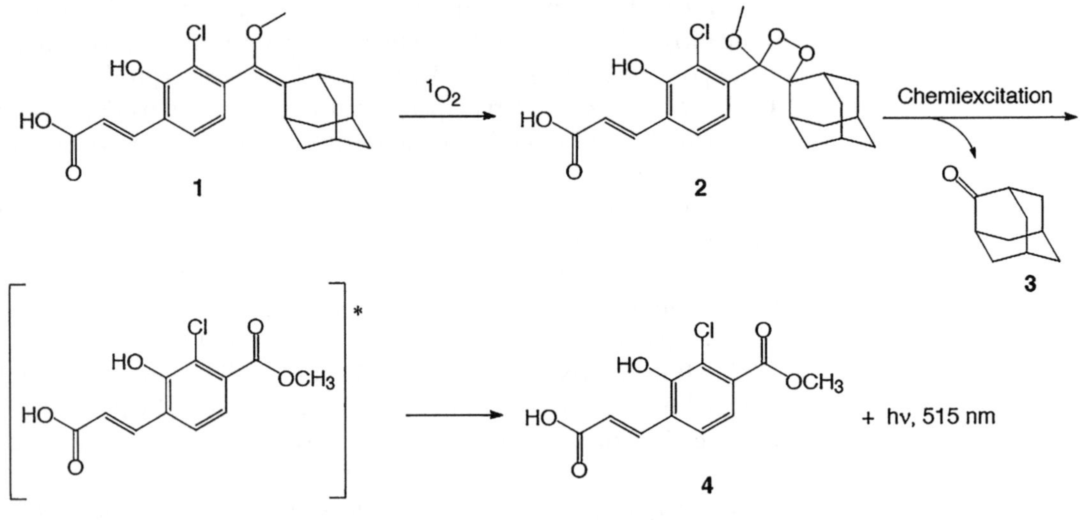

Fig. 1 Structure of the $^1O_2$ chemiluminescence probe (SOCL) and the reaction products of the chemiexcitation pathway upon the reaction of SOCL with $^1O_2$. (1) 3-(4-((adamantan-2-ylidene)(methoxy)methyl)-3-chloro-2-hydroxyphenyl)acrylic acid (SOCL), (2) 3-(3-chloro-2-hydroxy-4-(4′-methoxyspiro[adamantane-2,3′-[1,2]dioxetan]-4′-yl)phenyl)acrylic acid (SOCL phenoxy-dioxetane), (3) adamantan-2-one, and (4) 3-(3-chloro-2-hydroxy-4-(methoxycarbonyl)phenyl)acrylic acid. (Adapted from [33] with permission from the authors © 2019 and licensed by MDPI, Basel, Switzerland)

background emission from endogenous fluorescent molecules, which are often confronted in fluorescence-based assays, are avoided [34, 35]. Indeed, SOCL is found to be significantly more sensitive than the common fluorescence probes for $^1O_2$ sensing [32].

SOCL is suitable for detecting $^1O_2$ in vitro, for example, in cell extracts; however, to detect $^1O_2$ in living cells, a cell-permeable probe is required. A common strategy for providing molecules with cell permeability is equipping them with a CPP [36]. Therefore, SOCL-CPP was prepared by coupling SOCL with a nona-arginine peptide through a cysteine-maleimide bioconjugation reaction [37]. SOCL-CPP could successfully detect intracellular $^1O_2$ produced by a photosensitizer in HeLa cells during a photodynamic therapy (PDT) mode of action and is a promising $^1O_2$ probe for working within mammalian cells [35]. Furthermore, SOCL-CPP is expected to penetrate plant cells, as oligo-arginine CPPs were shown to provide molecules with high plant cell permeability [38, 39]. However, plant cell permeability must be validated experimentally prior to $^1O_2$ measurements in living plant cells.

2 Materials

Prepare all the buffers and stock solutions with ultrapure water, analytical-grade reagents, and biological-grade buffers. Adhere to all waste and biohazard disposal regulations while handling the chemical materials for proper and safe disposal.

2.1 Buffer and Stock Solutions for Singlet Oxygen Detection by Chemiluminescence

1. 1× phosphate buffered saline (PBS) containing methylene blue (MB): Dilute one-tenth PBS (supplied as 10× solution) with ultrapure water. Add a very small pinch of dry MB (mw 319.85 g mol^{-1}) powder to 1× PBS (hereafter simply PBS) and vortex the solution for a few tens of seconds several times (*see* **Note 1**). Read the absorbance of MB in PBS at 663 nm in a spectrophotometer and readjust the absorbance with fresh PBS until reaching a final absorbance value of 0.150. The final absorbance value is equivalent to a concentration of ~2 μM for MB in PBS (*see* **Note 2**).
2. Stock solution of furfuryl alcohol (FFA): 40 mM FFA (mw 98.10 g mol^{-1}) in ethanol. Dissolve 3.47 μL of FFA (d. 1.132 g cm^{-3}) in 1 mL of ethanol. Use a high-precision syringe to measure the volume of FFA (*see* **Note 3**).
3. 50% saturated KCl solution: ~2.3 M KCl. Dissolve 1.7 g of KCl (mw 74.55 g mol^{-1}) in ultrapure water and adjust the final volume to 10 mL.
4. Stock solution of bovine serum albumin (BSA) in PBS: ~30% (w/v) BSA (mw ~66 kDa, purity >96%) in PBS. Weigh 150 mg of lyophilized BSA powder in an analytical scale, mix it with 400 μL of PBS and gently vortex the mixture until full solubilization and homogenization. If needed, centrifuge the protein solution to remove bubbles and protein foam in a standard bench microfuge at the highest speed for 5 min at 4 °C (*see* **Note 4**).
5. Stock solutions of SOCL and SOCL-CPP: 50 mM SOCL (mw 374.13 g mol^{-1}) in DMSO and 10 mM SOCL-CPP (mw 2079.13 g mol^{-1}) in DMSO (*see* **Note 5**). Calculate the total volume of the stock solution required for the number of experiments planned to be carried out. Weigh the dry powder of SOCL (*see* **Note 6**) or SOCL-CPP (*see* **Note 7**) in an analytical scale, mix it with the corresponding volume of DMSO, and gently vortex the solution until full homogenization.
6. Stock solution of meta-tetra(hydroxyphenyl)chlorin (mTHPC) in DMSO: 10 mM mTHPC (mw 680.74 g mol^{-1}, purity >90%) in DMSO. Proceed similarly as indicated for the preparation of the stock solutions of SOCL and SOCL-CPP.

2.2 HeLa Culture

1. Human cervical adenocarcinoma HeLa cells: Purchase from Sigma Aldrich.
2. Growth medium: Dulbecco's Modified Eagle's Medium (DMEM) supplemented with 10% fetal bovine serum (FBS),

100 μg mL^{-1} streptomycin, 100 IU mL^{-1} penicillin, 12.5 IU mL^{-1} nystatin and 2 mM *L*-glutamine. A humidified atmosphere of 95% air and 5% CO_2 at 37 °C was set in the cell culture incubator.

3. Wash buffer: Idem as the solution for cell growth medium.

Keep standard chemicals and solvents at room temperature; lyophilized BSA, FFA, DMEM, FBS and antibiotics at 4 °C; and dry SOCL powder and SOCL-CPP powder at −20 °C in the dark until use. Use an oxygen electrode type Clark to monitor the molecular oxygen consumption rate [Chlorolab 2 system with optical ports for illumination via LH11/2R high-intensity red LED probes (Hansatech Instruments Ltd., Norfolk, England, UK) or equivalent] and a spectrofluorometer to monitor the chemiluminescence emission from SOCL phenoxy-dioxetane [Model QM-2000-4 (Photon Technology International Inc., Lawrenceville, NJ, USA) or equivalent]. Monitor the chemiluminescence emission from SOCL-CPP phenoxy-dioxetane with a BioSpace Lab Photon IMAGER™ (Nesles-la-Vallée, France).

3 Methods

3.1 Oxygen Consumption Rate Measurements

1. Measure molecular oxygen consumption rate in a PC-operated oxygen electrode type Clark equipped with a concentric water jacket to keep the temperature constant and optical ports to hold red LED light sources projecting a light irradiance of 2 mE m^{-2} s^{-1} at the centre of the cylindrical chamber. Place a piece of a paper spacer on the epoxy resin disc of the oxygen electrode, deposit a few drops of 2.3 M KCl on the paper spacer and trap it beneath an oxygen-permeable polytetrafluoroethylene (PTFE) membrane. Seal the PTFE membrane with an O-ring rubber. Screw the epoxy resin disc to the electrode chamber and place it on top of the oxygen electrode control unit. Fill the electrode chamber with approximately 1 mL of ultrapure water and leave the electrode chamber open for 30 min while maintaining the water under stirring with a tiny magnetic bar to reach a rapid gas exchange equilibrium with room air. Calibrate the oxygen electrode at 20 °C and 1 atm, first with air-saturated water and then with nitrogen-purged water for several minutes. Empty the oxygen electrode chamber with a plastic pipette avoiding any damage to the PTFE membrane and deposit approximately 1 mL of PBS until use.
2. Measure molecular oxygen consumption rate in a final volume of 1 mL at a constant temperature selected between 10 °C and 37 °C and atmospheric pressure (*see* **Note 8**). Deposit a volume of 990 μL of PBS containing 2 μM MB inside the oxygen

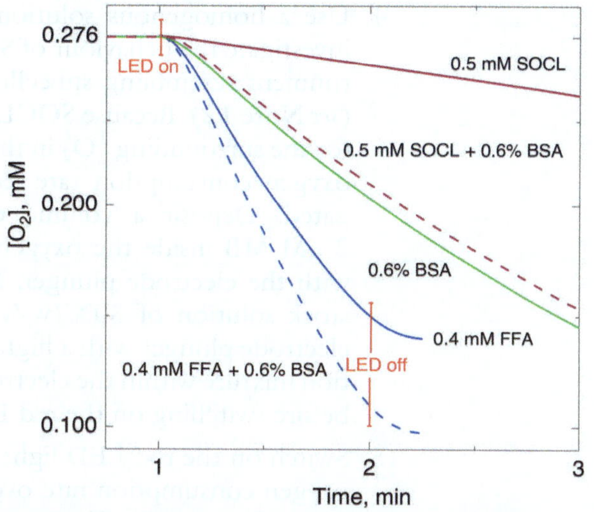

Fig. 2 Molecular oxygen uptake in the oxygen electrode chamber containing 2 μM MB as $^1O_2$ photosensitizer and 0.5 mM SOCL, 0.4 mM FFA, 0.3% (w/v) BSA (coloured solid lines) as $^1O_2$ quenchers in PBS pH 7.4 at 20 °C temperatures. The effect of 0.3% (w/v) BSA on the molecular oxygen uptake by 0.5 mM SOCL or 0.4 mM FFA is also shown for comparison reasons (coloured dashed lines). Samples were illuminated with a red LED source with an intensity of 2 mE m^{-2} s^{-1} for 1 min or 2 min depending on whether either FFA or SOCL was present in the reaction mixture. The photooxygenation reactions were performed under atmospheric pressure conditions. (Adapted from [33] with permission from the authors © 2019 and licensed by MDPI, Basel, Switzerland)

electrode chamber and close it with the electrode plunger. Inject a volume of 10 μL of the stock solutions of 40 mM FFA (or 50 mM SOCL) through the hole of the electrode plunger with a high-precision syringe. Keep the reaction mixture within the electrode chamber in the dark for 1 min before switching on the red LED light source (*see* **Note 9**).

3. Switch on the red LED light source to reach a steady state concentration of $^1O_2$ in the electrode chamber. $^1O_2$ photosensitization occurs after the molecular collision of ground-state molecular oxygen with triplet-excited-state MB, the latter formed through intersystem crossing after singlet excitation of the ground state MB. Record the molecular oxygen consumption rate over 1 min when using FFA or 2 min when using SOCL (*see* **Note 10**), then switch off the red LED light source. The traces of the consumption rate of molecular oxygen in the electrode as the result of the chemical reaction of $^1O_2$ with FFA (or SOCL) are directly displayed on-screen using the electrode oxygen control PC software (Fig. 2) (*see* **Note 11**). Repeat the measure of molecular oxygen consumption rate in triplicate for FFA (and SOCL) at any temperature under study.

4. Use a homogenous solution of 30% (w/v) BSA in PBS to investigate the behaviour of SOCL in a protein-crowding environment resembling subcellular compartments of living cells (*see* **Note 12**). Because SOCL and BSA can chemically compete for the surrounding $^1O_2$ in the reaction mixture, the molecular oxygen consumption rate for BSA is first separately investigated. Deposit a volume of 980 μL of PBS containing 2 μM MB inside the oxygen electrode chamber and close it with the electrode plunger. Inject a volume of 20 μL of the stock solution of 30% (w/v) BSA through the hole of the electrode plunger with a high-precision syringe. Keep the reaction mixture within the electrode chamber in the dark for 1 min before switching on the red LED light source.
5. Switch on the red LED light source and record the molecular oxygen consumption rate over 2 min, then switch off the red LED light source. The traces of the molecular oxygen consumption rate are directly displayed on-screen using the electrode oxygen control PC software (Fig. 2). Repeat the measure of molecular oxygen consumption rate for BSA in triplicate at any temperature under study.
6. Deposit a volume of 970 μL of PBS containing 2 μM MB inside the oxygen electrode chamber and close it with the electrode plunger. Inject a volume of 20 μL of the stock solution of 30% (w/v) BSA followed by a volume of 10 μL of the stock solution of 50 mM SOCL (or 40 mM FFA) through the hole of the electrode plunger with a high-precision syringe for each chemical. Keep the reaction mixture within the electrode chamber in the dark for 1 min before switching on the red LED light source.
7. Switch on the red LED light source and record the molecular oxygen consumption rate over 2 min when mixing SOCL and BSA (or 1 min when mixing FFA and BSA) in the oxygen electrode chamber (*see* **Note 13**), then switch off the red LED light source. The traces of the molecular oxygen consumption rate are directly displayed on-screen using the electrode oxygen control PC software (Fig. 2). Repeat in triplicate as indicated above.

3.2 Chemiluminescence Emission Measurements

1. Measure the time-dependent chemiluminescence emission of the SOCL phenoxy-dioxetane derivative at 515 nm (*see* **Note 14**) in a spectrofluorometer with a Peltier-based temperature-controlled cuvette holder, in which the optical path of the beam coming from the light source is blocked and the aperture of the emission slits is adjusted to maximize the signal-to-noise ratio.
2. Pre-incubate the spectrofluorometer cuvette in the holder at the temperature selected in the oxygen electrode.

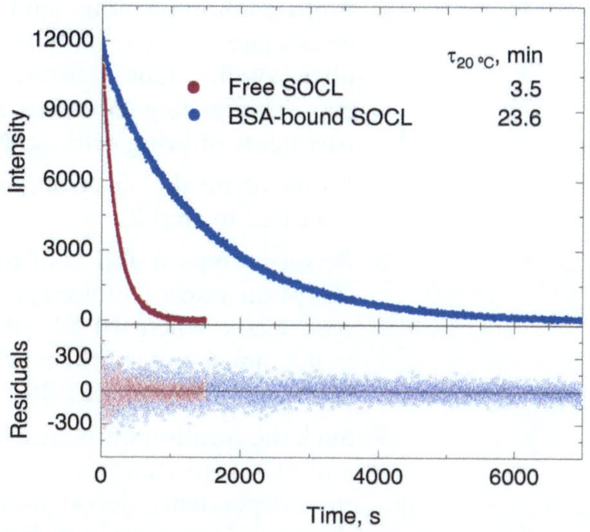

Fig. 3 Time-dependent variation of the chemiluminescence traces of the SOCL phenoxy-dioxetane derivative at 20 °C after the photooxygenation reaction of SOCL with $^1O_2$ in the absence or presence of 0.6% (w/v) BSA in PBS pH 7.4. Before monitoring the chemiluminescence traces, the reaction mixtures were set in the oxygen electrode chamber under atmospheric pressure, where 2 μM MB was present as $^1O_2$ photosensitizer and 0.5 mM SOCL as $^1O_2$ quencher. The reaction mixtures were illuminated with a red LED source with an intensity of 2 mE m^{-2} s^{-1} for 2 min at 20 °C. The data was fitted to a monoexponential equation, $I_{515nm} = \alpha_1 \exp(-t/\tau_1)$; also shown are the corresponding residuals. The standard error for the decay rate constants is <5%. (Adapted from [33] with permission from the authors © 2019 and licensed by MDPI, Basel, Switzerland)

3. Perform **steps 2** and **3** of the protocol to measure the consumption rate of molecular oxygen in the oxygen electrode type Clark, when SOCL is dissolved in PBS containing 2 μL MB, to accumulate the SOCL phenoxy-dioxetane derivative in solution.
4. Suck the solution of the reaction mixture with a plastic pipette from the electrode chamber immediately after the 2-min illumination with the red LED light source and place it in the spectrofluorometer cuvette. Click immediately the start button of the time-based acquisition mode of the spectrofluorometer to record the time-dependent decomposition of the SOCL phenoxy-dioxetane derivative (*see* **Note 15**).
5. Record the chemiluminescence emission traces at 515 nm until the chemiluminescence emission vanishes (*see* **Note 16**). The traces are displayed on-screen using the spectrofluorometer control PC software (Fig. 3).

6. Prepare a homogenous solution of 30% (w/v) BSA in PBS to investigate the chemiluminescence emission of the SOCL phenoxy-dioxetane derivative and its thermostability in a protein-crowding environment resembling subcellular compartments of living cells (*see* **Note 17**).
7. Pre-incubate the spectrofluorometer cuvette in the holder as indicated in **step 2**.
8. Perform **steps 6** and 7 of the protocol to measure the consumption rate of molecular oxygen in the oxygen electrode type Clark when 10 μL of the stock solution of 50 mM SOCL and 20 μL of the stock solution of 30% (w/v) BSA are dissolved in 970 μL of PBS containing 2 μM MB.
9. Suck the solution of the reaction mixture containing MB, BSA and SOCL in PBS in a way similar to **step 4** and record the time-dependent decomposition of the SOCL phenoxy-dioxetane derivative as indicated in **step 5** (*see* **Note 18**). The chemiluminescence emission traces of SOCL phenoxy-dioxetane are displayed on-screen using the spectrofluorometer control PC software (Fig. 3).

3.3 Detection of Singlet Oxygen in Living HeLa Using SOCL-CPP

1. Seed HeLa cells in 96 well-corning clear bottom plates (25,000 cells per well) 1 day before treatment.
2. Remove the growing medium using a pipette (cells are not in suspension) and add a fresh one.
3. Incubate the samples with SOCL-CPP or the photosensitizer mTHPC (*see* **Note 19**), or both, for 1 h at a final concentration for each compound of 10 μM and 1% DMSO (Fig. 4).
4. Wash the cells with the culture medium three times and add a fresh medium.
5. Irradiate the 96 plates with the cells for 1 min using an LED lamp (19 W, 2700 K, 1380 lumen). Keep control cells in the dark before imaging.
6. Measure the chemiluminescent signal by collecting the photons in all wavelengths (*see* **Note 20**) using a photon imager.

4 Notes

1. It is highly recommended to measure the absorbance of MB at 663 nm in PBS rather than weighing the dry MB powder on an analytical scale. The extinction coefficient of MB in pure water is 7.41×10^4 M^{-1} cm^{-1} at 664 nm. Prepare a large volume of ~2 μM MB in PBS (approximately 100 mL) if several technical repetitions are planned. Incubate the solution in a water bath at

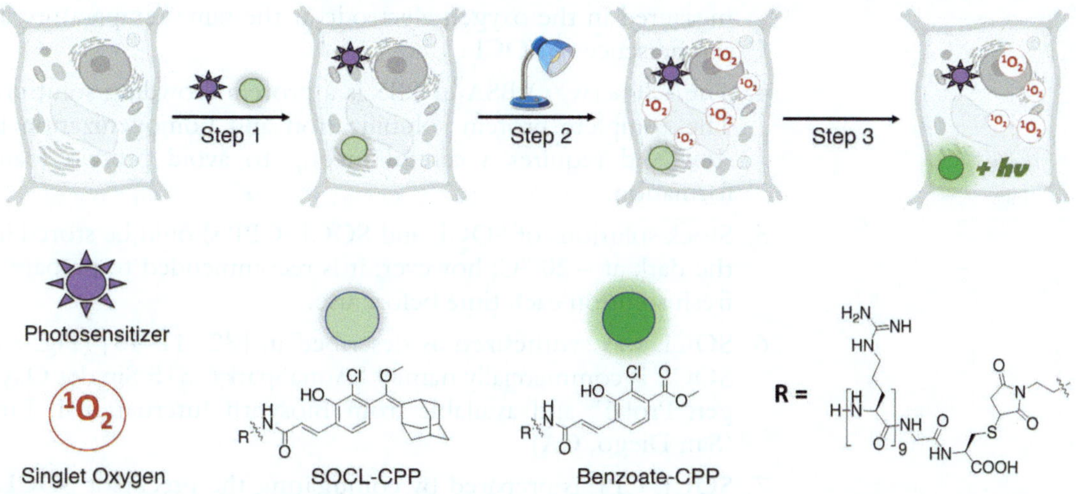

Fig. 4 General workflow for measuring the chemiluminescence generated by SOCL-CPP and $^1O_2$ in HeLa cells. **Step 1**—Treatment of the cells with SOCL-CPP and a photosensitizer. **Step 2**—Irradiation of the cells with a light source to allow generation of $^1O_2$ by the photosensitizer. **Step 3**—Collection of the chemiluminescence light emission that is generated by the reaction of SOCL-CPP with $^1O_2$

the temperature of the oxygen electrode chamber. Use the solution to wash the oxygen electrode between measurements.

2. MB is a water-soluble, $^1O_2$ photosensitizer with a relatively high quantum yield ($\Phi_\Delta \sim 0.52$) that can be substituted by other $^1O_2$ photosensitizers. $^1O_2$ photosensitization by thylakoid membranes or photosynthetic complexes such as PSII RC can be investigated with SOCL. If this is the case, use MB-free PBS and dissolve or disperse the photosynthetic complex preparations in the appropriate buffers and quantify their concentration by measuring Chl or protein in their stock solutions. Because the Φ_Δ of the photosynthetic/biological preparations can vary notably and, particularly, in some cases can be very small, the substitution of water-containing buffers by deuterium oxide–containing buffers is recommended if no significant changes in molecular oxygen consumption rate are observed in the oxygen electrode type Clark.

3. FFA selectively reacts with (photo)generated $^1O_2$ with a reaction efficiency of one and a bimolecular rate constant of $1.0 \times 10^8\ M^{-1}\ s^{-1}$ at 20 °C. The temperature-dependent function for the bimolecular rate constant for the reaction of FFA with $^1O_2$ was determined [40]. This temperature-dependent function can be used to indirectly estimate the value of the bimolecular rate constant for the quenching of $^1O_2$ by SOCL (or other $^1O_2$ quenchers) at an established temperature when the molecular oxygen consumption rate is

measured in the oxygen electrode at the same temperature in the presence of SOCL.

4. The ~30% (w/v) BSA in PBS is a protein-crowding solution. The complete protein solubilization and homogenization is slow and requires a careful mixing to avoid protein foam formation.
5. Stock solutions of SOCL and SOCL-CPP should be stored in the dark at −20 °C; however, it is recommended to prepare a fresh solution each time before use.
6. SOCL was synthesized as described in [32, 41–43] (Fig. 5). SOCL is commercially named "AquaSpark® 515 Singlet Oxygen Probe" and available from Biosynth International, Inc. (San Diego, CA).
7. SOCL-CPP is prepared by conjugating the precursor SOCL-maleimide and the ligand CPP-SH (Fig. 6). Importantly, CPP-SH is only an example of one potential possibility. Practically, any desired ligand can be installed as long as it has a thiol moiety. Therefore, SOCL-maleimide can potentially be used to generate a broad range of chemiluminescent $^1O_2$ probes with diverse biodistribution. As examples, other targeting peptides can be employed to target the chloroplast [44]. This will allow monitoring $^1O_2$ in this subcellular organelle.
8. Molecular oxygen consumption rates were monitored at different temperatures. The temperature of the water circulating through the concentric jacket was fixed using a programmable water bath. Temperature inside the oxygen electrode chamber could also easily be double-checked with a standard liquid-in-glass thermometer.
9. One-minute dark incubation in the oxygen electrode chamber is required to check, first, whether there are potential dark reactions in the reaction mixture consuming molecular oxygen and, second, to equilibrate the temperature of the reaction mixture with the bath-programmed temperature of water circulating through the concentric jacket of the oxygen electrode.
10. Illumination for 2 min when using SOCL in the reaction mixture increases the time-dependent concentration of SOCL phenoxy-dioxetane, product of the bimolecular reaction of SOCL with $^1O_2$.
11. Initial concentrations were chosen to ensure linear time-dependent molecular oxygen consumption rates (i.e. zero-order rates) for FFA and SOCL in the oxygen electrode experiments. This facilitates the analysis of the bimolecular rate constant for the reaction of SOCL with $^1O_2$ when the molecular oxygen consumption rates for FFA and SOCL are compared under identical experimental conditions.

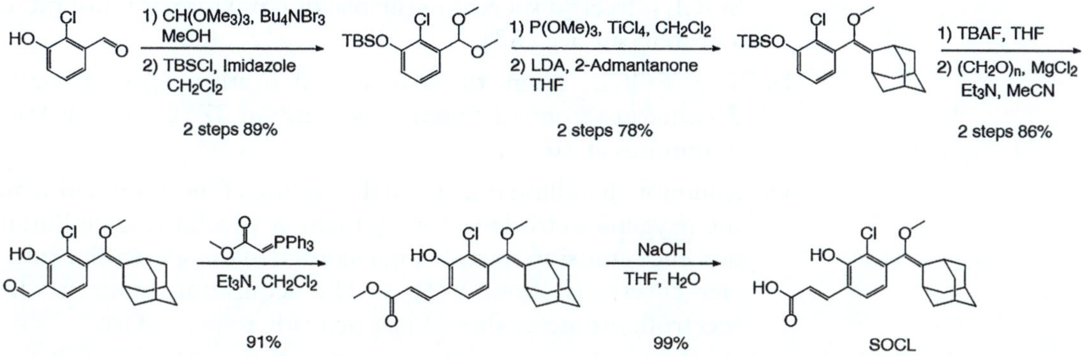

Fig. 5 Synthetic strategy and step reaction yields for SOCL production. Abbreviations for the chemicals used in the reaction steps stand for as follows: *$CH(OMe_3)_3$* trimethyl orthoformate, *Bu_4NBr_3* tetra-n-butylammonium tribromide, *TBSCl* *tert*-butyldimethylsilyl chloride, *$P(OMe)_3$* trimethyl phosphite, *LDA* lithium diisopropylamide, *THF* tetrahydrofuran, *TBAF* tetrabutylammonium fluoride, *$(CH_2O)n$* paraformaldehyde, *Et_3N* triethylamine, *MeCN* acetonitrile, *$Ph3P{=}CHCO_2CH_3$* methyl (triphenylphosphoranylidene)acetate

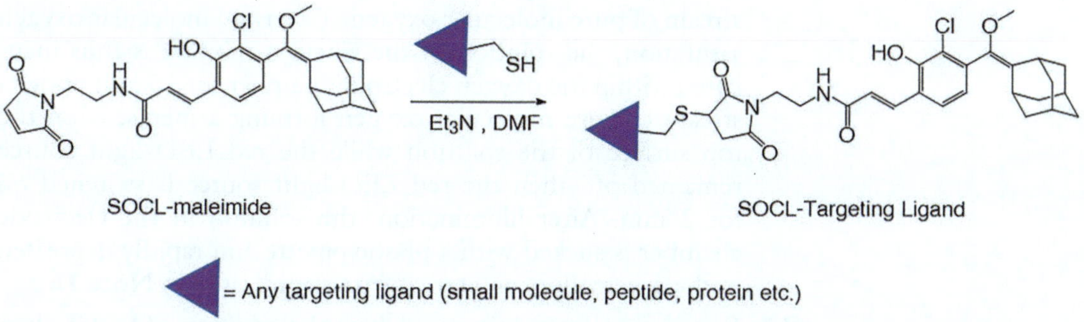

Fig. 6 General synthetic route for the preparation of SOCL conjugates with targeting ligand for subcellular sensing of $^1O_2$ [32]. As example, the purple triangle can be the guiding peptide sequence MASSMLS-SATMVGGC of the small subunit of rubisco that was used for nanoparticle targeting and translocation across the chloroplast envelope [44]

12. BSA is a rather inexpensive protein easily found in research laboratories and was used in this study as a representative water-soluble protein that could be found in a protein-crowding system. It can be replaced by photosynthetic complex preparations (thylakoids, PSII or PSII RC) if endogenous $^1O_2$ production is under investigation.

13. When combining two chemicals such as SOCL and BSA with high avidity for $^1O_2$, the overall molecular oxygen consumption rate does not necessarily correspond with the sum of the individual molecular oxygen consumption rates for each chemical while keeping the rest of the experimental conditions identical. The estimated bimolecular reaction rate constant between SOCL and $^1O_2$ might change depending on whether

SOCL is freely dispersed in the reaction mixture or is bound to a protein surface [33].

14. The SOCL phenoxy-dioxetane derivative spontaneously decomposes from a few tens of seconds at 37 °C to a few tens of minutes at 10 °C.
15. Minimum handling time from the ending of the illumination in the oxygen electrode to the moment at which the chemiluminescence emission starts being recorded in the spectrofluorometer cuvette is about 8–10 s. The acquisition mode of the spectrofluorometer should just be ready to press start.
16. Optimal accumulation of the SOCL phenoxy dioxetane derivative in the oxygen electrode chamber can be reached if illumination through the optical ports with the red LED light source is maintained for 2 min in air-saturated solutions at the temperature under study. The accumulation of the SOCL phenoxy dioxetane can be improved and even double if the dissolved air in PBS containing 2 μM MB and SOCL is displaced with a stream of pure molecular oxygen. To ensure molecular oxygen saturation, the solution of the reaction mixture is thus incubated within the oxygen electrode chamber for 15 min under a stream of pure molecular oxygen forming a meniscus on the top surface of the solution while the red LED light source remained off, then the red LED light source is switched on for 2 min. After illumination, the solution in the electrode chamber is sucked with a plastic pipette and rapidly deposited in the spectrofluorometer cuvette as indicated in **Note 15**.
17. Experimental conditions in which the addition of SOCL does not significantly modify the molecular oxygen consumption rate or even diminishes it do not exclude the formation of the SOCL phenoxy-dioxetane derivative. To unambiguously establish whether SOCL can compete for and react with $^1O_2$ in a protein-crowding system, it is necessary to record the traces of the chemiluminescence emission at 515 nm.
18. Changes in the chemiluminescence lifetime of the SOCL phenoxy-dioxetane derivative in protein-crowding systems might be associated with changes in its conformational freedom after its binding to protein pockets. An increase in the molecular rigidification of the SOCL phenoxy-dioxetane derivative can be a physicochemical property of interest to enhance the sensitivity of $^1O_2$ detection in cellular compartments of living cells.
19. mTHPC is a well-investigated photosensitizer that has been used vastly for photodynamic therapy [45].
20. Both SOCL-CPP and the product generated by its reaction with $^1O_2$ (phenoxy-benzoate) exhibit intrinsic green fluorescence. Therefore, it is possible to visualize them using

fluorescent confocal microscopy. For such experiments, incubate the cells with 10 μM SOCL-CPP for 1 h. Wash the cells three times with the growth medium, fixed with 4% paraformaldehyde for 30 min at room temperature, and wash again with PBS. Mount the cover glasses on microscope glass slides using ProLong™ Gold Antifade Mountant (Life technologies, USA) and monitor the cellular uptake of the probes utilizing Leica TCS SP8 confocal imaging systems with the 60× oil objective (Leica Microsystems, Wetzlar Germany).

Acknowledgements

The research was funded by MCIN/AEI/10.13039/501100011033 (Project n° PID2019-107154RB-100) and the regional government of Castilla y León (Project n° CSI260P20). The Project 'CLU-2019–05—IRNASA/CSIC Unit of Excellence' funded by the Junta de Castilla y León and co-financed by the European Union (ERDF 'Europe Drives Our Growth') and the CSIC Interdisciplinary Thematic Platform (PTI) Optimization of Agricultural and Forestry Systems (PTI-AGROFOR) are also acknowledged.

References

1. Arellano JB, Naqvi KR (2016) Endogenous singlet oxygen photosensitizers in plants. In: Nonell S, Flors C (eds) Singlet oxygen: applications in biosciences and nanosciences, 1st edn. Royal Society of Chemistry, London, pp 239–269
2. Fischer BB, Hideg E, Krieger-Liszkay A (2013) Production, detection, and signaling of singlet oxygen in photosynthetic organisms. Antioxid Redox Signal 18:2145–2162. https://doi.org/10.1089/ars.2012.5124
3. Chen T, Fluhr R (2018) Singlet oxygen plays an essential role in the root's response to osmotic stress. Plant Physiol 177:1717–1727. https://doi.org/10.1104/PP.18.00634
4. Khorobrykh S, Havurinne V, Mattila H, Tyystjärvi E (2020) Oxygen and ROS in photosynthesis. Plants 9:91. https://doi.org/10.3390/plants9010091
5. Pattanayak GK, Venkataramani S, Hortensteiner S et al (2012) ACCELERATED CELL DEATH 2 suppresses mitochondrial oxidative bursts and modulates cell death in Arabidopsis. Plant J 69:589–600. https://doi.org/10.1111/j.1365-313X.2011.04814.x
6. Triantaphylides C, Krischke M, Hoeberichts FA et al (2008) Singlet oxygen is the major reactive oxygen species involved in photooxidative damage to plants. Plant Physiol 148: 960–968. https://doi.org/10.1104/pp.108.125690
7. Triantaphylides C, Havaux M (2009) Singlet oxygen in plants: production, detoxification and signaling. Trends Plant Sci 14:219–228. https://doi.org/10.1016/j.tplants.2009.01.008
8. González-Pérez S, Gutiérrez J, García-García F et al (2011) Early transcriptional defense responses in Arabidopsis cell suspension culture under high-light conditions. Plant Physiol 156:1439–1456. https://doi.org/10.1104/pp.111.177766
9. Lemke MD, Woodson JD (2022) Targeted for destruction: degradation of singlet oxygen-damaged chloroplasts. Plant Signal Behav 17: e2084955. https://doi.org/10.1080/15592324.2022.2084955
10. Zhang S, Apel K, Kim C (2014) Singlet oxygen-mediated and EXECUTER-dependent signalling and acclimation of *Arabidopsis thaliana* exposed to light stress. Philos Trans R Soc Lond Ser B Biol Sci 369:20130227. https://doi.org/10.1098/rstb.2013.0227

11. Sánchez-Corrionero Á, Sánchez-Vicente I, González-Pérez S et al (2017) Singlet oxygen triggers chloroplast rupture and cell death in the zeaxanthin epoxidase defective mutant *aba1* of *Arabidopsis thaliana* under high light stress. J Plant Physiol 216:188–196. https://doi.org/10.1016/j.jplph.2017.06.009
12. Li M, Kim C (2023) Singlet oxygen in plants: from genesis to signaling. In: Van Breusegem F, Mittler R (eds) Advances in botanical research, 1st edn. Academic Press, Cambridge, MA, pp 1–42
13. Gútierrez J, Gónzalez-Pérez S, García-García F et al (2014) Programmed cell death activated by rose Bengal in *Arabidopsis thaliana* cell suspension cultures requires functional chloroplasts. J Exp Bot 65:3081–3095. https://doi.org/10.1093/jxb/eru151
14. Talaat NB (2019) Role of reactive oxygen species signaling in plant growth and development. In: Hasanuzzaman M, Fotopoulos V, Nahar K, Fujita M (eds) Reactive oxygen, nitrogen and sulfur species: production, metabolism, signaling and defense mechanisms, 1st edn. Wiley, Hoboken, pp 231–232
15. Li H, Melø TB, Arellano JB, Naqvi KR (2012) Temporal profile of the singlet oxygen emission endogenously produced by photosystem II reaction centre in an aqueous buffer. Photosynth Res 112:75–79. https://doi.org/10.1007/s11120-012-9739-4
16. Skovsen E, Snyder JW, Lambert JDC, Ogilby PR (2005) Lifetime and diffusion of singlet oxygen in a cell. J Phys Chem B 109:8570–8573. https://doi.org/10.1021/jp051163i
17. Ramel F, Birtic S, Cuine S et al (2012) Chemical quenching of singlet oxygen by carotenoids in plants. Plant Physiol 158:1267–1278. https://doi.org/10.1104/pp.111.182394
18. Dogra V, Singh RM, Li M et al (2022) EXECUTER2 modulates the EXECUTER1 signalosome through its singlet oxygen-dependent oxidation. Mol Plant 15:438–453. https://doi.org/10.1016/j.molp.2021.12.016
19. Danon A, Miersch O, Felix G et al (2005) Concurrent activation of cell death-regulating signaling pathways by singlet oxygen in *Arabidopsis thaliana*. Plant J 41:68–80. https://doi.org/10.1111/j.1365-313X.2004.02276.x
20. Kim C, Meskauskiene R, Zhang SR et al (2012) Chloroplasts of Arabidopsis are the source and a primary target of a plant-specific programmed cell death signaling pathway. Plant Cell 24: 3026–3039. https://doi.org/10.1105/tpc.112.100479
21. Mano J, Biswas MS, Sugimoto K (2019) Reactive carbonyl species: a missing link in ROS signaling. Plants 8:391. https://doi.org/10.3390/plants8100391
22. Ramel F, Birtic S, Ginies C et al (2012) Carotenoid oxidation products are stress signals that mediate gene responses to singlet oxygen in plants. Proc Natl Acad Sci USA 109:5535–5540. https://doi.org/10.1073/pnas.1115982109
23. Shumbe L, Chevalier A, Legeret B et al (2016) Singlet oxygen-induced cell death in Arabidopsis under high-light stress is controlled by OXI1 kinase. Plant Physiol 170:1757–1771. https://doi.org/10.1104/pp.15.01546
24. Havaux M (2020) β-Cyclocitral and derivatives: emerging molecular signals serving multiple biological functions. Plant Physiol Biochem 155:35–41. https://doi.org/10.1016/j.plaphy.2020.07.032
25. Woodson JD, Joens MS, Sinson AB et al (2015) Ubiquitin facilitates a quality-control pathway that removes damaged chloroplasts. Science 350:450–454. https://doi.org/10.1126/science.aac7444
26. Tano DW, Kozlowska MA, Easter RA, Woodson JD (2023) Multiple pathways mediate chloroplast singlet oxygen stress signaling. Plant Mol Biol 111:167–187. https://doi.org/10.1007/s11103-022-01319-z
27. Geng R, Pang X, Li X et al (2023) PROGRAMMED CELL DEATH8 interacts with tetrapyrrole biosynthesis enzymes and ClpC1 to maintain homeostasis of tetrapyrrole metabolites in Arabidopsis. New Phytol 238:2545–2560. https://doi.org/10.1111/NPH.18906
28. Nardi G, Manet I, Monti S et al (2014) Scope and limitations of the TEMPO/EPR method for singlet oxygen detection: the misleading role of electron transfer. Free Radic Biol Med 77:64–70. https://doi.org/10.1016/j.freeradbiomed.2014.08.020
29. Hideg E (2008) A comparative study of fluorescent singlet oxygen probes in plant leaves. Cent Eur J Biol 3:273–284. https://doi.org/10.2478/s11535-008-0018-5
30. Nagano T (2009) Bioimaging probes for reactive oxygen species and reactive nitrogen species. J Clin Biochem Nutr 45:111–124. https://doi.org/10.3164/jcbn.R09-66
31. Kim S, Tachikawa T, Fujitsuka M, Majima T (2014) Far-red fluorescence probe for monitoring singlet oxygen during photodynamic therapy. J Am Chem Soc 136:11707–11715. https://doi.org/10.1021/ja504279r
32. Hananya N, Green O, Blau R et al (2017) A highly efficient chemiluminescence probe for

the detection of singlet oxygen in living cells. Angew Chem 56:11793–11796. https://doi.org/10.1002/anie.201705803

33. Miranda-Apodaca J, Hananya N, Velázquez-Campoy A et al (2019) Emissive enhancement of the singlet oxygen chemiluminescence probe after binding to bovine serum albumin. Molecules 24:2422. https://doi.org/10.3390/molecules24132422
34. Hananya N, Shabat D (2019) Recent advances and challenges in luminescent imaging: bright outlook for chemiluminescence of dioxetanes in water. ACS Cent Sci 5:949–959. https://doi.org/10.1021/acscentsci.9b00372
35. Hananya N, Shabat D (2017) A glowing trajectory between bio- and chemiluminescence: from luciferin-based probes to triggerable dioxetanes. Angew Chem 56:16454–16463. https://doi.org/10.1002/anie.201706969
36. Copolovici DM, Langel K, Eriste E, Langel Ü (2014) Cell-penetrating peptides: design, synthesis, and applications. ACS Nano 8:1972–1994. https://doi.org/10.1021/nn4057269
37. Futaki S, Suzuki T, Ohashi W et al (2001) Arginine-rich peptides: an abundant source of membrane-permeable peptides having potential as carriers for intracellular protein delivery. J Biol Chem 276:5836–5840. https://doi.org/10.1074/JBC.M007540200
38. Numata K, Horii Y, Oikawa K et al (2018) Library screening of cell-penetrating peptide for BY-2 cells, leaves of Arabidopsis, tobacco, tomato, poplar, and rice callus. Sci Rep 8:1–17. https://doi.org/10.1038/s41598-018-29298-6
39. Wang JW, Goh N, Lien E, et al (2022) Quantification of cell penetrating peptide mediated delivery of proteins in plant leaves. bioRxiv 2022.05.03.490515. https://doi.org/10.1101/2022.05.03.490515
40. Appiani E, Ossola R, Latch DE et al (2017) Aqueous singlet oxygen reaction kinetics of furfuryl alcohol: effect of temperature, pH, and salt content. Environ Sci Process Impacts 19:507–516. https://doi.org/10.1039/c6em00646a
41. Hananya N, Boock AE, Bauer CR et al (2016) Remarkable enhancement of chemiluminescent signal by dioxetane-fluorophore conjugates: turn-on chemiluminescence probes with color modulation for sensing and imaging. J Am Chem Soc 138:13438–13446. https://doi.org/10.1021/jacs.6b09173
42. Bruemmer KJ, Green O, Su A et al (2018) Chemiluminescent probes for activity-based sensing of formaldehyde released from folate degradation in living mice. Angew Chem 130:7630–7634. https://doi.org/10.1002/ANGE.201802143
43. Son S, Won M, Green O et al (2019) Chemiluminescent probe for the in vitro and in vivo imaging of cancers over-expressing NQO1. Angew Chem Int Ed 58:1739–1743. https://doi.org/10.1002/ANIE.201813032
44. Santana I, Wu H, Hu P, Giraldo JP (2020) Targeted delivery of nanomaterials with chemical cargoes in plants enabled by a biorecognition motif. Nat Commun 11:1–12. https://doi.org/10.1038/s41467-020-15731-w
45. Senge MO (2012) MTHPC—a drug on its way from second to third generation photosensitizer? Photodiagnosis Photodyn Ther 9:170–179. https://doi.org/10.1016/J.PDPDT.2011.10.001

Chapter 4

Noninvasive *In Planta* Live Measurements of H_2O_2 and Glutathione Redox Potential with Fluorescent roGFPs-Based Sensors

Stefano Buratti, Matteo Grenzi, Giorgia Tortora, Sara Paola Nastasi, Elisa Dell'Aglio, Andrea Bassi, and Alex Costa

Abstract

In this protocol, we present a noninvasive *in planta* bioimaging technique for the analysis of hydrogen peroxide (H_2O_2) and glutathione redox potential in adult *Arabidopsis thaliana* plants. The technique is based on the use of stereo fluorescence microscopy to image *A. thaliana* plants expressing the two genetically encoded fluorescent sensors roGFP2-Orp1 and Grx1-roGFP2. We provide a detailed step-by-step protocol for performing low magnification imaging with mature plants grown in soil or hydroponic systems. This protocol aims to serve the scientific community by providing an accessible approach to noninvasive *in planta* bioimaging and data analysis.

Key words Noninvasive bioimaging, Fluorescent biosensors, Hydrogen peroxide, Glutathione redox status, Image analysis, Data analysis pipeline, Salt stress

1 Introduction

Plants, being sessile organisms, have evolved extensive mechanisms for acclimation and adaptation. Whereas the perception of external stimuli is mainly mediated by a battery of receptors, downstream, different second messengers are involved in the transduction of the signal and its amplification [1]. In plants, among second messengers, the role of Ca^{2+} and reactive oxygen species (ROS) has been largely established [2]. In particular, ROS were initially considered only as by-products of aerobic metabolism but their role in signaling is now recognized [3]. Among ROS, hydrogen peroxide (H_2O_2), due to its chemical properties such as its relatively mild

The original version of this chapter was previously published without open access. A correction to this chapter is available at https://doi.org/10.1007/978-1-0716-3826-2_19

Francisco J. Corpas and José M. Palma (eds.), *ROS Signaling in Plants: Methods and Protocols*, Methods in Molecular Biology, vol. 2798, https://doi.org/10.1007/978-1-0716-3826-2_4,

reactivity with a relatively long lifetime (from ms to s), plays a prominent role in signaling [3–5]. The recent discovery of the leucine-rich-repeat receptor kinase, hydrogen-peroxide-induced Ca^{2+} increases (HPCA1), and the cytosolic thiol peroxidase PRXIIB as respectively extracellular and intracellular H_2O_2 receptors has further certified the role of this molecule in both extracellular and intracellular signaling in plants [6, 7].

The photosynthetic process and photorespiration are the main sources of ROS and H_2O_2 in physiological conditions [3, 8], and a battery of ROS scavengers keep these molecules under a strict control [5]. However, when plants are challenged with both biotic and abiotic stimuli, ROS generators like the plasma membrane NADPH oxidases of the RBOH family are activated and a ROS burst can occur [9]. To cope with such a ROS burst, plants use ROS scavengers whose activity relies on the availability of reducing power. The plant antioxidant machinery includes enzymatic components such as superoxide dismutase (SOD), catalase (CAT), ascorbate peroxidase (APX), guaiacol peroxidase (GPX), glutathione reductase (GR), monodehydroascorbate reductase (MDHAR), dehydroascorbate reductase (DHAR), and nonenzymatic antioxidants like ascorbic acid (AA), reduced glutathione (GSH), α tocopherol, carotenoids, flavonoids, and the osmolyte proline [5]. Many antioxidant compounds are simultaneously present in oxidized and reduced states in the form of active redox couples [5]. One of the main redox couples of a plant cell is the oxidized/reduced glutathione (GSSG/2GSH), which is kept reduced by the electrons donated by Nicotinamide adenine dinucleotide phosphate (NADPH) and in turn, for example, can reduce the dehydroascorbate to ascorbate (DHA/ASC) [5]. NADPH is generated mainly via photosynthesis and by different types of NADP-dehydrogenases such as NADP-glyceraldehyde-3-phosphate dehydrogenase, NADP-isocitrate dehydrogenase, NADP malic enzyme, and the enzymes of the oxidative part of the pentose phosphate pathway, glucose-6-phosphate dehydrogenase, and 6-phosphogluconate dehydrogenase (reviewed in [10]), When plants are challenged with stress, ROS are thus efficiently scavenged by the antioxidant batteries that will be consequently oxidized. However, if the antioxidant batteries are not sufficient to scavenge entirely the produced ROS their cellular level can increase.

Based on these premises it becomes evident that measuring ROS, and specifically H_2O_2, as well as the effects on the redox status of the redox couples are important parameters to gather information about the physiology of the plants in control and stress conditions. Traditionally, biochemical analytical analyses have provided tools to quantify ROS (including H_2O_2) production, redox status of different redox couples, and their total amount [11]. However, these techniques are destructive, require tissue manipulation and homogenization, and thus, possibly also the introduction of artifacts. Moreover, these techniques hardly provide cellular and subcellular resolution [8, 11]. In the last 15 years

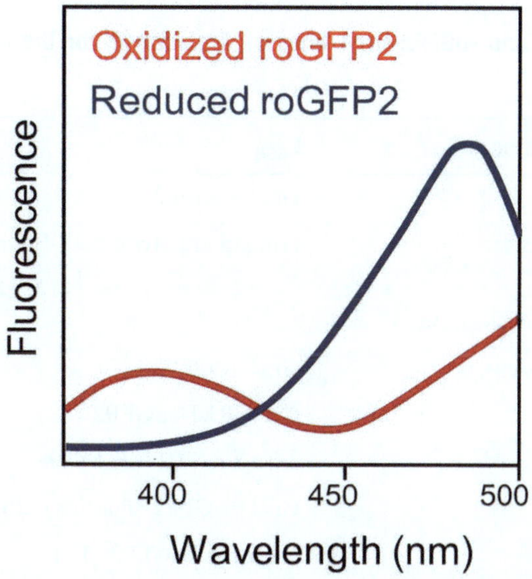

Fig. 1 The spectral properties of purified roGFP2-Orp1. The figure is based on original traces reported in [25], showing the excitation spectra of a purified roGFP2-Orp1 in its fully oxidized (red trace) and fully reduced (dark blue). The isosbestic point is at 422 nm and represents the redox insensitive point [24]

in vivo imaging based on the use of plants expressing genetically encoded ROS and redox sensors has been established as a powerful and noninvasive technique [11–13]. In particular, redox green fluorescent protein (roGFP)-based sensors have been designed with the intention of measuring both the redox potential of glutathione pool and the H_2O_2 synthesis [12, 14, 15]. Basically, these sensors are based on a modified GFP, where two cysteine residues have been inserted in adjacent β-strands on the surface of the protein β-barrel [11, 16, 17]. When oxidized, these two cysteines form a disulfide bond that induces a structural conformational change that influences the chromophore fluorescence and affects the quantum yields (QY) of the two main absorption peaks (at 405 and 488 nm, respectively) in an opposite way, leading to a ratiometric response [12]. Importantly, the emission spectrum of the sensor remains the same (Fig. 1).

In the cellular environment, roGFPs are in a redox potential equilibrium with the GSSG/2GSH couple, in a reaction catalyzed by glutaredoxin (Grx) (reviewed in [11]). Both the GSH pool and Grx are present in the cytosol and subcellular compartments [11]; however, since in vivo, the roGFP equilibrium with the GSSG/2GSH couple is mediated by Grx, to overcome any possible limitation of Grx availability, the human Grx1 was fused to the roGFP2, making a 1:1 proportion between the roGFP2 sensor and the catalyst [14]. *Arabidopsis thaliana* (hereinafter *Arabidopsis*) plants expressing the Grx1-roGFP2 targeted at the cytosol and different subcellular compartments are available (Table 1) [18–22]. In

Table 1
Grx1-roGFP2 and roGFP2-Orp1 lines of *Arabidopsis* for the analysis of the glutathione redox status and H_2O_2

Available sensors	E_{GSH}	H_2O_2
Sensor	Grx1-roGFP2	roGFP2-Orp1
Operator/sensor	Human glutaredoxin 1 (hGrx1)	Yeast glutathione peroxidases like 3 (Gpx3/Orp1)
Subcellular localization		
Cytosol/nucleus	Grx1-roGFP2 (17)	roGFP2-Orp1 (14)
Cytosol	Cyt-GRX1-roGFP2 (18)	roGFP2-Orp1 (Cyt) (21)
Mitochondria	mito-roGFP2-Grx1 (19)	mt-roGFP2-Orp1 (14)
Plastids	Grx1-roGFP2 (Plastids) (20)	roGFP2-Orp1 (plastids) (20)
Peroxisomes	per-Grx1-roGFP2 (18)	roGFP2-Orp1 (Per) (21)
Nucleus		roGFP2-Orp1 (Nuc) (21)
Apoplast		roGFP2-Orp1 (Apo) (21)
Optical characteristics	*Grx1-roGFP2*	*roGFP2-Orp1*
Fluorescent protein	EGFP	EGFP
Excitation maximum	395 nm, 490 nm	395 nm, 490 nm
Suggested excitation wavelength	405 nm, 488 nm	405 nm, 488 nm
Excitation setup	400 nm, 460 nm	400 nm, 460 nm
Emission maximum	511 nm	511 nm
Emission filter	525/36 nm	525/36 nm
Ratio calculation	$400_{exitation}/460_{excitation}$	$400_{exitation}/460_{excitation}$
In vitro dynamic range	12 (390 nm/480 nm)	8 (400 nm/482 nm)
Plant dynamic range	5 (400 nm/ 460 nm)	6 (400 nm/460 nm)

conclusion, the Grx1-roGFP2 is a sensor for the in vivo determination of the redox potential of the glutathione pool.

It has been shown that the Yeast Orp1 peroxidase has an intrinsic capacity to act as a specific H_2O_2-dependent protein thiol oxidase when it is recruited into proximity of oxidizable target proteins [23]. By leveraging this property, Gutscher et al. developed the roGFP2-Orp1 where the roGFP2 is subjected to an H_2O_2-dependent oxidation mediated by the Orp1. In this way, the roGFP2 becomes oxidized (formation of the disulfide bridge) specifically in response to H_2O_2 and no other oxidants. Thus, roGFP2-Orp1 is a bona fide H_2O_2 sensor. However, a key aspect that needs to be considered is that the roGFP2-Orp1 oxidized form is reversibly reduced in vivo, most probably by the Grx and/or the thioredoxins (Trx) systems that directly reduce Orp1 [15]. Because roGFP2-Orp1 undergoes a specific H_2O_2-dependent oxidation, but can also be reduced, its redox state is influenced either by the level of the oxidant (H_2O_2) or by the reductants themselves, such as GSH and thioredoxin [15]. This makes the use of this sensor not completely straightforward for the definition of the H_2O_2 levels [11]. Nonetheless, roGFP2-Orp1 has been successfully expressed in *Arabidopsis* plants, allowing to monitor the H_2O_2 dynamics in different cellular compartments (Table 1) [15, 21, 22]. In conclusion, the roGFP2-Orp1 is a sensor for the in vivo determination of the H_2O_2-dependent oxidation of the roGFP2 sensor.

So far, imaging of *Arabidopsis* plants expressing either Grx1-roGFP2 or roGFP2-Orp1 has been mainly carried out using young seedlings grown in plates or leaf sections from mature soil-grown plants imaged by wide-field and/or confocal fluorescence microscopy as well as by the plate reader. Detailed protocols describing these procedures are available [24, 25]. Remarkably, such microscopy approaches require a certain grade of sample manipulation, provide only a view of a few numbers of cells, and do not provide information at the organismic level. However, Hipsch and colleagues have imaged entire *Arabidopsis* and *Solanum tuberosum* (potato) plants expressing the roGFP2 localized into chloroplasts in physiological conditions using an Advanced Molecular Imager HT (Spectral Ami-HT, Spectral Instruments Imaging, LLC., USA) [26–28].

Here, we present a noninvasive methodological approach based on the use of a stereo fluorescence microscope to study cytosolic glutathione redox potential and H_2O_2 level in adult *Arabidopsis* plant leaves expressing Grx1-roGFP2 and roGFP2-Orp1 sensors. We provide a guideline to perform experiments with plants grown in standard conditions in soil and hydroponics and in response to salt stress treatment. The principles of the approach described here can be generalized to other fluorescent protein sensors such as for pH and Ca^{2+}.

2 Materials

2.1 Sensors, Plant Lines, and Growth Material

1. For the Grx1-roGFP2 and roGFP2-Orp1 sensors, a comprehensive in vitro characterization is available, including a reliable account of their respective spectroscopic characteristics, specificities, and sensitivity ranges [12, 15]. Different constructs for *Agrobacterium*-based plant transformation for stable transgenic lines in *Arabidopsis* Col-0 background are available (*see* **Note 1**) (Table 1 adapted from [25]). Barley seeds expressing the Grx1-roGFP2 have also been obtained [29].

2. Mature *Arabidopsis* Col-0 plants (3–4 weeks) expressing the Grx1-roGFP2 and roGFP2-Orp1 are used for treatments.
3. Plant growth chamber with temperature, lights, and humidity control.
4. MS/2 plates (0.5× Murashige and Skoog medium), pH 5.8 (adjusted with KOH), solidified with 0.8% plant agar.
5. 1 M hydrogen peroxide (H_2O_2) stock solution in deionized water.
6. 1 M dithiothreitol (DTT) stock solution in deionized water.

2.2 Hydroponic Cultures

1. Seed-holders (Araponics, https://www.araponics.com/) mounted on tip boxes.
2. Rockwool.
3. Composition of the hydroponic solution: 1.5 mM $Ca(NO_3)_2$, 1.5 mM KNO_3, 1.5 mM $Mg(SO_4)$, 1.5 mM KH_2PO_4, 1.5 mM $(NH_4)_2SO_4$, 0.1 mM $C_{10}H_{12}FeN_2NaO_8$, 0.15 mM $Na_2O_3Si \times 9H_2O$, 0.05 mM KCl, 0.01 mM $MnSO_4$, 0.0015 mM $CuSO_4$, 0.002 mM $ZnSO_4 \times 7H_2O$, 0.05 mM H_3BO_3, 75 nM $(NH_4)_6Mo_7O_{24}$.
4. As a final concentration, 100 mM NaCl was added to the hydroponic solution.

2.3 Microscopy

1. A stereo fluorescence microscope fits the requirement to carry out the measurements described here. The method described is for Nikon SMZ18 (Nikon Instruments Europe B.V, https://www.microscope.healthcare.nikon.com/en_EU/products/stereomicroscopes-macroscopes/smz25-smz18) (Fig. 2d), with motorized Zoom and filter wheel, but it can be applied to other setups. The microscope was equipped with standard GFP-B/GFP-L/RFP filters and with a custom-built set for roGFP2 ratiometric imaging (Excitation filter 405/20×, dichroic mirror T505lpxr and Emission filter 525/36 nm) (*see* **Note 2**).
2. Excitation light for the sensor is produced by a pE-300ultra CoolLED (CoolLED, https://www.coolled.com/products/pe-300ultra/) with a 400, 460, and 550 nm wavebands configuration to match filter sets within the microscope. The roGFP2 sensors are excited sequentially with two different wavelengths and specifically switching between 460 nm and 400 nm by motorized filter wheel (*see* **Notes 3** and **4**).
3. Images are collected with a Hamamatsu ORCA-Flash4.0 LT Digital CMOS camera (Hamamtsu, https://www.hamamatsu.com/eu/en/product/cameras/cmos-cameras/C11440-42U40.html) (*see* **Note 5**).

Fig. 2 Experimental setup for in vivo mesoscopic imaging. (**a**, **b**) Plant-holders for plant growth in hydroponic. (**c**) 3D-printed hydroponic system. (**d**) Nikon SMZ18 fluorescence stereo microscope. (**e**) Whole plant imaging for unperturbed in vivo acquisition

4. For roGFP2s analysis, images are collected using a 505/530 nm bandpass filter (GFP) at both excitation wavelengths (460 and 400 nm). The two collected images used for ratio calculation are named 460_{em} (emission 1, 505/530 nm for roGFP2 excited with 460 nm) and 400_{em} (emission 2, 525/36 nm for roGFP2 excited with 400 nm) (*see* **Note 6**).
5. For the largest image acquisition, a 0.5× objective numerical aperture (NA) 0.075 or equivalent is used and zooming can be used (from 0.75 to 13.5×) (*see* **Note 7**).
6. For higher magnifications a 1× objective NA 0.15 can be selected.
7. The NIS-Elements (Nikon Instruments Europe B.V) is used as a platform to control the microscope, illuminator, and camera, and for post-acquisition analyses (*see* **Note 8**).

8. Fiji (https://fiji.sc/) image processing package is installed on PC or Mac (*see* **Note 9**).
9. Ratiometric image analysis is critical to extract qualitative and quantitative data. We use the NIS-Elements (Nikon Instruments Europe B.V), the Fiji software for pixel intensity quantification of 400_{em} and 460_{em} signals in the acquired images (*see* Subheading 3.5).

3 Methods

A procedure is presented here to quantify the glutathione redox potential and the H_2O_2 levels, based on measurements of the $400_{em}/460_{em}$ ratio for Grx1-roGFP2 and roGFP2-Orp1 in non-detached rosette leaves of adult *Arabidopsis* plants in three conditions: soil grown plants, plants grown hydroponically in standard conditions, and plants grown hydroponically in the presence of 100 mM NaCl.

The aim is to illustrate the main steps and considerations for routine measurements.

3.1 Plant Growth and Sample Treatment

1. For soil grown plants, *Arabidopsis* sterile seeds are germinated in MS/2 medium, checked for the presence of fluorescence by using the GFP filter set, and then transferred to pots filled with soil. For homozygous lines, seeds can be directly sown in soil.
2. For hydroponically grown plants, *Arabidopsis* sterile seeds are germinated in MS/2 medium, checked for the presence of fluorescence by using the GFP filter set, and then transferred on Araponics seed-holders filled with Rockwool, and placed in the pipette tip rack located in the tip box filled with the hydroponic solution (Fig. 2a) (*see* **Note 10**).
3. Pots and boxes with seeds are transferred to the growth chamber under long-day conditions (16 h light/8 h dark cycles, 100 Me m^{-2} s^{-1} of cool white neon lamps) at 22 °C and 75% relative humidity and imaged after 3 weeks of growth (*see* **Note 11**). Long-day-grown plants are well-suited to whole-plant imaging due to the relatively small size of the rosette (Fig. 2b, c).
4. Plants grown on soil are watered every 2 days, whereas for hydroponically grown plants solution is replaced weekly (*see* **Note 12**).
5. Live imaging of plants in physiological conditions would require minimal perturbation to the tissues during measurement and performing *in planta* imaging with a stereo microscope (Fig. 2d, e) guarantees to avoid mechanical stress helping to reduce sample manipulation artifacts (*see* **Note 13**).

6. For salt stress treatment experiment, the hydroponic standard solution is replaced with the same solution supplemented with 100 mM NaCl. Plants are then kept for up to 96 h with or without NaCl in the same growth conditions and images are acquired every 24 h.

3.2 Plants Handling

1. Soil- or hydroponically grown plants at the desired time (e.g., 72 h posttreatment with salt) are removed from the growth chamber and directly placed under the stereo macro lens (Fig. 2e).
2. For large images, a 0.5× objective is chosen which allows the full coverage of the plant. A higher magnification of 1× can be also used, but this enables to capture the image of only single leaves.
3. The sensor expression is checked using fluorescent light in combination with 460 nm excitation and GFP filter.
4. The first specimen is used to set up the microscope and to optimize the instrument settings (zooming, focus, and exposure time, *see* Subheading 3.3). Data collection then begins with a separate set of individuals.

3.3 Adjustments for Sensor Imaging

1. Exposure time is routinely settled to 100–1000 ms with 4 × 4 Camera binning for each fluorescence emission (*see* **Notes 14** and **15**). The exposure time is then chosen based on the expression level of the sensor and the zoom factor. With different excitations, different exposure times can be settled, but the same camera binning must be used.
2. Based on our experience the measured pixel intensities expressed as Fluorescence Arbitrary Units (FAU) should provide at least a value of 3000 (considering the maximum of 65,000 grey levels for a 16-bit camera) for the 400_{em} of both Grx1-roGFP2 and roGFP2-Orp1. Keeping exposure time as short as possible avoids sensor bleaching and minimizes photo-oxidative stress of plants (*see* **Note 16**).
3. Pixel saturation (pixel intensities close to 65,000 FAU) must be avoided as this prevents a correct analysis. This latter case can potentially occur for roGFP2 emissions using higher magnification and zooming. With the 0.5× objective, the amount of emitted light is generally in the linear range of the CMOS sensor.
4. Once set, the settings must be maintained strictly constant for a given experiment (the different tested treatments). Any change of settings will directly impact the results rendering the measured ratio meaningless.

5. Images are acquired every 5 s for a total of 90 s (*see* **Notes 17** and **18**).
6. The experiment is stored in a dedicated folder using the default software extension (*.nd2 with the NIS software).

3.4 Evaluation of the Proper Sensors' Functionality

1. Before performing any experiment, it is important to evaluate the proper ratiometric response of the sensors. This is necessary to check that the hardware is properly configured.
2. Place an adult plant under the stereo microscope and image a single leaf (Fig. 3a).
3. Start performing the image acquisitions as reported in Subheading 3.3 for a maximum of 5–10 min.
4. With a micropipette directly applied to the leaf adaxial side 5 μL of H_2O_2 solutions at different concentrations (10 mM, 100 mM, 1 M) (Fig. 3a).
5. Collect the images for both fluorescence emissions for any given time and save the data.
6. Perform the image and data analysis as reported in Subheading 3.5.

3.5 Data Collection and Quantitative Image Analysis

1. The simplest type of experiment is to monitor for up to 90 s the steady state H_2O_2 and glutathione redox potential in *Arabidopsis* rosette leaves in standard growth conditions using the fixed acquisition settings. Typically, single images are collected for ≥3 individual plants *per* treatment and or conditions ($N = 5$ recommended, from independent plants).
2. Of the analyzed Regions of Interest (ROIs), 400_{em} and 460_{em} corresponding to the entire leaf area or entire rosette are used for the ratio (R) calculation ($400_{em}/460_{em}$) (yellow polygons in Figs. 3, 4, and 5). The 400_{em} and 460_{em} fluorescence emissions are background subtracted before ratios are calculated. The background is mainly based on CMOS noise and autofluorescence from soil or hydroponic support. Background fluorescence must be taken for an ROI chosen outside of the leaf but close to the plant (*see* **Note 19**) (red squares in Figs. 3, 4, and 5).
3. The procedure to perform the image analysis for the $400_{em}/460_{em}$ ratio calculation using the Open-Source Fiji software is briefly described. Open a given acquired experiment with its original file extension (e.g., *nd2 of NIS) with the Fiji platform. Fiji will automatically open two series of images, one for the 400_{em} and one for the 460_{em}. Click on the "Analyze" button and in the drop menu select "Tool," and then press the "ROI Manager" button. A new window will appear. By using the drawing tools of the main bar (square, circle, or

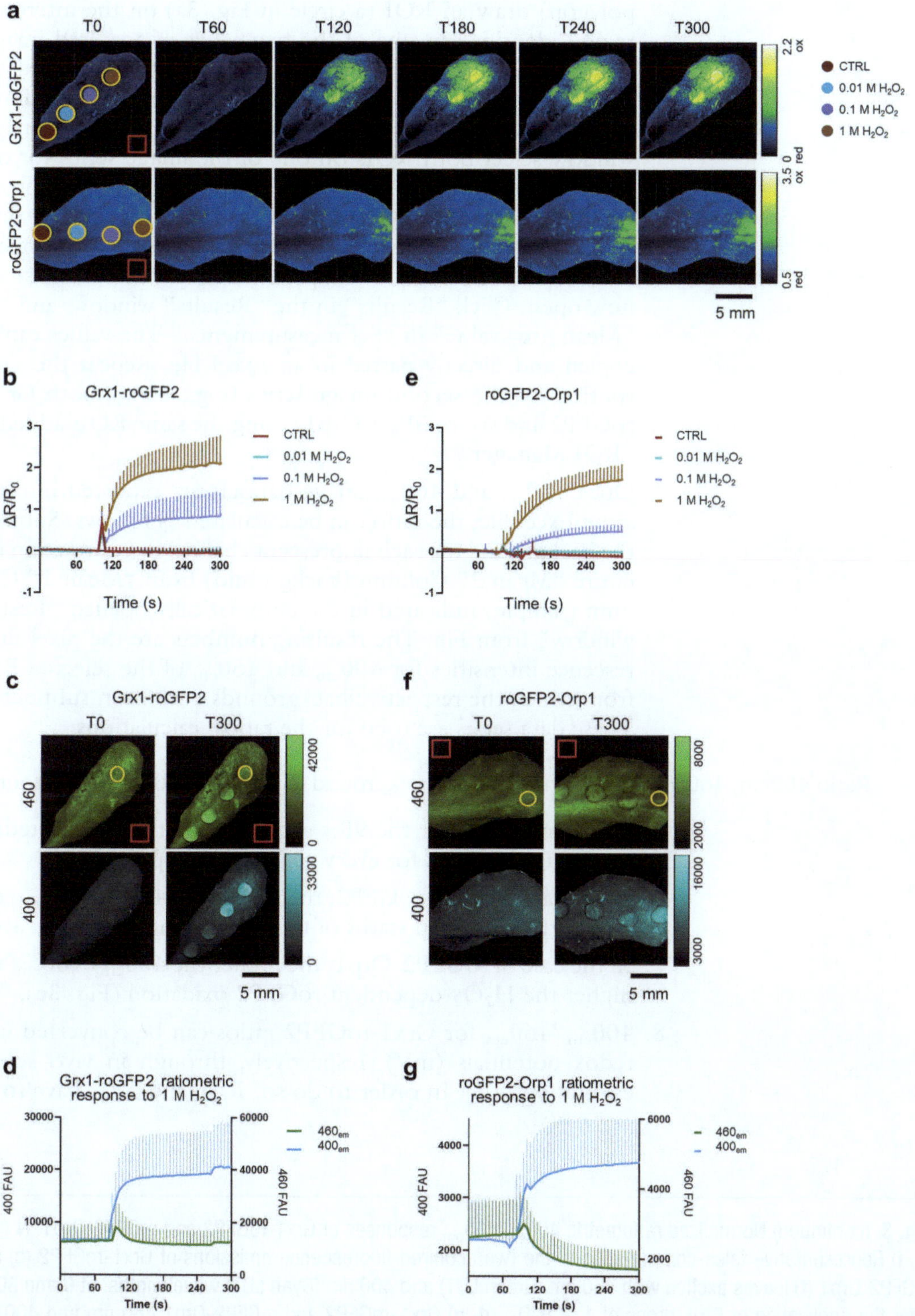

Fig. 3 In vivo H_2O_2-dependent oxidation of roGFP2-based sensors and confirmation of ratiometric response. (**a**) Representative $400_{em}/460_{em}$ ratio images [Green Fire Blue Look Up Table (LUT)] of Grx1-roGFP2 and roGFP2-Orp1 *Arabidopsis* leaves at 0, 60, 120, 180, 240, and 300 s after the start of the acquisition. Treatments (5 μL) from left to right are the following: CTRL, 0.01 M H_2O_2, 0.1 M H_2O_2, 1 M H_2O_2. (**b, e**)

polygon) draw an ROI (a circle in Fig. 3a) on the interested region identified in one of the two series of acquired images then on the ROI Manager window press "Add." Repeat the same operation for a background ROI which should be drawn in a region outside the leaf sample. In the ROI Manager window select both ROIs on one of the image series for one of the sensors (e.g., 460_{em} for both Grx1-roGFP2 and/or roGFP2-Orp1) and press the "More" button. In the drop menu select "Multi Measure." A new window will appear, tick the first two options and press "OK." A "Results" list will now open. Click "Results" in the "Results" window, and tick "Mean grey value" in "Set measurements." The values can be copied and directly pasted in an Excel file. Repeat the same routine for the second image series (e.g., 400_{em} both Grx1-roGFP2 and/or roGFP2-Orp1) using the same ROIs added to "ROI Manager."

4. Once 400_{em} and 460_{em} image datasets are reported into the same Excel file, the ratio can be calculated as follows. Subtract the background for each fluorescent channel by subtracting the entire "Mean 2" Column (background) from "Mean 1" Column (sample) indicated in the automatically opened "Results window" from Fiji. The resulting numbers are the pixel fluorescence intensities for 400_{em} and 460_{em} of the selected ROI from which the respective backgrounds have been subtracted. These data series are used for the ratios' calculations:

$$\text{Ratio } 400\text{em}/460\text{em} = (400\text{em} - 400\text{em background})/(460\text{em} - 460\text{em background})$$

5. An average ratio for the 90-s acquisition is then calculated by averaging the ratios for every acquired time point.
6. In the case of Grx1-roGFP2, the higher the $400_{em}/460_{em}$, the higher the oxidation status of the glutathione pool (Fig. 3b).
7. In the case of roGFP2-Orp1, the higher the $400_{em}/460_{em}$, the higher the H_2O_2-dependent roGFP2 oxidation (Fig. 3e).
8. $400_{em}/460_{em}$ for Grx1-roGFP2 ratios can be converted into redox potentials (mV), respectively, through in vivo sensor calibration [12]. In order to do so, R_{max} and R_{min} have to be

Fig. 3 (continued) Normalized ratiometric $400_{em}/460_{em}$ responses of Grx1-roGFP2 and roGFP2-Orp1. $N \geq 5$. (**c**, **f**) Representative false-color images of the two acquired fluorescence emissions of Grx1-roGFP2 (**c**) and roGFP2-Orp1 (**f**) leaves excited with 460 nm (Green LUT) and 400 nm (Cyan LUT) wavelengths, at 0 and 300 s after the application of 5 μL drops of 1 M H_2O_2. (**d**, **g**) Grx1-roGFP2 and roGFP2-Orp1 460 nm and 400 nm emissions changes after 1 M H_2O_2 treatment applied at 100 s. $N \geq 5$. Analyzed Regions of Interest (ROIs) are highlighted by yellow circles in (**c**) and (**f**), background ROIs are highlighted in red. Images were acquired using 0.5× objective lens and 1.5× zoom

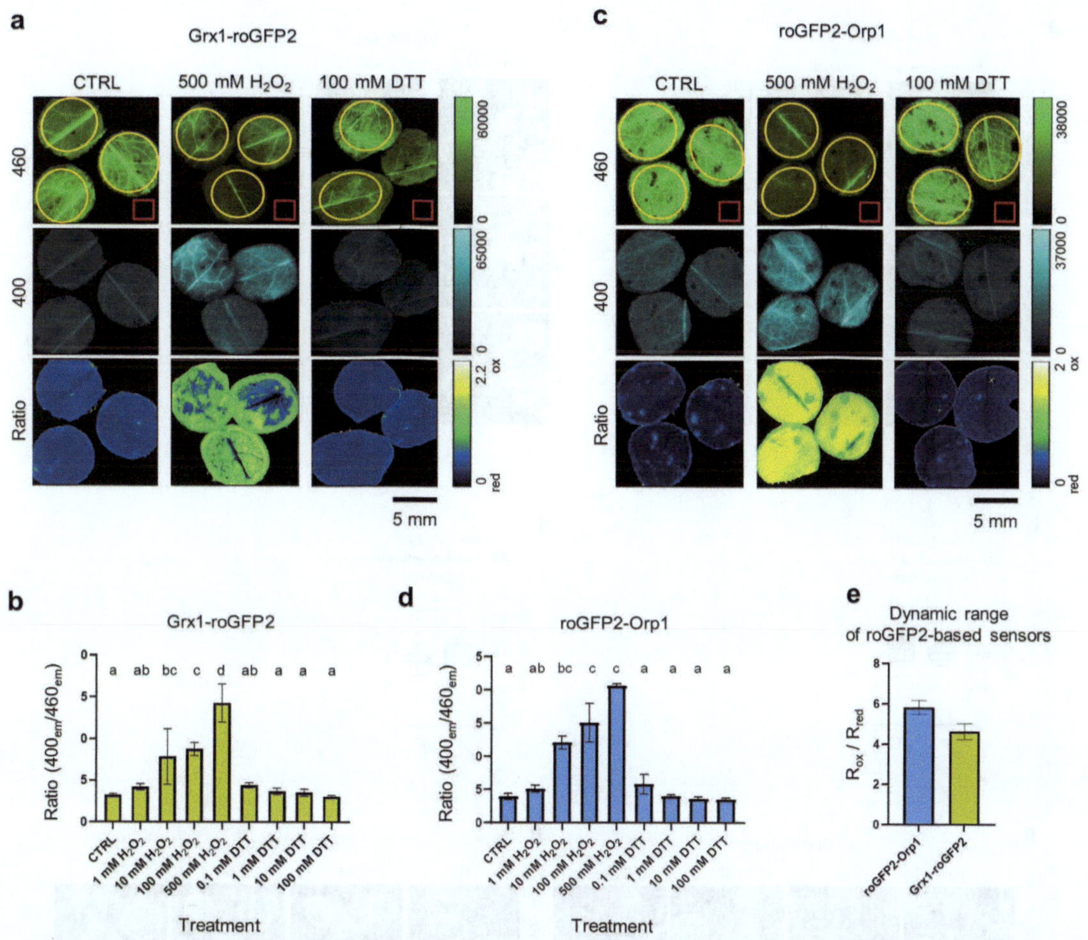

Fig. 4 Determination of in vivo dynamic range of roGFP2-based sensors. (**a**) Representative false-color images of Grx1-roGFP2 leaf disks treated with control (CTRL), 500 mM H_2O_2, and 100 mM DTT solutions for the two acquired fluorescence emissions excited with 460 nm (Green LUT) and 400 nm (Cyan LUT) excitations, and the resulting $400_{em}/460_{em}$ ratio (Green Fire Blue LUT). (**b**) $400_{em}/460_{em}$ ratios of Grx1-roGFP2 leaf disks incubated for 10 min in solutions with different concentrations of H_2O_2 and DTT. (**c**) Representative false-color images of roGFP2-Orp1 leaf disks treated with CTRL, 500 mM H_2O_2, and 100 mM DTT solutions for the two acquired fluorescence emissions excited with 460 nm (Green LUT) and 400 nm (Cyan LUT) excitations, and the resulting $400_{em}/460_{em}$ ratio (Green Fire Blue LUT). (**d**) $400_{em}/460_{em}$ ratios of roGFP2-Orp1 leaf disks incubated for 10 min in solutions with different concentrations of H_2O_2 and DTT. (**e**) In vivo dynamic ranges of roGFP2-based sensors calculated as the ratio between maximum oxidation, obtained by treating the leaf disks with 500 mM H_2O_2, and maximum reduction, obtained by treating the leaf disks with 100 mM DTT. Analyzed regions of interest (ROIs) are highlighted in yellow; background ROIs are highlighted in red. Images were acquired using 0.5× objective lens and 2× zoom. Data are plotted as column bar graphs. Bars represent means and the whiskers indicate the standard deviation. One-way ANOVA with post hoc Tukey's test. $N = 3$. *CTRL* control

measured by treating entire leaves with a concentration of H_2O_2 sufficiently high to fully oxidize the sensor (up to 500 mM) and DTT to fully reduce the sensor (up to 100 mM) (Fig. 4a, c). To improve the efficacy of the treatment

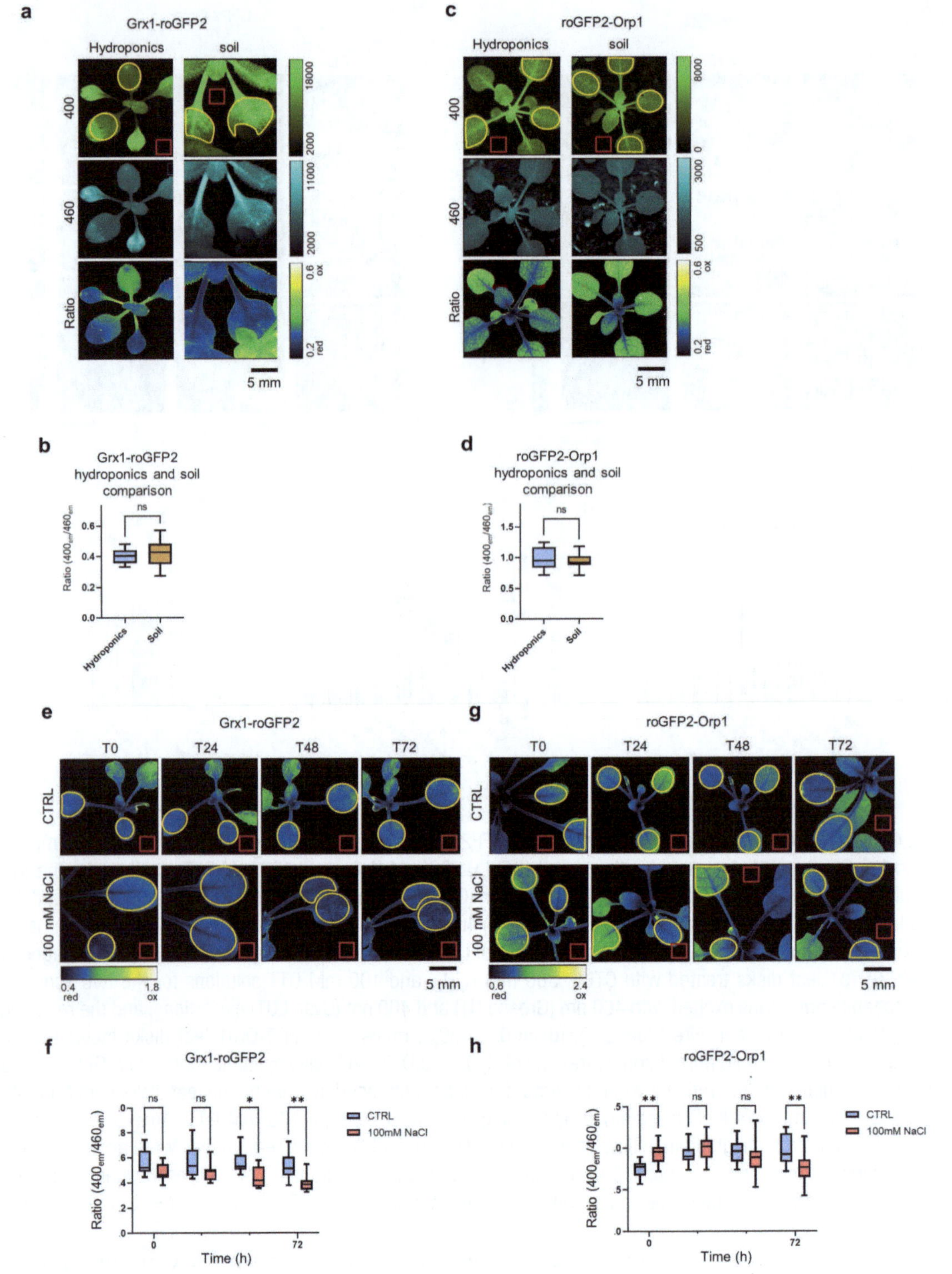

Fig. 5 Case studies: Redox status of Grx1-roGFP2 and roGFP2-Orp1 sensors in adult plants grown in different conditions. (**a**) Representative false-color images for the two acquired fluorescence emissions of Grx1-roGFP2 excited with 460 nm (Green LUT) and 400 nm (Cyan LUT) excitations, and the resulting $400_{em}/460_{em}$ ratio (Green Fire Blue LUT) of plants grown in hydroponics and soil. (**b**) $400_{em}/460_{em}$ ratios of Grx1-roGFP2 plants

leaf discs are obtained from plant sensor lines and dipped for 10 min in solutions containing different concentration of H_2O_2 (1 mM, 10 mM, 100 mM, 500 mM) or the reducing agent DTT (0.1 mM, 1 mM, 10 mM, 100 mM) and then imaged (Fig. 4b, d).

9. Ratio images such as those reported in Figs. 3a, 4a, c, 5a, c, e, and g, for both roGFP2-Orp1 and Grx1-roGFP2 are generated using the "Ratio Plus Plugin" that can be freely downloaded from https://imagej.nih.gov/ij/plugins/ratio-plus.html and installed accordingly to the indications reported in the webpage. Shortly, once the two series of images are open (e.g., 400_{em} and 460_{em}), press the "Plugin" button and select, in the drop menu, the "Ratio Plus" button. A new window will open, and here select as "Image 1" the 400_{em} and select as "Image 2" the 460_{em}, and then press "OK." A new series of images (the ratio images) will open, and now by selecting the Look Up Table (LUT), different colors can be chosen (e.g., Green Fire Blue for $400_{em}/460_{em}$ Figs. 3, 4, and 5). To allow visual comparisons among ratio images, corresponding to the different treatments, it is necessary to uniform their "minimum" and "maximum" ratio values. This is carried out by opening a ratio images with Fiji and then pressing the "Adjust" button which is found in the "Image" drop menu. Then press the "Brightness/Contrast" button and press the "Set" button. Here, the values of the minimum and maximum ratios can be indicated. Repeat the same routine and use the same minimum and maximum values for all the ratio images which need to be compared (*see* **Note 20**).
10. Ratio (R) values at different time points can be normalized to the initial (R_0) by applying the following formula:

$$\Delta R/R_0 = (R - R_0)/R_0$$

Fig. 5 (continued) grown in hydroponics and soil. $N \geq 14$. (**c**) Representative false-color images for the two acquired fluorescence emissions of roGFP2-Orp1 excited with 460 nm (Green LUT) and 400 nm (Cyan LUT) excitations, and the resulting $400_{em}/460_{em}$ ratio (Green Fire Blue LUT) of plants grown in hydroponics and soil. (**d**) $400_{em}/460_{em}$ ratios of roGFP2-Orp1 plants grown in hydroponics and soil. $N \geq 14$. (**e**) Representative $400_{em}/460_{em}$ ratio images (Green Fire Blue LUT) of Grx1-roGFP2 hydroponically grown plants at 0, 24, 48, and 72 h in control conditions and treated with 100 mM NaCl. (**f**) $400_{em}/460_{em}$ ratios of Grx1-roGFP2 plants at 0, 24, 48, and 72 h in control conditions and treated with 100 mM NaCl. $N \geq 9$. (**g**) Representative $400_{em}/460_{em}$ ratio images (Green Fire Blue LUT) of roGFP2-Orp1 hydroponically grown plants at 0, 24, 48, and 72 h in control conditions and treated with 100 mM NaCl. (**h**) $400_{em}/460_{em}$ ratios of roGFP2-Orp1 plants at 0, 24, 48, and 72 h in control conditions and treated with 100 mM NaCl. $N \geq 9$. Analyzed regions of interest (ROIs) are highlighted in yellow; background ROIs are highlighted in red. Images were acquired using 0.5× objective lens and 1.5× zoom. Data are plotted as boxplots, with whiskers indicating minimum and maximum values. Upper and lower boxplot's boundaries define the 75th and 25th percentiles. The line in the middle of the box corresponds to the median. *ns* not significant, $^{*}p < 0.05$, $^{**}p < 0.005$ (one-way ANOVA with post hoc Tukey's test). $N = 3$. *CTRL* control

and plotted versus time (Fig. 3b, e). R_0 is the ratio measured before treatment administration.

3.6 Analyses of Steady-State Glutathione Redox Potential and H_2O_2-Dependent roGFP2 Oxidation in Soil or Hydroponics

1. Grow plants in soil and hydroponics for 3 weeks as described in Subheading 3.1.
2. Replace the hydroponic solution with a fresh solution weekly and water soil-grown plants every 2 days.
3. Move plants to the stereo microscope and adjust the microscope settings.
4. Record the fluorescence of the channels with 460 and 400 nm excitation lights as described in Subheading 3.3 for 90 s and place the plants back in the growth chamber.
5. Collect data and perform image analysis as described in Subheading 3.5.

3.7 Analyses of Steady-State Glutathione Redox Potential and H_2O_2-Dependent roGFP2 Oxidation in Control and Salt-Stressed Plants

1. Grow plants in hydroponics for 3 weeks as described in Subheading 3.1.
2. Replace the hydroponic solution with a fresh solution with 100 mM NaCl. Control plants are kept with standard solution.
3. At time 0, which corresponds to the shift to 100 mM NaCl, move the plants to the stereo microscope and adjust the microscope settings.
4. Record the fluorescence of the channels with 460 and 400 nm excitation as described in Subheading 3.3 for 90 s and place the plants back in the growth chamber.
5. Repeat the image acquisition at 24, 48, and 96 h posttreatment.
6. Collect data and perform image analysis as described in Subheading 3.5.

4 Notes

1. Both binary vectors and *Arabidopsis* plants expressing the Grx1-roGFP2 and roGFP2-Orp1 localized at different compartments are available. Specifically, to what concern plants where Grx1-roGFP2 and roGFP2-Orp1 are localized to cytosol, nucleus, chloroplasts, mitochondria, and peroxisomes have been recently reported by Arnaud et al. [22].
2. This filter set is needed to excite the sensors with 400 nm and collect the fluorescence emitted in the GFP emission window.
3. Based on our direct experience, we chose to excite the sensor at every acquisition first with the 460 nm and then with 400 nm.

The 400 nm excitation might induce a photosensitization of the sensor as previously reported for HyPer sensor [30].

4. The use of other light sources instead of light-emitting diode (LED) can be used. Mercury or xenon incandescent light sources can be used. In this latter case, there is however the need to have proper excitation filters, a filter wheel (to switch the excitation lights) and a shutter. Nevertheless, LEDs have several advantages over incandescent light sources including lower energy consumption, longer lifetime, improved physical robustness, smaller size, and faster switching since they do not require a shutter.
5. CMOS cameras from other companies can be chosen.
6. Different excitation wavelengths are admissible (e.g., 405 and 488 nm) if they are before or after the isosbestic point of roGFP2 (~ 422 nm), respectively [25].
7. Higher magnifications can be used by using the Zoom knob or other lenses.
8. The Open-Source Microscopy Software Micro-Manager (https://micro-manager.org/) can be used instead of a commercial software as a platform for controlling the microscope, illuminator, and camera.
9. Napari (https://napari.org/) can be used as an alternative to Fiji with the ROI Registration plugin (https://www.napari-hub.org/plugins/napari-roi-registration).
10. Make sure that the rockwool was well soaked with the solution.
11. To uniform the treatment, pots can be randomly distributed among standard greenhouse flats.
12. The water status of plants is very important. For hydroponic cultures the level of the solution must be carefully checked. With old plants the high rate of transpiration can alter the level of the solution.
13. Plants subjected to imaging analyses can be transferred to the microscopy facility 1 day ahead the analyses and possibly maintained in condition like the normal growth conditions. A home-made growth chamber using LED light sources can be easily built.
14. The use of low magnification objectives has the disadvantage of allowing a limited amount of light to illuminate the specimen which requires an increased camera exposure time as well as the need to set the Camera binning to 4 × 4 which entails a lower resolution compared with the use of a 2 × 2 camera binning. 2 × 2 camera binning can be used with specimens having a high expression level of the sensor or when imaged at higher magnification.

15. For a precise and reliable quantification of 400_{em} and 460_{em} fluorescent signal intensities (quantified as pixel intensity), signal should be at least three times higher than the background noise. Due to the different quantum yield of the two roGFP2 excitation peaks [12, 15] it is better to use a higher exposure time for the 400_{em} with respect to the 460_{em} fluorescent channels.
16. Note that blue light can induce plant responses mediated by phototropins such as transient cytosolic calcium concentration increase.
17. For the proper measurement of the resting $400_{em}/460_{em}$ ratios, a period of 90 s acquisition is strongly suggested since it allows to verify if the measured ratios are constant or undergo changes/fluctuations during the analysis. Instantaneous and early ratio changes can be due to a transient touch response, or even by the light used to illuminate the sensor. Fluctuations reduce the reliability of ratio measurements.
18. Longer experiments can be performed. The sampling time can be adjusted depending on the experiment. For long-term treatments, it can be changed.
19. To reduce the background signal, the support for the Araponics holder should be black and opaque. The use of 3D-printed support with polylactic acid (PLA) material is an affordable and reliable solution (Fig. 2c).
20. If ratio images have to be compared as shown in Fig. 3a, the minimum and maximum ratio intervals must be the same for all ratio images.

Acknowledgments

This work was carried out with the support of the NOLIMITS Center of Excellence for Plant Biology and Other Life Sciences established by the University of Milan. This study was carried out within the Agritech National Research Center and received funding from the European Union Next-GenerationEU (PIANO NAZIONALE DI RIPRESA E RESILIENZA (PNRR)—MISSIONE 4 COMPONENTE 2, INVESTIMENTO 1.4–D.D. 1032 17/06/2022, CN00000022). This study reflects only the authors' views and opinions; neither the European Union nor the European Commission can be considered responsible for them. This work was also supported by a PhD fellowship from the University of Milan to S.B. and M.G., by the European Union's HORIZON-TMA-MSCA-PF-E to E.D.A. (proposal number 101059695).

References

1. Zhu JK (2016) Abiotic stress signaling and responses in plants. Cell 167(2):313–324
2. Ravi B, Foyer CH, Pandey GK (2023) The integration of reactive oxygen species (ROS) and calcium signalling in abiotic stress responses. Plant Cell Environ 46(7): 1985–2006
3. Waszczak C, Carmody M, Kangasjärvi J (2018) Reactive oxygen species in plant signaling. Annu Rev Plant Biol 69:209–236
4. Apel K, Hirt H (2004) Reactive oxygen species: metabolism, oxidative stress, and signal transduction. Annu Rev Plant Biol 55:373–399
5. Das K, Roychoudhury A (2014) Reactive oxygen species (ROS) and response of antioxidants as ROS-scavengers during environmental stress in plants. Front Environ Sci 2:2014
6. Wu F, Chi Y, Jiang Z et al (2020) Hydrogen peroxide sensor HPCA1 is an LRR receptor kinase in Arabidopsis. Nature 578(7796): 577–581
7. Bi G, Hu M, Fu L (2022) The cytosolic thiol peroxidase PRXIIB is an intracellular sensor for H_2O_2 that regulates plant immunity through a redox relay. Nat Plants 8(10):1160–1175
8. Zaffagnini M, Fermani S, Marchand CH (2019) Redox homeostasis in photosynthetic organisms: novel and established thiol-based molecular mechanisms. Antioxid Redox Signal 31(3):155–210
9. Sagi M, Fluhr R (2006) Production of reactive oxygen species by plant NADPH oxidases. Plant Physiol 141(2):336–340
10. Corpas FJ, González-Gordo S, Palma JM (2021) Nitric oxide and hydrogen sulfide modulate the NADPH-generating enzymatic system in higher plants. J Exp Bot 72(3):830–847
11. Schwarzländer M, Dick TP, Meyer AJ et al (2016) Dissecting redox biology using fluorescent protein sensors. Antioxid Redox Signal 24(13):680–712
12. Meyer AJ, Brach T, Marty L et al (2007) Redox-sensitive GFP in *Arabidopsis thaliana* is a quantitative biosensor for the redox potential of the cellular glutathione redox buffer. Plant J 52(5):973–986
13. Schwarzländer M, Fricker MD, Müller C et al (2008) Confocal imaging of glutathione redox potential in living plant cells. J Microsc 231(2): 299–316
14. Gutscher M, Pauleau AL, Marty L et al (2008) Real-time imaging of the intracellular glutathione redox potential. Nat Methods 5(6): 553–559
15. Nietzel T, Elsässer M, Ruberti C et al (2019) The fluorescent protein sensor roGFP2-Orp1 monitors *in vivo* H_2O_2 and thiol redox integration and elucidates intracellular H_2O_2 dynamics during elicitor-induced oxidative burst in Arabidopsis. New Phytol 221(3):1649–1664
16. Dooley CT, Dore TM, Hanson GT et al (2004) Imaging dynamic redox changes in mammalian cells with green fluorescent protein indicators. J Biol Chem 279(21):22284–22293
17. Meyer AJ, Dick TP (2010) Fluorescent protein-based redox probes. Antioxid Redox Signal 13(5):621–650
18. Marty L, Siala W, Schwarzländer M et al (2009) The NADPH-dependent thioredoxin system constitutes a functional backup for cytosolic glutathione reductase in Arabidopsis. Proc Natl Acad Sci USA 106(22):9109–9114
19. Rosenwasser S, Rot I, Sollner E et al (2011) Organelles contribute differentially to reactive oxygen species-related events during extended darkness. Plant Physiol 156(1):185–201
20. Albrecht SC, Sobotta MC, Bausewein D et al (2014) Redesign of genetically encoded biosensors for monitoring mitochondrial redox status in a broad range of model eukaryotes. J Biomol Screen 19(3):379–386
21. Ugalde JM, Fuchs P, Nietzel T et al (2021) Chloroplast-derived photo-oxidative stress causes changes in H_2O_2 and EGSH in other subcellular compartments. Plant Physiol 186(1):125–141
22. Arnaud D, Deeks MJ, Smirnoff N (2023) Organelle-targeted biosensors reveal distinct oxidative events during pattern-triggered immune responses. Plant Physiol 191(4): 2551–2569
23. Gutscher M, Sobotta MC, Wabnitz GH et al (2009) Proximity-based protein thiol oxidation by H_2O_2-scavenging peroxidases. J Biol Chem 284(46):31532–31540
24. Doccula FG, Luoni L, Behera S et al (2018) *In vivo* analysis of calcium levels and glutathione redox status in Arabidopsis epidermal leaf cells infected with the hypersensitive response-inducing bacteria *Pseudomonas syringae* pv. tomato AvrB (PstAvrB). Methods Mol Biol 1743:125–141
25. Ugalde JM, Fecker L, Schwarzländer M et al (2022) Live monitoring of ROS-induced cytosolic redox changes with roGFP2-based sensors in plants. Methods Mol Biol 2526:65–85
26. Haber Z, Lampl N, Meyer AJ et al (2021) Resolving diurnal dynamics of the chloroplastic glutathione redox state in Arabidopsis reveals

its photosynthetically derived oxidation. Plant Cell 33(5):1828–1844

27. Hipsch M, Lampl N, Zelinger E et al (2021) Sensing stress responses in potato with whole-plant redox imaging. Plant Physiol 187(2): 618–631
28. Hipsch M, Michael Y, Lampl N et al (2023) Early detection of late blight in potato by whole-plant redox imaging. Plant J 113(4): 649–664
29. Bohle F, Klaus A, Tegethof H et al (2022) High robustness of cytosolic glutathione redox potential under combined salt and osmotic stress in barley as revealed by the biosensor Grx1-roGFP2. bioRxiv 2022.12.22.521445
30. Belousov VV, Fradkov AF, Lukyanov KA et al (2006) Genetically encoded fluorescent indicator for intracellular hydrogen peroxide. Nat Methods 3(4):281–286

Chapter 5

Determination of ROS Generated by *Arabidopsis* Xanthine Dehydrogenase1 (AtXDH1) Using Nitroblue Tetrazolium (NBT) and 3,3′-Diaminobenzidine (DAP)

Aigerim Soltabayeva and Moshe Sagi

Abstract

Plants generate reactive oxygen species (ROS) during different metabolic processes, which play an essential role in coordinating growth and response. ROS levels are sensitive to environmental stresses and are often used as a marker for stress in plants. While various methods can detect ROS changes, histochemical staining with nitroblue tetrazolium (NBT) and 3,3′-diaminobenzidine (DAB) is a popular method, though it has faced criticism. This staining method is advantageous as it enables both the quantification and localization of ROS and the identification of the enzymatic origin of ROS in plants, cellular compartments, or gels. In this protocol, we describe the use of NBT and DAP staining to detect ROS generation under different stresses such as nitrogen starvation, wounding, or UV-C. Additionally, we describe the use of NBT staining for detecting enzymatic generation of ROS in native and native SDS PAGE gels. Our protocol also outlines the separation and comparison of the origin of ROS generated by xanthine dehydrogenase1 (XDH1) using different substrates.

Key words *Arabidopsis*, ROS, XDH1, NBT and DAP staining, Native and native-SDS PAGE in gel assay, Nitrogen starvation, Wounding, or UV-C

1 Introduction

Reactive oxygen species as (ROS) byproduct of metabolic activity in lower levels play a role in signaling [1–3], while in higher levels they are known as toxic molecules which cause cell death in plant [4, 5]. Environmental stressors, both biotic and abiotic, induce ROS generation in plants, leading to ROS accumulation, which is a sensitive indicator of stress. Compared to other stress markers, ROS changes in plants provide critical information for stress identification [6]. The detection of the ROS changes, their location in plants, in organs, or at the cellular level, and identification of ROS origin, or global ROS dynamics are important parameters in understanding ROS signaling conducted through developing different

Francisco J. Corpas and José M. Palma (eds.), *ROS Signaling in Plants: Methods and Protocols*,
Methods in Molecular Biology, vol. 2798, https://doi.org/10.1007/978-1-0716-3826-2_5,

ROS measurement methods. While, spectrophotometric methods using epinephrine and Amplex red have been popular in the past, the focus is on in vivo monitoring of ROS in plant cells using fluorescent dyes such as 2′-7′-dichloro-dihydrofluorescein diacetate (DCFH2-DA) and/or Amplex Red, as well as electron paramagnetic resonance (EPR) spectroscopy to increase sensitivity of ROS detection [7–9]. However, the histochemical staining method using nitroblue tetrazolium (NBT) and 3,3′-diaminobenzidine (DAB) has been criticized for requiring quantification of removed chlorophyll and low sensitivity [8, 9], although it can provide additional information on the origin of ROS molecules by using substrates and/or inhibitors [10–12]. This method also allows for the visualization of ROS location at the cellular level using microscope facilities [13, 14]. Additionally, the antioxidant properties of different metabolites such as allantoin/allontoate [10] or nanoparticles or drugs [15] in plants were mainly evidenced by DAB or NBT staining.

Moreover, NBT [or MTT, 3-(4,5-dimethylthiazol-2-yl)-2,5-diphenyltetrazolium bromide, a cheap analog of NBT] staining can be used to analyze ROS generation activity of enzymes such as XDH (xanthine dehydrogenase1), which is involved in xanthine degradation but can also generate superoxide and/or act as NADH oxidase [11, 16]. Notably, XDH contributes significantly (30–60%) to the plant redox milieu under normal and stressed growth conditions such as dark, drought, nitrogen starvation, wounding, and UV-C [10, 11, 17, 18]. In addition, in-gel NBT staining of ROS generation by enzymes enables the differentiation of the origin of ROS and how different substrates affect ROS generation properties of enzymes. This information can be used to differentiate the superoxide-generating XOR form from the XDH form in *Arabidopsis* and pea, for example [12, 18, 19].

Thus, we have described the ROS detection method by NBT or DAP staining used in a recent publication for detection and accumulation of ROS after stresses UV-C, wound, and nitrogen starvation in plants. Furthermore, we focused on identification of the origin of ROS generated after stress plant tissue or leaf disks. In addition, we are describing the ROS generation activity by XDH in-gel used for separation origin of ROS source in XDH: either xanthine and/or NADH.

2 Materials

To ensure consistency, all solutions used in the experiments should be prepared freshly prepared and using Milli-Q grade water. Additionally, all experiments should be conducted at room temperature.

2.1 Detection of ROS in Plant Tissue

1. Nitroblue tetrazolium (NBT) solution: 4 mg of NBT dissolved in 10 mL of 50 mM Tris–HCl solution at pH 7.5, covered with aluminum foil.
2. 3,3′-Diaminobenzidine (DAB) solution: freshly prepared by dissolving 1 mg of DAB in 1 mL of distilled water, falcon tube covered with aluminum foil.
3. 95% Ethanol solution: a solution composed of ethanol and water, with a volumetric ratio of 95% ethanol (pure or 99%) to 5% water.
4. Ethyl alcohol with glycerol solution: ethyl alcohol and glycerol mixed in a 1:2 volume ratio.
5. Plastic file (use folder for A4 paper), where to put stained material and safely scan the stained material.
6. Diphenylene iodonium (DPI) stock solution: 10 millimolar (mM) concentration in solvent (DMSO, Dimethyl sulfoxide).
7. Allopurinol (1 M) stock solution: dissolve 0.6805 g of allopurinol powder in water to a final volume of 5 mL.

2.2 Protein Extraction and Evaluation

1. Protein extraction buffer composition (10 mL):

- Tris–HCl: 250 mM, pH 8.48, 10 mL.
- EDTA: 1.84 mg.
- GSH (glutathione): 22 mg.
- DTT (dithiothreitol): 3 mg.
- L-cysteine (6 mg).
- Sodium molybdate (Na_2MoO_4) solution: 0.1 M, 25 μL.
- Protease inhibitors: PMSF (phenylmethylsulfonyl fluoride) (10 mM, 5 μL), aprotinin (10 μg/mL, 5 μL), pepstatin (10 μg/mL, 5 μL), leupeptin (10 μg/mL, 5 μL).
- Sucrose: 856 mg.

The protein extraction buffer was prepared by combining the above-listed components in a total volume of 10 mL of Tris–HCl (250 mM) pH 8.48 solution.

2.3 Native PAGE

1. Resolving buffer (#3 solution): 947 mM Tris–HCl and 0.289 N HCl at pH 8.48.
2. Stacking buffer (#4 solution): 158 mM Tris–HCl and 0.256 N H_3PO4 at pH 6.9.
3. 40%T, 5%C Monomer solution (#5 solution): dissolve 38 g of acrylamide and 2 g of bis-acrylamide in water to a total volume of 100 mL.

4. 6.25%T, 20%C Monomer solution (#6 solution): dissolve 5 g of acrylamide and 1.25 g of bis-acrylamide in water to a total volume of 100 mL.
5. Ammonium persulfate (APS) solution: dissolve 1 g of APS in 10 mL water.
6. TEMED (N,N,N,N′-Tetramethyl-ethylenediamine).
7. 70% Ethanol (EtOH).
8. Lower tank buffer: 63 mM Tris and 50 N HCl at pH 7.47.
9. Upper tank buffer: 37.6 mM Tris and 40 mM glycine at pH 8.89.
10. Sample buffer (Loading Dye #1) composition:
    - Sucrose 50% solution in water.
    - Bromophenol blue 0.002% (w/v).

To prepare the sample buffer (loading dye #1), combine a 50% sucrose solution in water with bromophenol blue at a concentration of 0.002% (w/v).

2.4 SDS Page

1. Resolving buffer ×4 SDS: 1.5 M Tris–HCl pH 8.48 with 0.4% SDS.
2. Prepare stacking buffer ×4 SDS: 0.5 M Tris–HCl pH 6.8 with 0.4% SDS.
3. 30% Acrylamide/bis-acrylamide solution: dissolve 29.2 g acrylamide and 0.8 g N,N′-methylenebisacrylamide (bis-acrylamide) in 100 mL distilled water.
4. APS 10% in water.
5. TEMED.
6. EtOH 70%.
7. TGS (Tris/Glycine/SDS) buffer: 25 mM Tris–HCl, 192 mM glycine, 0.1% SDS at pH 8.3.
8. Sample buffer (loading dye#2): 47 mM Tris–HCl (pH 7.5), 2% (w/v) SDS, 7.5% (v/v) glycerol, and 40 mM 1,4-dithio-DL-threitol (DTT) as the thiol-reducing agent, and 0.002% (w/v) bromophenol blue [12, 20].
9. Washing buffer removing SDS: 7.5 mL Tris 1 M pH 7.8, add 7.5 mL TritonX-100, add Milli-Q water till 750 mL and add 0.438 g EDTA (ethylenediaminetetraacetic acid).

2.5 In-Gel Assay

1. Staining solution for general XDH activity: In 0.1 mM Tris–HCl buffer (pH 8.5), add 0.1 mM phenazine methosulfate (PMS), 1 mM 3-(4,5-dimethylthiazol-2-yl)-2,5-diphenyltetrazolium bromide (MTT) or nitro blue tetrazolium (NBT), and 0.5 mM xanthine along with 1 mM hypoxanthine. Cover the

solution with aluminum foil to protect it from light. Thoroughly mix the solution using a vortex. It is essential to prepare a fresh solution each time.

2. Staining solution for xanthine depending of XDH activity: In 0.1 mM Tris buffer (pH 8.5), add 1 mM MTT (or NBT), 0.5 mM xanthine mixed with 1 mM hypoxanthine. Cover the solution with aluminum foil to protect it from light. Mix well by vortex. Every time prepare a fresh solution.
3. Staining solution for NADH depending of XDH activity: In 0.1 mM Tris buffer (pH 8.5), combine 1 mM MTT and 0.25 mM NADH (nicotinamideadenine dinucleotide). Cover the solution with aluminum foil to protect it from light. Mix well by vortex. Every time prepare a fresh solution.

3 Methods

3.1 Stress Application of Plants

1. To apply UV-C stress, select plants that are under 25 days old and have not yet bolted. Irradiate the plants with 0.150 Joule/m^2 UV-C ($\lambda = 254$ nM) using a CL 508G UV cross-linker. After irradiation, promptly transfer the plants to a growth room for recovery, allowing a recovery period of 1 h to 3 days.
2. For the wounding stress application, choose plants that are under 25 days old and have not yet bolted. Induce mechanical wounding by applying pressure to the middle of the leaves perpendicular to the main vein using forceps. Immediately after the treatment, place the plants in a growth room for recovery, allowing a recovery period of 1 h to 3 days.
3. To induce nitrogen starvation stress, transplant 6-day-old seedlings into 0.128 L pots filled with a 1:1 mixture of perlite and nutrient-free soil. Irrigate the plants twice per week using 0.5× Hoagland solution [21], which should be modified to contain either 1 or 5 mM $NaNO_3$ as the sole nitrogen source. Ensure balanced sodium levels across all treatments by adding NaCl. To prevent salinization, perform irrigation to leach out 50% of the applied nutrient solution. Sample the plants before they reach the bolting stage.

3.2 Detection of ROS Generation in Leaves and Leaf Disks

1. Cut the leaves from plants (*see* **Note 1**). For leaf disks, immediately detach the leaves and cut them into disks with an 8 mM diameter. Ensure that the middle section of the leaf is used for all samples. The leaves used for analysis must be of the same age.
2. To measure total superoxide and hydrogen peroxide (H_2O_2) generation in leaves, transfer the leaf or leaf disk into a 2 mL Eppendorf tube containing 1 mL NBT/DAP solution (*see*

Note 2). Label all tubes before use and ensure that the leaves are fully covered by the solution. Incubate the tube with NBT solution in the dark at room temperature for 2 h. Incubate the tube with DAP solution in light at room temperature for 2 h.

(a) For xanthine oxidase-dependent ROS generation in leaves, place the leaf or leaf disk onto filter paper soaked with a 0.5 mM allopurinol solution for 30 min in the growth room conditions. Ensure that the abaxial surface of the leaf is facing the filter paper side. Then transfer the soaked leaf or leaf disk into a 2 mL Eppendorf tube containing 1 mL NBT/DAP solution and incubate for 2 h in the dark or light, respectively, as done for total ROS detection in leaves.

(b) For NADPH oxidase-dependent ROS generation in leaves, transfer the leaf or leaf disk onto filter paper soaked with a 100 μM DPI solution for 30 min in the growth room conditions. Then transfer the leaf or leaf disk soaked in DPI into a 2 mL Eppendorf tube containing 1 mL NBT/DAP solution and incubate for 2 h in the dark or light, respectively, as done for total ROS detection in leaves.

3. After 2 h of staining with NBT/DAP, transfer the leaves or leaf disks into another tube containing 1 mL of 95% ethanol.
4. Place the tube with the stained material on a thermal block set to 80 °C for approximately 20 min. Gently vortex the tube every 5 min (*see* **Note 3**).
5. Once the chlorophyll is fully removed from the plant material, place the stained leaves without chlorophyll in a plastic folder. Apply approximately 100 μL of ethyl alcohol solution (1:2 v/v) onto the leaves or leaf disks to minimize damage (*see* **Note 4**).
6. Scan the plastic folder where the leaves are organized in order and properly labeled.
7. Use ImageJ software (http://imagej.nih.gov/ij/) to analyze the intensity of the blue color (reduced NBT) and brown color (DAP) in the leaves or leaf disks. Calculate the relative intensity of reduced NBT and precipitated DAP using one of the leaves as a control (*see* **Notes 5** and **6**). Figure 1a illustrates H_2O_2 production in leaves of wild type (WT) and *Atxdh1* mutant subjected to nitrogen starvation stress, while Fig. 1b, d shows H_2O_2 production in leaf disks under UV-C exposure and wounding stress, respectively. Figure 1c presents superoxide production in leaf disks of both genotypes subjected to wounding stress.

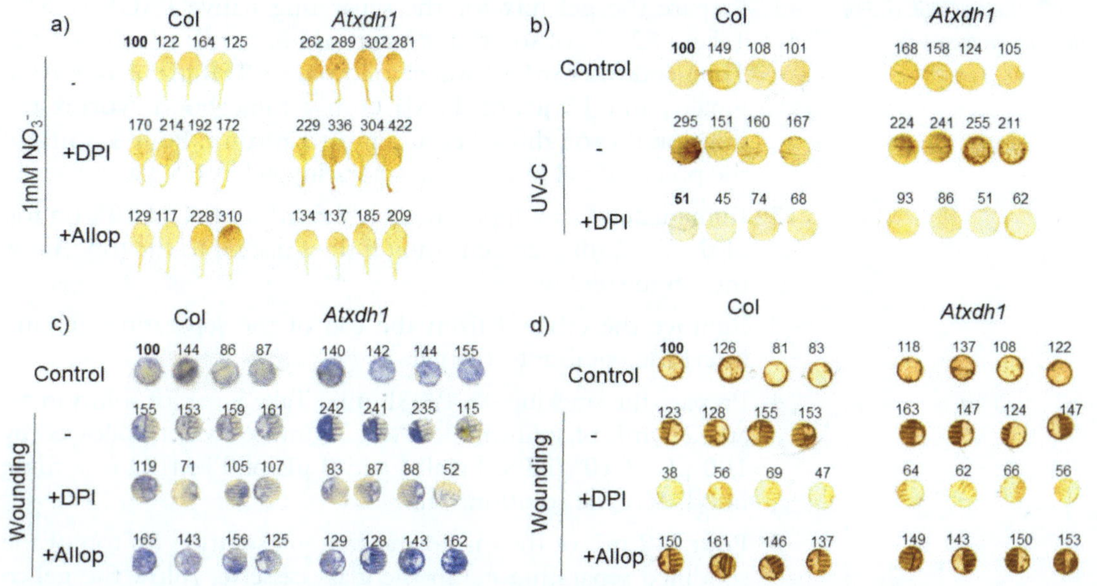

Fig. 1 Visualization of ROS generation in leaves and leaf disks stressed by nitrogen starvation (**a**) UV-C (**b**) and wound stress (**c**) in the absence or presence of allopurinol (+Allo) or diphenylene iodonium (+DPI). H_2O_2 production in leaf and or leaf discs was detected by staining with 3,3′-diaminobenzidine (DAB), and the superoxide levels were visualized by NBT staining of leaf discs. Col is wild type and *Atxdh1* mutant is defective in xanthine dehydrogenase1 (AtXDH1). (Soltabayeva et al. [17, 18]). The intensity of the staining was determined using the ImageJ software, with the value of the Col-0 control set as 100

3.3 Detection ROS Generation by XDH in-Gel Assay

3.3.1 Sample Preparation for In-Gel Assay

1. Collect plant material samples of the same fresh weight (100 mg) from leaves located at the same position. Immediately place the samples in liquid nitrogen and store them at − 80 °C.
2. Extract the proteins in protein extraction buffer (1:4 w/v) by crushing 100 mg plant tissue into powder using mortar and pestle with liquid nitrogen and adding 400 μL of extraction solution.
3. Centrifuge for 10 min at 13,000 g at +4C and transfer the supernatant into a new tube.
4. Determine the total protein concentration in the supernatant using the Bradford method. Dilute 5 μL of the supernatant with 495 μL of Milli-Q water. Add 20 μL of the diluted protein to 220 μL of diluted Bradford Assay reagent. Measure the protein concentration at 595 nM. Prepare a standard curve following a similar procedure.
5. Then equalize the protein concentration in the extracted sample supernatant to 0.1 in a volume of 0.1 mL.

3.3.2 Native PAGE (7.5%) Running Protocol

1. Prepare the gel mix for the separating native PAGE (7.5%): Take 2.82 mL of solution #5 and add 3.75 mL of solution #3. Then add 8.44 mL of water followed by 150 μL of 10% APS. Finally, add 15 μL of TEMED in a fume hood. Vortex the solution thoroughly to ensure proper mixing. Pour 4.5 mL of the prepared gel mix for the separating gel into a glass cassette.
2. Immediately pour approximately 0.5 mL of EtOH 70% on top of the separating gel, ensuring it does not fill to the top. Allow the gel to solidify.
3. Remove the ethanol from the top of the separating gel and insert the comb into the gel.
4. Prepare the stacking 4% PAGE mix: Take 5 mL of solution #4 and 2.5 mL of solution #6. Add 2.5 mL of water, followed by 100 μL of 10% APS. Finally, add 8 μL of TEMED in a fume hood. Mix the solution well.
5. Pour 1–2 mL of the mixed stacking gel solution on top of the solidified separating gel in the glass cassette. Allow the gel to solidify.
6. Place the prepared glass cassettes with the native gel into the Mini-PROTEAN electrophoresis chamber in the electrophoresis module. Fill the inside of the glass cassette with lower tank buffer (63 mM Tris–HCl, pH 7.47), and the chamber with upper tank buffer (37.6 mM Tris–HCl, 40 mM glycine, pH 8.89). Ensure that all buffers are pre-chilled.
7. Load 30–60 μL of the samples mixed with loading buffer #1 (in a 1:10 ratio of sample to dye) into the wells. Set the current to 15 mA for the electrophoresis run.
8. Run the native PAGE at 30 V for 30 min (until the dye has migrated approximately 0.5 to 1 cm above the bottom of the gel), then increase the voltage to 60 V. Place ice around the electrophoresis chamber to maintain a cool temperature.

3.3.3 Staining Normal XDH Activity in Native Gel Assay

1. After the separation of proteins, transfer the gel into a clean box containing 20 mL of the prepared solution for general XDH activity. The solution should consist of 1 mM MTT, 0.5 mM xanthine, and 1 mM hypoxanthine in 0.1 mM Tris–HCl buffer (pH 8.5).
2. Allow the gel to incubate overnight at room temperature in a dark place.
3. After the incubation, stop the reactions by immersing the gels in 7% acetic acid or rinsing them with tap water.
4. For further quantification of the intensity analysis, either scan the gels (placed in a plastic folder) using a scanner or take a photo of the gels. Alternatively, use a ChemiDoc Imaging System for image capture. The results of general in native gel

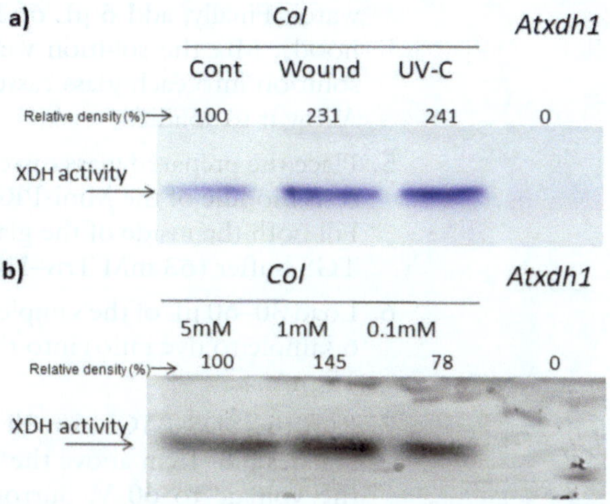

Fig. 2 The detection of ROS generation by xanthine dehydrogenase (XDH) activity in native PAGE gels after wounding and UV-C stress (**a**) and different nitrate level applications (**b**). The negative control used was the *Atxdh1* (GABI_049004) mutant. The gel staining was documented by scanning the gel photo (**a**) and ChemiDoc Imaging Systems (**b**). Equal amounts of protein were loaded into each lane, and the intensities were determined using the ImageJ software. (http:/imagej.nih.gov/ij/), with the value of Col without UV-C and wounding treatment in the leaves and Col-0 at 5 mM nitrate treatment set as 100

XDH activity in WT samples stressed by wound and UV-C treatment were shown in Fig. 2a, displayed through a photo taken by a scanner. In Fig. 2b, the XDH activity under nitrogen starvation conditions was depicted, as shown by the ChemiDoc Imaging System.

5. Analyze the scanned gel photo using image analysis software such as ImageJ (http://imagej.nih.gov/ij/). Calculate the relative intensity of the XDH activity compared to the control.

3.3.4 SDS PAGE (7.5%) Running Protocol

1. Prepare the separating SDS 7.5% PAGE gel mix: Take 3.87 mL of 30% acrylamide/bis, add 3.75 mL of resolving buffer (4× SDS), then add 150 μL of 10% APS. Add 7.2 mL of distilled water, followed by 15 μL of TEMED (toxic, handle in a fume hood). Mix the solution well.
2. Pour 4.5 mL of the prepared gel mix into a glass cassette, ensuring not to fill it to the top. Immediately pour about 0.5 mL of ethanol on top of the separating gel.
3. Allow the gels to solidify. Once solid, remove the ethanol from the top of the separating gel and insert the comb.
4. Prepare the stacking SDS 4% PAGE: Take 1.2 mL of 30% acrylamide/bis, add 1.5 mL of stacking buffer (4× SDS) (pH 6.8), then add 60 μL of 10% APS and 3.3 mL of distilled

water. Finally, add 6 μL of TEMED (toxic, handle in a fume hood). Mix the solution well. Pour 1–2 mL of the prepared solution into each glass cassette, ensuring it covers the comb. Allow it to solidify.

5. Place the prepared glass cassettes with gels into the electrophoresis module of the Mini-PROTEAN electrophoresis chamber. Fill both the inside of the glass cassettes and the chamber with TGS buffer (63 mM Tris–HCl, pH 7.47).
6. Load 30–60 μL of the samples mixed with loading buffer #2 (1: 6 sample to dye ratio) into the wells. Set the electrophoresis at 15 mA.
7. Run the SDS-PAGE for 30 min at 30 V (until the dye front reaches 0.5–1 cm above the bottom of the gel). Then increase the voltage to 60 V. Surround the electrophoresis chamber with ice to maintain a cool temperature throughout the run.

3.3.5 Staining the XDH Activity in Native SDS Gel

1. After separating the proteins in the SDS-PAGE gel, place the gel in a washing buffer on a shaker for 1 to 1.5 h. Use a pre-cold washing solution. For 1 gel (7 mL), you will need 455 mL of washing buffer.
2. After washing the gel to remove SDS, transfer the gel to a staining solution in a small box or petri dish.
   (a) For general XDH activity, immerse the gel in a solution containing 1 mM MTT, 0.1 mM PMS, 0.5 mM xanthine, and 1 mM hypoxanthine in 0.1 mM Tris–HCl buffer (pH 8.5). Keep the gel at room temperature in the dark overnight. The results showing general XDH activity in WT plants stressed by UV-C treatment (right panel) are presented in Fig. 3a.
   (b) For xanthine-dependent XDH activity, immerse the gel in a solution containing 1 mM MTT, 0.5 mM xanthine, and 1 mM hypoxanthine in 0.1 mM Tris–HCl buffer (pH 8.5). Keep the gel at room temperature in the dark overnight. *See* **Note 7** for additional details. The results illustrating xanthine-dependent XDH activity in WT plants stressed by UV-C treatment are shown in Fig. 3b.
   (c) For NADH-dependent XDH activity, immerse the gel in a solution containing 1 mM MTT and NADH in 0.1 mM Tris–HCl buffer (pH 8.5). Keep the gel at room temperature in the dark overnight. The results illustrating NADH-dependent XDH activity in WT plants stressed by UV-C treatment are presented in Fig. 3c.
3. After the incubation of the staining solutions, stop the reactions by immersing the gels either in 7% acetic acid or tap water.

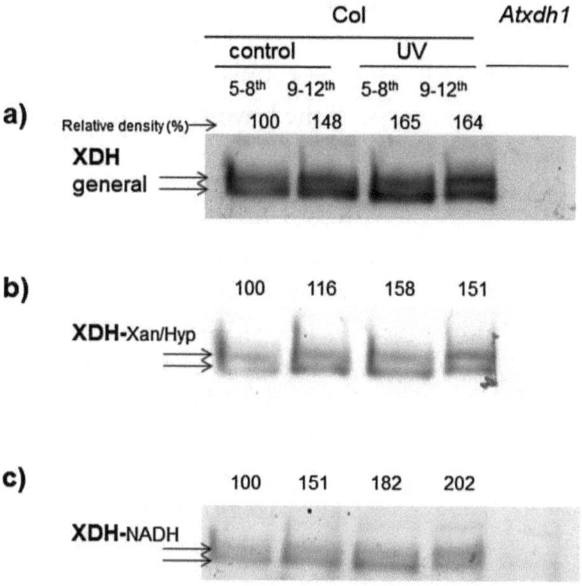

Fig. 3 The detection of ROS generation by xanthine dehydrogenase (XDH) activity in SDS-Native gel. Activities of (**a**) xanthine dehydrogenase (XDH-general), (**b**) xanthine-dependent XDH (XDH-xan), and (**c**) NADH-dependent XDH (XDH-NADH) were assessed after treatments with UV-C. For UV-C treatment analysis, old leaves from the fifth to eighth leaf position and young leaves from the top ninth to 12th position were used. The negative control employed was the *Atxdh1* (GABI_049004) mutant. The gel staining was documented using the ChemiDoc Imaging Systems. Equal amounts of protein were loaded into each lane, the gels were scanned, and the intensities were determined using the ImageJ software. (http:/imagej.nih.gov/ij/), with the value of Col without UV-C treatment in the fifth to eighth leaves set as 100

4. Place the stained gel in a transparent file for safekeeping and scan the results using a scanner or a ChemiDoc Imaging System.
5. Analyze the scanned gel photo using the ImageJ software (http://imagej.nih.gov/ij/) and calculate the relative intensity compared to the control. Refer to **Note 7** for further instructions.

4 Notes

1. The sampling should be performed quickly using a tube or forceps to avoid damaging the leaves or leaf disks and to minimize drought stress.
2. It is important to include a control group in each experiment, either using plants without stress or utilizing mutant lines as controls.

3. Only NBT should be used for the staining process as MTT dissolves in ethanol and will be washed out along with the chlorophyll.
4. If the leaves contain a significant amount of chlorophyll, it may be necessary to discard the ethanol solution and repeat the washing procedure to ensure thorough removal of chlorophyll. Use 1 mL of ethanol solution for each wash.
5. After boiling the leaves at +80 °C, they become fragile and prone to breakage. Adding ethyl alcohol to the solution helps to prevent damage to the leaves.
6. It is crucial to thoroughly wash out any chlorophyll from the plant materials. The presence of green pigments in the samples can interfere with result analysis and make it difficult to quantify using ImageJ (http://imagej.nih.gov/ij/).
7. The *Atxdh1* mutant can be used as a control to identify the activity generated by XDH.

References

1. Mittler R, Vanderauwera S, Suzuki N, Miller G, Tognetti VB, Vandepoele K, Gollery M, Shulaev V, Van Breusegem F (2011) ROS signaling: the new wave? Trends Plant Sci 16: 300–309. https://doi.org/10.1016/j.tplants.2011.03.007
2. Baxter A, Mittler R, Suzuki N (2014) ROS as key players in plant stress signalling. J Exp Bot 65:1229–1240. https://doi.org/10.1093/jxb/ert375
3. Sharma P, Jha AB, Dubey RS, Pessarakli M (2012) Reactive oxygen species, oxidative damage, and Antioxidative defense mechanism in plants under stressful conditions. J Bot 2012:1–26. https://doi.org/10.1155/2012/217037
4. Van Breusegem F, Dat JF (2006) Reactive oxygen species in plant cell death. Plant Physiol 141:384–390. https://doi.org/10.1104/pp.106.078295
5. Gechev TS, Van Breusegem F, Stone JM, Denev I, Laloi C (2006) Reactive oxygen species as signals that modulate plant stress responses and programmed cell death. BioEssays 28:1091–1101. https://doi.org/10.1002/bies.20493
6. Soltabayeva A, Ongaltay A, Omondi JO, Srivastava S (2021) Morphological, physiological and molecular markers for salt-stressed plants. Plan Theory 10:243. https://doi.org/10.3390/plants10020243
7. Ortega-Villasante C, Burén S, Barón-Sola Á, Martínez F, Hernández LE (2016) In vivo ROS and redox potential fluorescent detection in plants: present approaches and future perspectives. Methods 109:92–104. https://doi.org/10.1016/j.ymeth.2016.07.009
8. Steffens B, Steffen-Heins A, Sauter M (2013) Reactive oxygen species mediate growth and death in submerged plants. Front Plant Sci 4: 1–7. https://doi.org/10.3389/fpls.2013.00179
9. Anjum NA, Amreen TAY, Khan NA, Ahmad A (2020) Reactive oxygen species detection-approaches in plants: insights into genetically encoded FRET-based sensors. J Biotechnol 308:108–117. https://doi.org/10.1016/j.jbiotec.2019.12.003
10. Brychkova G, Alikulov Z, Fluhr R, Sagi M (2008) A critical role for ureides in dark and senescence-induced purine remobilization is unmasked in the *Atxdh1* Arabidopsis mutant. Plant J 54:496–509. https://doi.org/10.1111/j.1365-313X.2008.03440.x
11. Yesbergenova Z, Yang G, Oron E, Soffer D, Fluhr R, Sagi M (2005) The plant Mo-hydroxylases aldehyde oxidase and xanthine dehydrogenase have distinct reactive oxygen species signatures and are induced by drought and abscisic acid. Plant J 42:862–876. https://doi.org/10.1111/j.1365-313X.2005.02422.x
12. Srivastava S, Brychkova G, Yarmolinsky D, Soltabayeva A, Samani T, Sagi M (2017) Aldehyde oxidase 4 plays a critical role in delaying silique senescence by catalyzing aldehyde

detoxification. Plant Physiol 173:1977–1997. https://doi.org/10.1104/pp.16.01939

13. Javvaji PK, Dhali A, Francis JR, Kolte AP, Mech A, Roy SC, Mishra A, Bhatta R (2020) An efficient Nitroblue tetrazolium staining and bright-field microscopy based method for detecting and quantifying intracellular reactive oxygen species in oocytes, cumulus cells and embryos. Front Cell Dev Biol 8:1–12. https://doi.org/10.3389/fcell.2020.00764
14. Romero-Puertas MC, Rodríguez-Serrano M, Corpas FJ, Gómez M, Del Río LA, Sandalio LM (2004) Cadmium-induced subcellular accumulation of $O_2^{\cdot-}$ and H_2O_2 in pea leaves. Plant Cell Environ 27:1122–1134. https://doi.org/10.1111/j.1365-3040.2004.01217.x
15. Singh A, Singh NB, Hussain I, Singh H (2017) Effect of biologically synthesized copper oxide nanoparticles on metabolism and antioxidant activity to the crop plants *Solanum lycopersicum* and *Brassica oleracea* var. botrytis. J Biotechnol 262:11–27. https://doi.org/10.1016/j.jbiotec.2017.09.016
16. Zarepour M, Kaspari K, Stagge S, Rethmeier R, Mendel RR, Bittner F (2010) Xanthine dehydrogenase AtXDH1 from Arabidopsis thaliana is a potent producer of superoxide anions via its NADH oxidase activity. Plant Mol Biol 72:301–310. https://doi.org/10.1007/s11103-009-9570-2
17. Soltabayeva A, Srivastava S, Kurmanbayeva A, Bekturova A, Fluhr R, Sagi M (2018) Early senescence in older leaves of low nitrate-grown atxdh1 uncovers a role for purine catabolism in n supply. Plant Physiol 178:1027–1044. https://doi.org/10.1104/pp.18.00795
18. Soltabayeva A, Bekturova A, Kurmanbayeva A, Oshanova D, Nurbekova Z, Srivastava S, Standing D, Sagi M (2022) Ureides are accumulated similarly in response to UV-C irradiation and wounding in Arabidopsis leaves but are remobilized differently during recovery. J Exp Bot 73:1016–1032. https://doi.org/10.1093/jxb/erab441
19. Corpas FJ, Palma JM, Sandalio LM, Valderrama R, Barroso JB, del Río LA (2008) Peroxisomal xanthine oxidoreductase: characterization of the enzyme from pea (*Pisum sativum* L.) leaves. J Plant Physiol 165:1319–1330. https://doi.org/10.1016/j.jplph.2008.04.004
20. Sagi M, Fluhr R (2001) Superoxide production by plant homologues of the gp91(phox) NADPH oxidase. Modulation of activity by calcium and by tobacco mosaic virus infection. Plant Physiol 126:1281–1290. https://doi.org/10.1104/pp.126.3.1281
21. Hoagland D, Arnon DI (1950) The water-culture method for growing plants without soil. Calif Agric Exp Stn Circ 347:1–32

Chapter 6

Malondialdehyde Assays in Higher Plants

Melanie Morales and Sergi Munné-Bosch

Abstract

Malondialdehyde is a three-carbon dialdehyde produced as a byproduct of polyunsaturated fatty acid peroxidation widely used as a marker of the extent of lipid peroxidation in plants. There are several methodological approaches to quantify malondialdehyde contents in higher plants, ranging from the simplest, cheapest, and quickest spectrophotometric approaches to the more complex ones using tandem mass spectrometry. This chapter summarizes the advantages and limitations of approaches followed and provides brief protocols with some tips to facilitate the selection of the best method for each experimental condition and application.

Key words TBARS, Aldehydes, Oxidative stress, Lipid peroxidation, Methodologies, Accuracy

Abbreviations

BHT	butyl hydroxytoluene
DNPH	2,4-dinitrophenylhydrazone
HPLC	high performance liquid chromatography
UHPLC	ultra-high performance liquid chromatography
GC	gas chromatography
mQH_2O	milli-Q water
MDA	malondialdehyde
MeCN	acetonitrile
MeOH	methanol
MS	mass spectrometry
PFPH	pentafluorophenylhydrazine
PTFE	polytetrafluoroethylene
ROS	reactive oxygen species
TBA	thiobarbituric acid
TCA	trichloroacetic acid
THF	tetrahydrofuran
UB	ultrasonic bath

Francisco J. Corpas and José M. Palma (eds.), *ROS Signaling in Plants: Methods and Protocols*,
Methods in Molecular Biology, vol. 2798, https://doi.org/10.1007/978-1-0716-3826-2_6,

1 Introduction

Malondialdehyde (MDA) is a highly reactive molecule produced nonenzymatically by reactive oxygen species (ROS) or enzymatically by lipoxygenase activity in plant cells [1, 2]. Although it is a small molecule formed by three carbons with two aldehyde groups at the carbon 1 and carbon 3 positions, its pH-dependent tautomeric chemical property marks its reactivity. At acidic pH, both protonated enol aldehyde and dialdehyde forms appear in equilibrium. MDA is known for the high reactivity of the α-β-unsaturated carbonyl groups resulting from the peroxidation of triunsaturated fatty acids, mainly linoleic acid, which is the in vivo source of up to 75% of MDA in *Arabidopsis thaliana* leaves [3]. Primary lipid hydroperoxide products are difficult to quantify (due to its high instability and reactivity) and, for that reason, lipid peroxidation is usually estimated by measuring the concentration of secondary oxidation products derived from these initial hydroperoxides [4], which are mostly aldehydes, such as MDA. Taking advantage of the electrophilic character of the MDA molecule, several methodologies have been developed to quantify MDA.

Malondialdehyde assays in higher plants include several methods of diverse specificity and accuracy. To choose the optimal methodology, we should consider the type of samples and available resources, while keeping the highest sensitivity, specificity, and accuracy. These methods include spectrophotometry, liquid chromatography (LC), gas chromatography (GC), and mass spectrometry (MS). Traditionally, the most used method to detect MDA contents in plants has been the thiobarbituric acid-reactive substances (TBARS) assay using UV-visible spectrophotometric detection [5]. This method relies on an easy and quick protocol based on the formation of MDA complex with TBA, but its contents could be overestimated when plant samples present other molecules rich in carbonyl groups (such as carbohydrates and anthocyanins) [6]. For that, it has been criticized for its low specificity and artifact formation. To solve this limitation, methods including a control assay without TBA can be used [7]. The latest is nowadays the most commonly spectrophotometric method used because of its simplicity, although it is not completely free of possible artifacts. More accurate techniques are required such as high-performance liquid chromatography (HPLC) to enhance the specificity of MDA quantification across plant species and tissues [8–10]. Reversed-phase high-performance liquid chromatography (HPLC) coupled with UV detection reduces some limitations of the spectrophotometric methods, but it also shows specificity problems since other adducts formed could have the same retention time and be indistinguishable from MDA-$(TBA)_2$ complexes. To overcome these adducts overestimation, here we recommend using mass spectrometry to identify peaks (e.g., GC-MS) after derivatization with specific reagents [3]. However, these more precise techniques imply enhanced cost and, in consequence, its access is more restricted, at least for part of the scientific community. More recently, an ultra-high-performance liquid chromatography coupled to positive electrospray-tandem mass spectrometry (UHPLC-(+)-ESI-MS/MS) technique has proved to be simple, rapid, reliable, sensitive, selective, and reproducible for the determination of MDA and its variations, without needing a derivatization procedure [11]. This work compared the MDA levels determined by UHPLC-MS/MS versus traditional TBARS assay using spectrophotometry and demonstrated the unspecific response and overestimation of the spectrophotometric approach as well as its inadequacy to discriminate significant differences in MDA contents. Again, a more precise technique implies higher costs and many more requirements of equipment for the use of UHPLC-(+)-ESI-MS/MS which restricts this optimal MDA quantification method as the main option in most studies.

Despite the specificity and accuracy, MDA determination assays using UHPLC-MS/MS are laborious and costly, due to the equipment used. Therefore, spectrophotometric determination of TBARS is still the most frequently used method for estimation of lipid peroxidation. If the use of UHPLC-MS/MS is not possible, we strongly recommend validating spectrophotometric methods particularly in the comparison of groups of samples obtained from the same species, as has been shown previously [12]. Another good alternative is the use of HPLC, which strongly reduces costs and the need for sophisticated MS/MS equipment, while keeping a relatively good accuracy. Here, we describe the various methods mentioned thus far for the estimation of MDA in plant samples, from more unspecific but cheap spectrophotometric methods to the most accurate methodology using UHPLC-MS/MS.

2 Materials

Chemicals can all be purchased from Sigma-Aldrich (Poole, UK) and HPLC columns from Teknokroma (St. Cugat, Spain), but other suppliers are also available worldwide.

2.1 Reagents (See Note 1)

1. Acetonitrile (MeCN).
2. Butylated hydroxytoluene (BHT).
3. 2,4-dinitrophenylhydrazine (DNPH).
4. DNPH-FBz (internal standard).
5. Desferoxamine mesylate.
6. Ethanol 80% or methanol.

7. Formic acid.
8. [$^2H_2$]-MDA (internal standard).
9. Isooctane.
10. MiliQ water (mQ H_2O).
11. Pentafluorophenylhydrazine (PFPH).
12. Potassium phosphate buffer (pH = 7).
13. Phosphate-citrate buffer.
14. Sodium tungstate.
15. Sulfuric acid.
16. 1,1,3,3-tetraethoxypropane (internal standard).
17. Tetrahydrofuran (THF).
18. Thiobarbituric acid (TBA).
19. Trichloroacetic acid (TCA) ACS reagent $\geq$99.0%.

2.2 Equipment

1. Mortar and pestle or mixer mills including grinding balls.
2. Laboratory balance.
3. Microcentrifuge tubes (1 mL and 2 mL).
4. Pyrex® glass tubes with tap.
5. Vortex mixer.
6. Microcentrifuge.
7. Micropipette (100–1000 and 1000–5000 μL).
8. Oven.
9. Ultrasonic bath.
10. Glass beads.
11. Spectrophotometer:
    (a) Microplate spectrophotometer or spectrophotometer.
    (b) Semi-micro cuvettes or 96-well plates.
12. HPLC:
    (a) HPLC equipment.
    (b) Vials suitable for HPLC, plastic syringes (5 mL), and PTFE filters.
    (c) Photodiode-array detector controlled by respective software.
    (d) Hypersyl ODS 5 μm column (250 × 4.6 mm).
13. LC-MS.
    (a) LC-MS equipment.
    (b) Vials suitable for LC, plastic syringe (5 mL), and PTFE filters.

   (c) Hypersyl 5 µm C-18 reversed-phase column (250 × 2.0 mm, Phenomenex, Macclesfield, UK).

14. UHPLC-(+)-ESI-MS/MS.
   (a) UHPLC-(+)-ESI-MS/MS.
   (b) Vials suitable for UHPLC, plastic syringes (5 mL), and PTFE filters.
   (c) ACQUITY UHPLC® BEH C18 (Waters, Milford, USA) column with 2.1 mm internal diameter, 50 mm length, and 1.7 μm particle size).
15. GC-MS.
   (a) GC-MS equipment.
   (b) Vials suitable for GC.
   (c) HP-5MS capillary column (30 m × 0.25 mm).

3 Methods

3.1 General Recommendations to Prevent Hazards

1. Wear gloves always while handling acid reagents and materials that have been in touch with acids.
2. Always prepare solutions including TBA and TCA reagents in open bottles and manipulate them in a laboratory fume hood.
3. Before using each material make sure they are clean and fully dried.

3.2 The Thiobarbituric Acid-Reactive Substances Assay by Spectrophotometry

The first method of assessing the TBA–MDA complex was described five decades ago and it is still the most used method worldwide for both biological (plant and animal) samples and clinical studies. This traditional method [5] originally described a simple and quick assay to estimate MDA generated by the polyunsaturated fatty acid (PUFA)-photoperoxidation process in isolated chloroplasts. It is based on the electrophilic character of MDA, which quickly binds, at low pH and elevated temperature, to the nucleophilic site of thiobarbituric acid (TBA). After extraction, plant extracts containing MDA are incubated with TBA at high temperatures, yielding an MDA-$(TBA)_2$ adduct of reddish color with an absorbance maximum at 532 nm. The main limitation of this rapid approach is analyzing plant samples that contain carbohydrates and anthocyanins (or other molecules rich in carbonyl groups). These other secondary compounds are also susceptible to nucleophilic attack by TBA and produce interferences with the absorbance measurements at 532 nm, resulting in an overestimation of MDA content [11]. An alternative spectrophotometrically method was proposed to solve the problems with the traditional TBARS method [5], which consists mainly with a subtraction of the non-specific absorbance by spectrophotometry in the TBARS assay [7]. Both spectrophotometric methods are detailed in the following sections.

3.3 Method for Isolated Chloroplasts or Tissues without Interfering Compounds

3.3.1 Preparation of the Chemical Solutions

Firstly, calculate the milliliters needed based on the number of samples. This example is based on 100 samples, so calculations are presented to prepare 100 mL of TBA+.

TBA+ solution (it includes trichloroacetic acid containing the above plus 0.65% thiobarbituric acid):

1. Use half of the TBA− solution (previously prepared) inside a glass tube (maximum 100 mL).
2. Weigh 0.65 g of TBA and add it to the glass tube.
3. Dissolve: Stir at low velocity, heat, and cover with aluminum foil. Do not close it tight and do not let it boil.
4. Once it is dissolved, let it cool down.
5. Fill until 100 mL with the TBA− solution. Once it is finished, cover it with aluminum foil.

3.3.2 Extraction Steps

1. Each sample (Eppendorf tube) contains 50–100 mg of fresh weight (FW) (plant tissues, for example, leaves) that must be ground in liquid nitrogen. Ensure that all plant samples must be completely powdered.
2. Add 0.5–1 mL cold ethanol 80% (+ 0.01% BHT). From this step onward, it is ALWAYS required to keep it (ice) cold and protect it from direct light exposure.
3. Use vortex mixer to totally homogenize the extraction medium (in this case methanol) with the plant samples (solid phase). Thirty seconds per sample is recommended.
4. Put the samples in a cold ultrasonic bath (UB) for 30 min. It is recommended to add a vortex mixer hallway to this process, e.g., vortex +15 min UB + vortex +15 min UB + vortex.
5. Centrifuge conditions: 13,000 rpm, 15 min, 4 °C, Ac 9, Br 1.
6. Recover supernatant into falcon tubes.
7. Repeat **steps 2–6** twice more (be sure to recover extremely clean supernatant in the last extraction).

3.3.3 Incubation Procedure

It just includes the TBA incubation (Fig. 1).

1. Centrifuge (13,000 rpm, 10 min, 4 °C) pull of supernatant to clarify the liquid phase.
2. Prepare each sample in Pyrex glass tubes:
   (a) 1 mL of sample with 1 mL of TBA+ solution (TBA+).
   (b) It is suggested to change the volumes but always keep sample proportions (1/1; v/v).

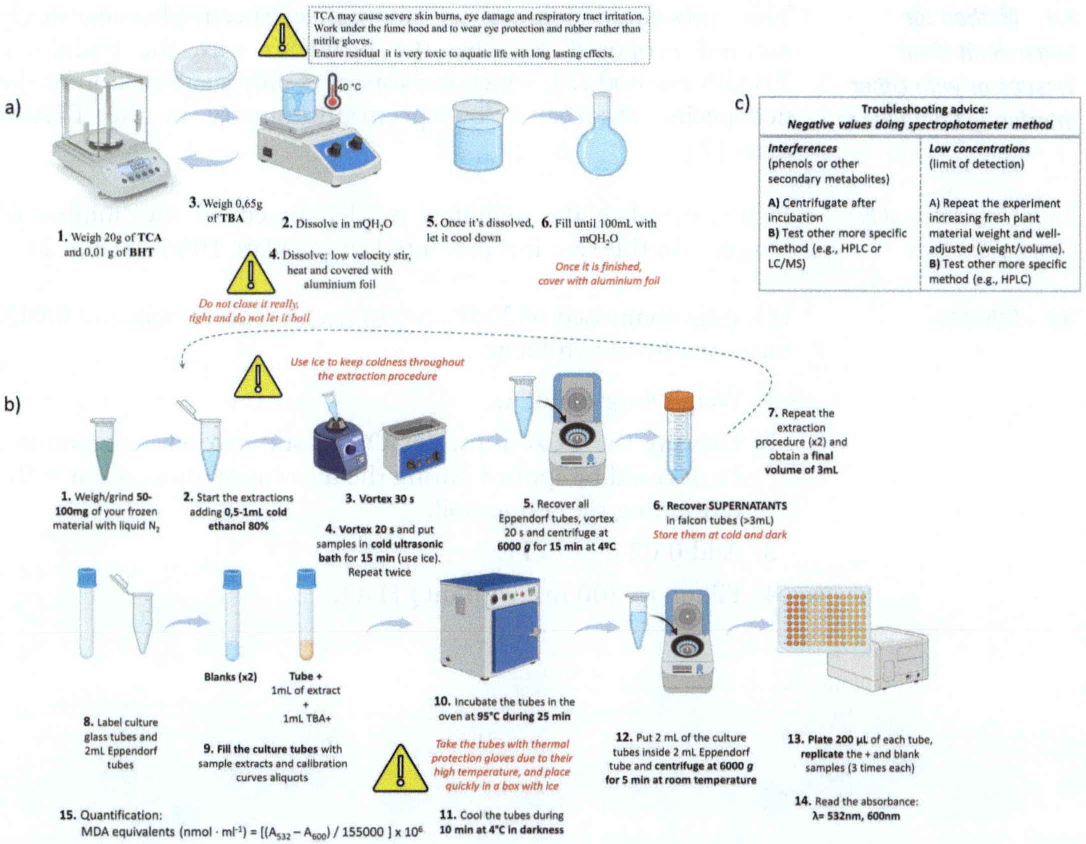

Fig. 1 Malondialdehyde assay using spectrophotometry for plant samples that do not contain interfering compounds in the TBARS assay, such as isolated chloroplasts. (**a**) Preparation of the TBA solution; (**b**) spectrophotometric assay; (**c**) troubleshooting advice

3. Incubate samples with TBA+ solution for 25 min at 95 °C (*see* **Note 2**).
4. Cool samples in ice box and keep in cold chamber and darkness (to stop the reaction).
5. Centrifuge for 1 min to clarify the solution.
6. Plate 200 μL of each tube and replicate the positive samples and blanks thrice for each one..
7. Using a spectrophotometer read each sample at the following absorbances:

$$\lambda = 532 \text{ nm and } 600 \text{ nm}.$$

3.3.4 Quantification

Quantification is performed using the following equation [5]:

$$\text{MDA equivalents } (\text{nmol} \cdot \text{mL}^{-1}) = [(\text{A}_{532}\text{—A}_{600})/155{,}000] \cdot 10^6$$

where 532 nm represents the maximum absorbance of the TBA–MDA complex, 600 nm is the correction for nonspecific turbidity, and 155,000 is the molar extinction coefficient for MDA (Fig. 1).

3.4 Method for Sugar-Rich Plant Tissues or with Other Interfering Compounds

This procedure follows an alternative spectrophotometrically method proposed to solve the problems with the traditional TBARS method [5], which consists of mainly a subtraction of the nonspecific absorbance by spectrophotometry in the TBARS assay [7].

3.4.1 Preparation of the Chemical Solutions

Firstly, calculate the milliliters needed based on the number of samples. In this case it is presented to calculate 100 mL (Fig. 2).

TBA– Solution

It is only comprised of 20.0% (w/v) trichloroacetic acid and 0.01% butylated hydroxytoluene:

1. Weigh 40 g of TCA.
2. Dissolve the TCA in mQ H_2O and stir at room temperature. Heat could be applied during the dissolution process but without letting the solute boil.
3. Add 0.02 g of BHT.
4. Fill up to 200 mL with mQ H_2O.

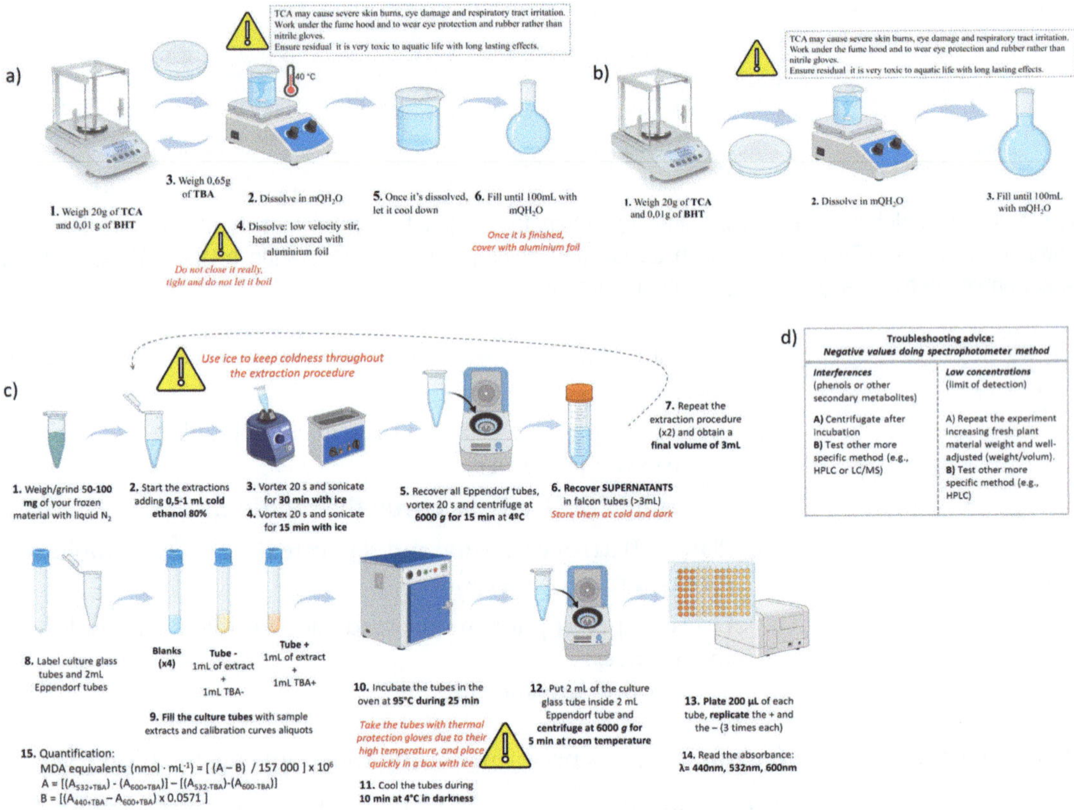

Fig. 2 Malondialdehyde assay using spectrophotometry for plant samples that contain interfering compounds in the TBARS assay, such as leaves. (**a**) Preparation of the TBA+ solution; (**b**) preparation of the TBA– solution; (**c**) spectrophotometric assay; (**d**) troubleshooting advice

TBA+ Solution

It includes trichloroacetic acid containing the above plus 0.65% thiobarbituric acid):

1. Use half of the TBA– solution (previously prepared) inside a glass tube (maximum 100 mL).
2. Weigh 0.65 g of TBA and add it to the glass tube.
3. Dissolve: Stir at low velocity, heat, and cover with aluminum foil. Do not close it tight and do not let it boil.
4. Once it is dissolved, let it cool down.
5. Fill until 100 mL with the TBA– solution. Once it is finished, cover with aluminum foil.

3.4.2 Extraction Steps for Plant Samples

1. Each sample (Eppendorf tube) contains 50–100 mg of fresh weight (plant tissues, for example, leaves) that must be ground in liquid nitrogen. Ensure that all plant samples must be completely powdered.
2. Add 0.5–1 mL cold ethanol 80% (+0.01% BHT). From this step onward, it is required ALWAYS to keep cold (ice) and protect from direct light exposure.
3. Use vortex mixer to totally homogenize the extraction medium (in this case methanol) with the plant samples (solid phase). Thirty seconds per sample is recommended.
4. Put the samples in a cold ultrasonic bath (UB) for 30 min. It is recommended to add a vortex mixer hallway to this process, e.g., vortex +15 min UB + vortex +15 min UB + vortex.
5. Centrifuge conditions: 13,000 rpm, 15 min, 4 °C, Ac 9, Br 1.
6. Recover supernatant into falcon tubes.
7. Repeat **steps 2–6** twice more (be sure to recover extremely clean supernatant in the last extraction).

3.4.3 Incubation Procedure (MDA Derivatization)

1. Centrifuge (13,000 rpm, 10 min, 4 °C) pull of supernatant to clarify the liquid phase.
2. Separate each sample into two aliquots and mix in Pyrex glass tubes:
   (a) 1 mL of sample with 1 mL of TBA– solution (TBA–).
   (b) 1 mL of sample with 1 mL of TBA+ solution (TBA+).

   It is suggested to change the volumes but always keep sample proportions (1/1; v/v).
3. Incubate samples with TBA –/+ solutions for 25 min at 95 °C.
4. Cool samples in ice box and keep in cold chamber and darkness (to stop the reaction).
5. Centrifuge for 1 min to clarify the solution.

6. Plate 200 μL of each tube, replicate the positive/negative samples and blanks thrice for each one.
7. Using a spectrophotometer read each sample at the following absorbances:

$$\lambda = 440 \text{ nm}, 532 \text{ nm}, \text{and } 600 \text{ nm}.$$

3.4.4 Quantification

Quantification is performed using following equations [7] summarized in Fig. 2.

3.5 Protocol by HPLC

MDA is also one of the commonly used biomarkers of oxidative stress for many research fields, particularly in biomedical and animal studies. Since most studies quantifying MDA by HPLC are performed in plasma or animal samples, an optimized protocol (Fig. 3) is presented for leaves [8]. It is based on following the first HPLC method developed by Iturbe-Ormaetxe et al. [10] with some modifications based on other studies [13, 14]. Briefly, the cold methanol 100% containing 1 ppm BHT was selected as the optimal extracting medium for MDA. It is recommended to test the

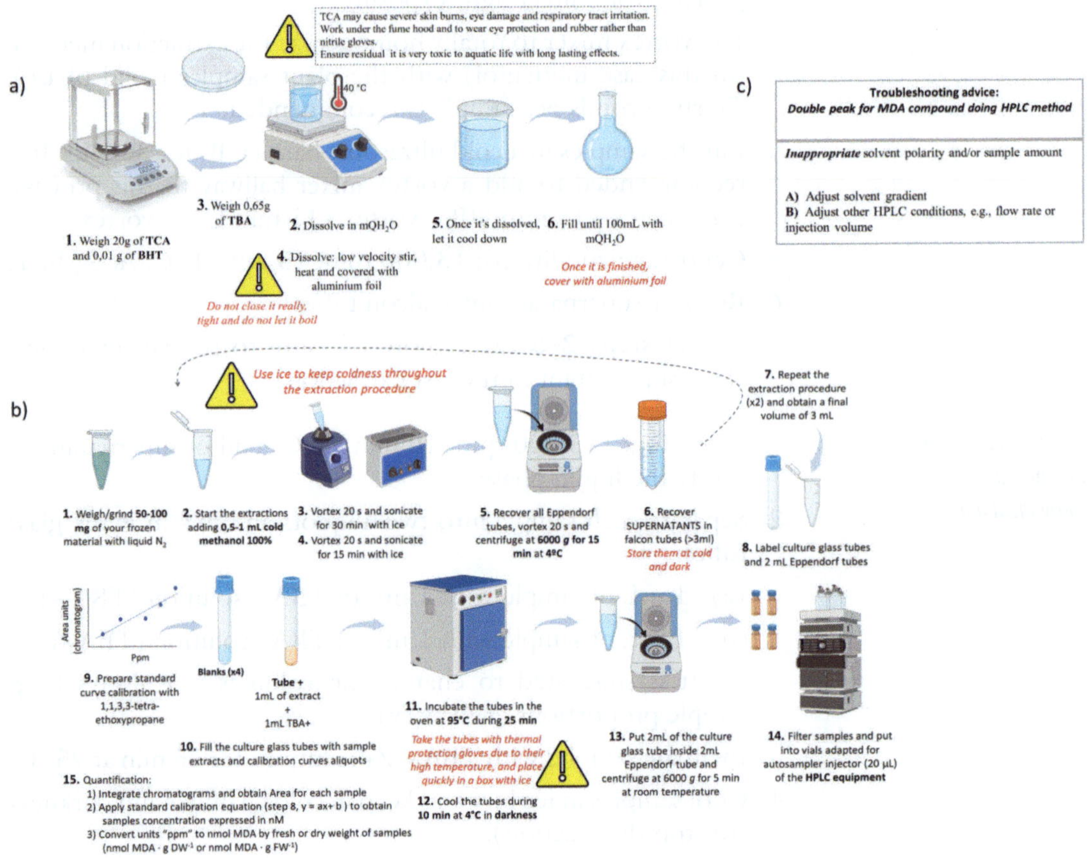

Fig. 3 Malondialdehyde assay using HPLC. (**a**) Preparation of the TBA solution; (**b**) HPLC assay; (**c**) troubleshooting advice

optimum organic solvent (methanol or ethanol) and proportion for each different plant tissue among species. The HPLC equipment (Agilent) included a photodiode-array detector controlled by Millennium software. The MDA-$(TBA)_2$ adduct was resolved on an Hypersyl ODS-5 μm column (250 × 4.6 mm, Teknokroma, St. Cugat, Spain) and eluted in the mobile phase with 5 mm potassium phosphate buffer (pH 7.0) containing 14% acetonitrile and 0.6% tetrahydrofuran. The flow rate was 1 mL/min and the injection volume 20 μL. The MDA-$(TBA)_2$ adduct was quantified by its absorbance at 532 nm (Diode array detector 1000S, Applied Biosystems), and was identified by its characteristic spectra and by coelution with an authentic standard of 1,1,3,3-tetraethoxypropane (Sigma, Steinheim, Germany), which is stoichiometrically converted into MDA during the acid-heating step of the assay.

3.5.1 Preparation of Chemical Solutions and Standard Calibration

1. *TBA+ solution* (containing 20 g TCA + 0.01 g BHT + 0.65 g TBA) in 100 mL mQ H_2O.
2. *Standard calibration* curve with 1,1,3,3-tetraethoxypropane (Sigma) in nM units. It is important to adjust the calibration curve depending on the concentration range in each type of sample. It could change among plant species and plant tissues.

3.5.2 Extraction Steps

1. Each sample (Eppendorf tube) contains 50–100 mg of FW that must be ground in liquid nitrogen. Plant samples must be completely powdered.
2. Add 0.5–1 mL cold methanol 100% (0.01% BHT). From this step onward, it is required ALWAYS to keep cold (ice) and protect from direct light exposure.
3. Use vortex to totally homogenize the extraction medium (in this case methanol) with the plant samples (solid phase). Thirty seconds per sample is recommended.
4. Put the samples in cold ultrasounds + vortex +20 min cold ultrasounds + vortex.
5. Centrifuge conditions: 13,000 rpm, 15 min, 4 °C, Ac 9, Br 1.
6. Recover supernatant into a new supernatant.
7. Repeat **steps 2–6** twice more (be sure to recover extremely clean supernatant in the last extraction).

3.5.3 Incubation Procedure (MDA Derivatization)

1. Centrifuge (13,000 rpm, 10 min, 4 °C) pull of supernatant to clarify the solution.
2. Prepare standard calibration curve with 1,1,3,3-tetraethoxypropane.
3. For each sample, mix in special glass tubes 1 mL of sample and standard calibration curve with 1 mL of TBA+ solution. In this

step, it can change volumes but always keep the same proportions (1:1 v/v).

4. Incubate the tubes/solution for 25 min at 95 °C.
5. Cool samples in ice box and keep in cold chamber and darkness (to stop the reaction).
6. Centrifuge for 1 min to clarify the solution.
7. Filter and place sample extractions in vials specific to HPLC equipment.

3.5.4 Quantification.

1. Step 1: Integrate chromatograms and obtain area for each sample.
2. Step 2: Apply standard calibration equation ($y = ax + b$) to obtain sample concentration expressed in nM.
3. Step 3: Convert units "ppm" to nmol MDA by fresh or dry weight of samples (nmol MDA · g/DW or nmol MDA · g/FW).

3.5.5 HPLC Conditions

1. Column: Hypersyl ODS-5 μm (250 × 4.6 mm, Teknokroma, St. Cugat, Spain)..
2. Absorbance at 532 nm (diode array detector 1000S).
3. Injection volume: 20 μL.
4. Flow rate: 1 mL·min^{-1}.
5. Mobile phase: 5 mM potassium phosphate buffer (pH = 7) containing 14% MeCN and 0.6% tetrahydrofuran.
6. Chromatogram time: 7 min.
7. Cleaning column and HPLC system: First cleaning: (1) 80% H_2OmQ, 20%MeCN (1 mL·min^{-1}, 15–20 min); (2) 20% H_2OmQ, 80%MeCN (1 mL·min^{-1}, 15–20 min); Pump A: change buffer to H_2OmQ and purge; Pump B: change MeCN to MeOH and purge; Pump C: change THF to MeOH and purge.

3.6 Protocol by LC-MS

A more rapid and sensitive protocol for the MDA quantification in plant samples is proposed here (Fig. 4), following the first method described [15] with latest modifications [16, 17]. Briefly, the plant sample was put into a 2 mL Eppendorf tube containing 1 mL methanol with BHT 0.01% (w/v). Glass beads (0.25 g of 1 mm and 0.25 g of 0.1 mm) were added to the tissue extract and the samples placed in a cell disrupter or ultrasonic bath (MiniBead Beater, Biospec Products, Bartlesville, OK, USA). The tissues were disrupted in three 30 s cycles (5000 cycles·min^{-1}) with cooling over ice between cycles. The homogenate was then centrifuged (5 min, 500 g) and the supernatant was transferred to a bijou bottle containing 1 mL 2,4-dinitrophenylhydrazine (DNPH; 350 mg/L) in 1 M HCl. After swirling, the reaction proceeded for 10 min at

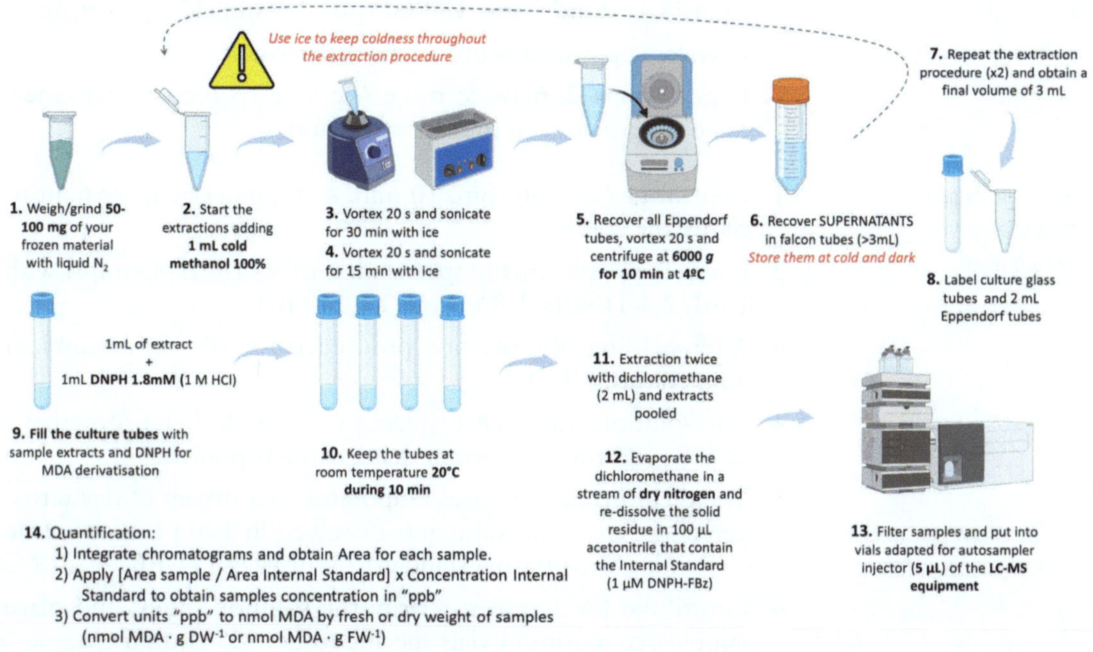

Fig. 4 Malondialdehyde assay using LC-MS

ambient temperature (20 °C). The solution was then extracted twice with dichloromethane (2 mL) and the dichloromethane extracts pooled. The dichloromethane was evaporated in a stream of dry nitrogen and the solid residue was dissolved in 100 μL acetonitrile containing the internal standard 1 μM DNPH-FBz (the 2,4-dinitrophenylhydrazone of 4-fluorobenzaldehyde). The analyses were performed in duplicate on a Finnigan MAT SSQ 710C single quadrupole mass spectrometer with APCI interface (Thermo Quest, Hemel Hempstead, UK).

3.6.1 Preparation of Chemical Solutions and Internal Standard

1. Solution with DNPH of 350 mg/mL.
2. Use of internal standard of 1 μM DNPH-FBz.

3.6.2 Extractions Steps

1. Each sample (Eppendorf tube) containing 50–100 mg FW must be ground in liquid nitrogen. Plant samples must be completely powdered.
2. Add 1 mL methanol containing 0.01% BHT. From this step onward, it is required ALWAYS to keep cold (ice) and protect from direct light exposure.
3. Use vortex to totally homogenize the extraction medium (in this case methanol) with the plant samples (solid phase). Thirty seconds per sample is recommended.
4. Put the samples in cold ultrasounds + vortex +20 min cold ultrasounds + vortex.

5. Centrifuge conditions: 13,000 rpm, 10 min, 4 °C, Ac 9, Br 1.
6. Recover supernatant into a new supernatant.
7. Repeat **steps 2–6** twice more (be sure to recover extremely clean supernatant in the last extraction).

3.6.3 Incubation Procedure (MDA Derivatization)

1. Centrifuge (13,000 rpm, 10 min, 4 °C) pull of supernatant to clarify the solution.
2. For each sample, mix in special glass tubes 1 mL of sample with 1 mL 2,4-DNPH 1.8 mM in 1.0 M HCl.
3. After swirling, the reaction proceeded for 10 min at ambient temperature (20 °C).
4. The solution was then extracted twice with dichloromethane (2 mL) and the dichloromethane extracts pooled.
5. The dichloromethane was evaporated in a stream of dry nitrogen and the solid residue was dissolved in 100 μL acetonitrile containing the internal standard 1 μM DNPH-FBz.
6. Centrifuge for 1 min to clarify the solution, filter, and place sample extractions in vials specific to LC-MS equipment.

3.6.4 Quantification

1. Step 1: Integrate chromatograms and obtain area for each sample.
2. Step 2: Apply standard calibration equation ($y = ax + b$) to obtain sample concentration expressed in nM.
3. Step 3: Convert units “ppm” to nmol MDA by fresh or dry weight of samples (nmol MDA · g DW^{-1} or nmol MDA · g FW^{-1}).

3.6.5 LC-MS Conditions

1. Column: Hypersyl 5 μm C-18 reversed-phase column (250 × 2.0 mm, Phenomenex, Macclesfield, UK).
2. HPLC pump: Constrametric 4100 MS HPLC pump, AS3000 autosampler and UV 1000 detector. Effluent from the UV detector was passed via a 1-m length of PEEK tubing (60 μm internal diameter) to the APCI interface.
3. UV detection: at 360 nm.
4. Injection volume: 5 μL.
5. Flow rate: 0.35 mL/min.
6. Solvent gradient elution: 5-min linear gradient from 41% to 75% acetonitrile in water, both solvents containing 40 mM ammonium acetate.
7. MS Detection: negative ion chemical ionization (NICI) mode at m/z 234 and 335 [M-H]- for the DNPH-derivatives of MDA 4-fluorobenzaldehyde (internal standard), respectively (Muckenschnabel et al. 2001).
8. Chromatogram time: 5 min.

3.7 Protocol by UHPLC-(+)-ESI-MS/MS

An UHPLC-MS/MS method has been developed to determine MDA in plant samples, particularly as a stress biomarker in fresh leaves of melon plants [11]. This method, UHPLC-(+)ESI-MS/MS, has proved to be simple, rapid, reliable, sensitive, selective, and reproducible for the determination of MDA and its variations, without needing a derivatization procedure. After the ESI source parameters were optimized (polarity, capillary voltage, source temperature, probe temperature, drying gas flow rate, and drying gas temperature), the proper values were achieved, as mentioned in the "mass spectrometry conditions" section of the original methodology [11], and demonstrated that the ESI source in positive mode is the best strategy for MDA ionization. The analytical response for the generation of the MDA protonated molecule [M + H] + was higher than the obtained for the generated [M − H] − ion when the negative ESI mode was assayed. Also, this study showed that the ESI source allowed to obtain an ionization efficiency at least an order of magnitude higher than the APCI source.

As summarized in Fig. 5, leaves were processed with a tissue homogenizer Bio-Gen PRO200 at 17,000 rpm, using 10 mL of 0.1% (v/v) formic acid as extraction solvent. Afterward, the extract was centrifuged at 11,500 rpm at 4 °C for 15 min. From each supernatant, 2 mL was filtered through a 0.2 μm PTFE (polytetrafluoroethylene) syringe filter into a glass vial prior to underivatized

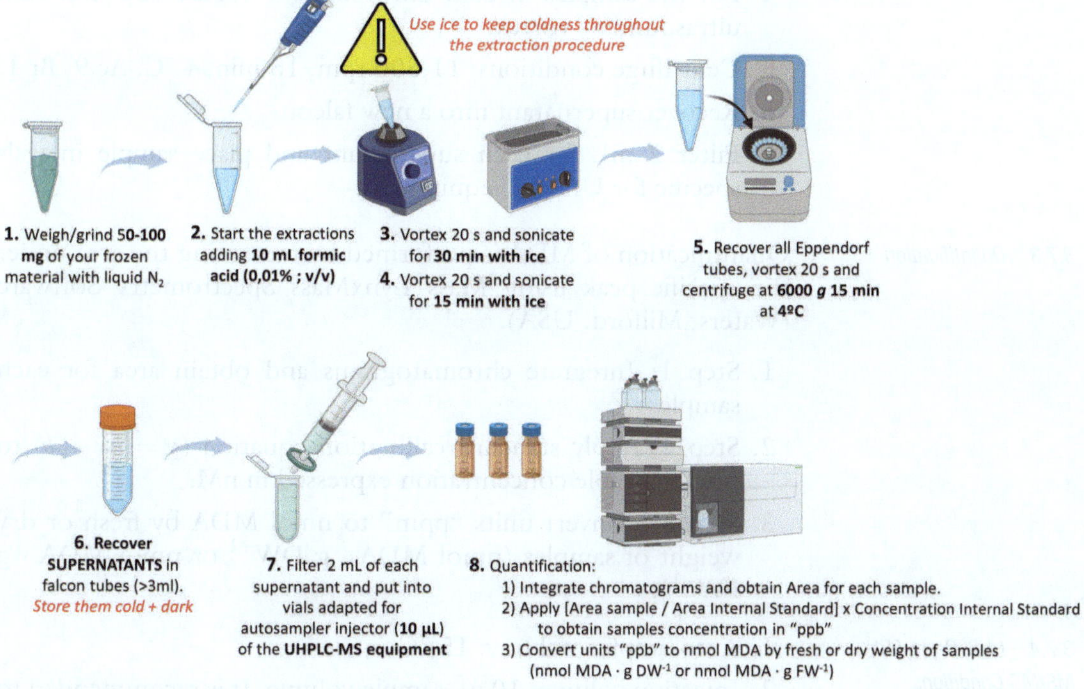

Fig. 5 Malondialdehyde assay using UHPLC-MS/MS

MDA analysis by UHPLC-(+)-ESI-MS/MS. To reduce the observed matrix effect on the MDA analytical response, a cleanup strategy through a solid-phase extraction (SPE) procedure was performed. For SPE, Oasis HLB reverse-phase cartridges were used. Aliquots of 2 mL sample were applied to the cartridges previously conditioned with 2 mL of methanol and equilibrated with 2 mL of 5% (v/v) methanol. The retained fraction was eluted with 2 mL of 10% (v/v) acetonitrile containing 0.1% (v/v) formic acid. Following this SPE procedure, the MDA recovery was above 90% and the matrix effect decreased by 10% after SPE clean-up. An Acquity® UHPLC (Waters, Milford) equipped with autosampler injection and pump systems (Waters, Milford) was used [11].

3.7.1 Preparation of Chemical Solutions and Standard Calibration

Prepare the formic acid 0.01% (v/v) solution.

3.7.2 Extractions Steps (Without Derivatization)

1. Each sample (Eppendorf tube) contains 50–100 mg FW and must be ground with liquid nitrogen. Plant samples must be completely powdered.
2. Add 10 mL formic acid 0.01% (v/).
3. Use vortex to totally homogenize the extraction medium (in this case methanol) with the plant samples (solid phase). Thirty seconds per sample is recommended.
4. Put the samples in cold ultrasounds + vortex +20 min cold ultrasounds + vortex.
5. Centrifuge conditions: 11,500 rpm, 15 min, 4 °C, Ac 9, Br 1.
6. Recover supernatant into a new falcon.
7. Filter 2 mL for each supernatant and place sample in vials specific for UHPLC equipment.

3.7.3 Quantification

Quantification of MDA is performed by measuring the area under the specific peak using Mass LynxMass Spectrometry Software (Waters, Milford, USA).

1. Step 1: Integrate chromatograms and obtain area for each sample.
2. Step 2: Apply standard calibration equation ($y = ax + b$) to obtain sample concentration expressed in nM.
3. Step 3: Convert units "ppm" to nmol MDA by fresh or dry weight of samples (nmol MDA · g DW^{-1} or nmol MDA · g FW^{-1}).

3.7.4 UHPLC-(+)-ESI-MS/MS Conditions

1. Autosampler vial tray: 15 °C.
2. Injection volume: 10 μL sample volume. It is recommended to wash the needle with proper mixtures of acetonitrile and water.

3. Analytical column: ACQUITY UHPLC® BEH C18 (Waters, Milford, USA) with 2.1 mm internal diameter, 50 mm length, and 1.7 μm particle size).
4. Column temperature: 25 °C.
5. Binary mobile phases: water with 0.1% (v/v) of formic acid (A) and acetonitrile with 0.1% (v/v) of formic acid (B). The elution was isocratic with a mobile phase composition of 70% A and 30% B.
6. Flow rate: 0.1 mL·min^{-1}.
7. Chromatogram time: The total run time, including conditioning time, was 3.0 min.
8. Mass spectrometry conditions: Quattro Premier® XE Micromass MS Technologies triple quadrupole mass spectrometer with a Z-Spray® electrospray ionization source (Waters, Milford, USA). The source was operated in the positive (ESI+) mode at 300 °C with N_2 as the nebulizer and the source temperature was kept at 150 °C. The capillary voltage was maintained at 1.9 kV and the extractor voltage was set at 4.0 kV. Ultrapure nitrogen was used as desolvation gas with a flow of 850 L/h.
9. MS Detection: MDA-protonated molecule [MDA-H] + of m/z 73 was performed in the selected ion monitoring mode (SIR). To choose the maximum sensitivity conditions, direct infusion (via syringe pump) into the MS of MDA standard solution in 0.1% (v/v) formic acid was performed and the product ion scan mass spectra were recorded over an m/z range of 50–150.

3.8 Protocol by GC-MS

This protocol is based on the quantification of MDA levels in *Arabidopsis* leaves by gas chromatography-mass spectrometry (GC-MS) with the latest modifications [3] based on preliminary studies [18, 19]. As summarized in Fig. 6, leaf material (100–150 mg) was collected from plants, frozen in liquid nitrogen, and ground to a fine powder. The frozen powder material was added phosphate buffered saline containing 1 mM desferoxamine mesylate, 0.08 mM butylated hydroxytoluene, and 0.5 nmol of the internal standard [$^2H_2$]-MDA. After rapid vortexing, 10 μL of 6.6 N H_2SO_4 was added immediately, and the sample was incubated for 10 min on ice. Sodium tungstate (0.3 M, 75 μL) was then added, and the samples were centrifuged for 5 min. The supernatant was combined with phosphate-citrate buffer (0.21 M citric acid, 0.58 M NA_2HPO_4) and 50 μL of pentafluorophenylhydrazine reagent (5 mg PFPH mL $-$ ^{1} H_2O) and incubated for 30 min at room temperature. It was added 9 μL of 9 N H_2SO_4 and the MDA-PFPH product was extracted with 125 μL of isooctane. The organic phase was recovered, and just 1 μL was analyzed by

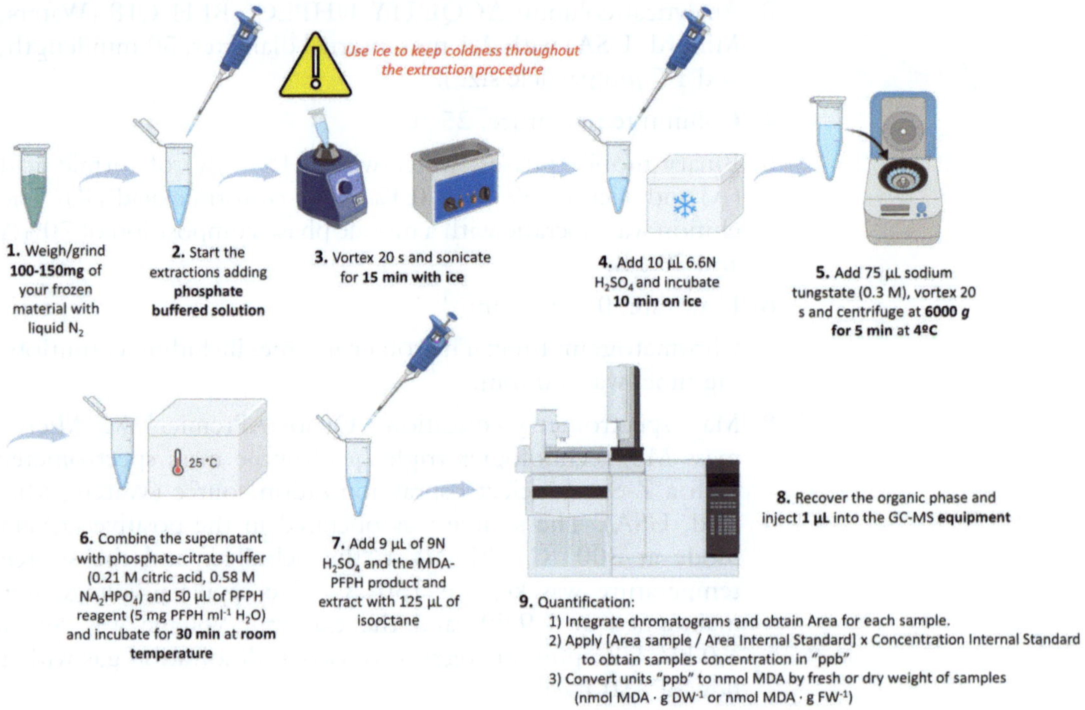

Fig. 6 Malondialdehyde assay using GC-MS/MS

GC-MS (Hewlett-Packard 5890 gas chromatograph, 5927 mass spectrometer, electron ionization mode, 70 eV electron potential, 11 psi column head pressure, HP-5MS capillary column 30 m × 0.25 mm). The temperature gradient was 50 °C for 1 min, 50–150 °C at 10 °C min,$^{-1}$ and 150–280 °C at 20 °C min^{-1}. Quantification was carried out using selective ion monitoring measuring ions m/z ¼ 234 for MDA and m/z ¼ 236 for the internal standard [$^2H_2$]-MDA.

3.8.1 Preparation of Chemical Solutions and Standard Calibration

1. Phosphate buffered saline containing 1 Mm desferoxamine mesylate, 0.08 Mm butylated hydroxytoluene, and 0.5 nmol of the internal standard [$^2H_2$]-MDA.
2. 6.6 N and 9 N H_2SO_4.
3. Sodium tungstate 0.3 M.
4. Phosphate-citrate buffer (0.21 M citric acid, 0.58 M Na_2HPO_4).
5. 5 mg PFPH mL $-$ ^{1} mQH_2O.

3.8.2 Extractions Steps

4. Each sample (Eppendorf tube) contains 100–150 mg of fresh plant material and it is completely powdered with liquid nitrogen.

5. The frozen powder was added to phosphate-buffered saline containing 1 mM desferoxamine mesylate, 0.08 mM butylated hydroxytoluene, and 0.5 nmol of the internal standard [$^2H_2$]-MDA.
6. Use vortex to totally homogenize the extraction medium with the plant samples. Always 30 s per sample is recommended.
7. After rapid vortexing, 10 μL of 6.6 N H_2SO_4 was added immediately, and the sample was incubated for 10 min on ice.
8. Add sodium tungstate (0.3 M, 75 μL) and centrifuge the samples for 5 min.
9. The supernatant was combined with phosphate-citrate buffer (0.21 M citric acid, 0.58 M NA_2HPO_4) and 50 μL of PFPH reagent (5 mg PFPH mL – ^{1} H_2O) and incubated for 30 min at room temperature.
10. Add 9 μL of 9 N H_2SO_4 and the MDA-PFPH product was extracted with 125 μL of isooctane.
11. Recover the organic phase and 1 μL was analyzed by GC/MS.

3.8.3 Quantification

1. Step 1: Integrate chromatograms and obtain area for each sample.
2. Step 2: Apply standard calibration equation ($y = ax + b$) to obtain sample concentration expressed in nM.
3. Step 3: Convert units "ppm" to nmol MDA by fresh or dry weight of samples (nmol MDA · g DW^{-1} or nmol MDA · g FW^{-1}).

3.8.4 GC-MS Conditions

1. Hewlett-Packard 5890 gas chromatograph.
2. 5927 mass spectrometer, electron ionization mode, 70 eV electron potential HP-5MS capillary column 30 m × 0.25 mm (11 psi column head pressure).
3. The temperature gradient was: 50 °C for 1 min, 50–150 °C at 10 °C min,$^{-1}$ and 150–280 °C at 20 °C min $^{-1}$.
4. Quantification was carried out using selective ion monitoring measuring ions m/z ¼ 234 for MDA and m/z ¼ 236 for the internal standard [$^2H_2$]-MDA.

4 Notes

1. TCA may cause severe skin burns, eye damage, and respiratory tract irritation. Therefore, it is recommended to work under the fume hood and to wear eye protection and rubber rather than nitrile gloves (as the latter can dissolve when exposed to

TCA). Also, it is very toxic to aquatic life with long-lasting effects.

2. After incubation take the tubes with thermal protection gloves due to their high temperature, and place quickly in a box with ice.

5 Discussion

The scientific community working in various disciplines within the plant sciences, from plant molecular biology to agrifood biotechnology, includes MDA quantification as a biomarker of lipid peroxidation. In this context, we have summarized diverse methodological approaches which can be used as a rapid guideline to choose the best MDA method for every plant sample. For cheap and quick analyses, traditional spectrophotometry methods are still recommended, but always using assays that consider possible interfering compounds, such as the method developed by Hodges et al. [7]. These assays advance significantly to the initial method described by Heath and Packer [5] that might be used for isolated chloroplasts or other samples that do not contain TBARS interfering compounds only. When a higher cost investment is possible, we strongly recommend the use of more precise techniques like HPLC [3, 8], LC-MS [4, 16], UHPLC-MS/MS [5, 11], or GC-MS [3, 6].

6 Conclusions

In conclusion, when choosing the most appropriate method for MDA quantification we should consider several relevant aspects. First, it is important to consider the type of experiment and plant material. For example, the traditional TBARS method could be a proxy for estimating MDA in isolated chloroplasts, but it would not be specific enough for quantification of MDA in complex plant samples leading to erroneous estimates and an overestimation of MDA contents. Second, it is very important to consider relevant limitations of each method for adequate interpretation of results. For instance, methods based on the TBARS assay are likely to measure compounds other than MDA that can also react with thiobarbituric acid, thus also leading to slight overestimations. Furthermore, methods based on plant extract derivatization are also prone to slightly erroneous results, in this case usually leading to an underestimation of MDA content. Although the available resources in every lab and the objectives of the study are determinants in choosing the most adequate MDA method, selecting the best option must be done with caution and in any case limitations of the approach used must be considered in the interpretation of results.

Acknowledgments

The authors are very grateful to the *Beatriu de Pinós* MSCA-COFUND postdoctoral grant (2018 BP-00106) and the ICREA Academia award, respectively, received from the Government of Catalonia.

References

1. Farmer EE, Mueller MJ (2013) ROS-mediated lipid peroxidation and RES activated signaling. Annu Rev Plant Biol 64:429–450
2. Sofo A, Dichio B, Xiloyannis C, Masia A (2004) Lipoxygenase activity and proline accumulation in leaves and roots of olive trees in response to drought stress. Physiol Plant 121: 58–65
3. Weber H, Chételat A, Reymond P, Farmer EE (2004) Selective and powerful stress gene expression in Arabidopsis in response to malondialdehyde. Plant J 37:877–888
4. Davey MW, Stals E, Panis B, Keulemans J, Swennen RL (2005) High-throughput determination of malondialdehyde in plant tissues. Anal Biochem 347:201–207
5. Heath RL, Packer L (1968) Photoperoxidation in isolated chloroplasts. I. Kinetics and stoichiometry of fatty acid peroxidation. Arch Biochem Biophys 125:189–198
6. Taulavuori E, Hellström E-K, Taulavuori K, Laine K (2001) Comparison of two methods used to analyse lipid peroxidation from *Vaccinium myrtillus* (L.) during snow removal, reacclimation and cold acclimation. J Exp Bot 52: 2375–2380
7. Hodges DM, DeLong JM, Forney CF, Prange RK (1999) Improving the thiobarbituric acid-reactive-substances assay for estimating lipid peroxidation in plant tissues containing anthocyanin and other interfering compounds. Planta 207:604–611
8. Morales M, Roach DA, Quarles BM, Cotado A, Salguero-Gómez R, Dwyer J, Munné-Bosch S (2021) Validity of photo-oxidative stress markers and stress-related phytohormones as predictive proxies of mortality risk in the perennial herb *Plantago lanceolata*. Environ Exp Bot 191:104598
9. Davey MW, Stals E, Panis B, Keulemans J, Swennen RL (2005) High throughput determination of malondialdehyde in plant tissues. Anal Biochem 347:201–207
10. Iturbe-Ormaetxe I, Escuredo PR, Arrese-Igor C, Becana M (1998) Oxidative damage in pea plants exposed to water deficit or paraquat. Plant Physiol 116:173–181
11. Yonny ME, Rodríguez-Torressi A, Nazareno MA, Cerutti S (2017) Development of a novel, sensitive, selective, and fast methodology to determine malondialdehyde in leaves of melon plants by ultra-high-performance liquid chromatography-tandem mass spectrometry. J Anal Methods Chem 2017:4327954
12. Lykkesfeldt J (2001) Determination of malondialdehyde as dithiobarbituric acid adduct in biological samples by HPLC with fluorescence detection: comparison with ultraviolet-visible spectrophotometry. Clin Chem 47:1725–1727
13. Munné-Bosch S, Alegre L (2002) Interplay between ascorbic acid and lipophilic antioxidant defences in chloroplasts of water-stressed Arabidopsis plants. FEBS Lett 524:145–148
14. Draper HH, Squires EJ, Mahmoodi H, Wu J, Agarwal S, Hadley M (1993) A comparative evaluation of thiobarbituric acid methods for the determination of malondialdehyde in biological materials. Free Radic Biol Med 15: 353–363
15. Deighton N, Magill WJ, Bremner DH, Benson EE (1997) Malondialdehyde and 4-hydroxy-2-nonenal in plant tissue cultures: LC-MS determination of 2,4-dinitrophenylhydrazone derivatives. Free Radic Res 27:255–265
16. Deighton N, Muckenschnabel I, Goodman BA, Williamson B (1999) Lipid peroxidation and the oxidative burst associated with infection of *Capsicum annuum* by *Botrytis cinerea*. Plant J 20:485–492

17. Muckenschnabel I, Williamson B, Goodman BA, Lyon GD, Stewart D, Deighton N (2001) Markers for oxidative stress associated with soft rots in French beans (*Phaseolus vulgaris*) infected by *Botrytis cinerea*. Planta 212: 376–381
18. Yeo HC, Helbock HJ, Chyu DW, Ames BN (1994) Assay of malondialdehyde in biological-fluids by gas-chromatography mass-spectrometry. Anal Biochem 220:391–396
19. Yeo HC, Liu J, Helbock H, Ames BN (1999) Assay of malonaldehyde and other alkanals in biological fluids by gas chromatography-mass spectrometry. Meth Enzymol 300:70–73

Chapter 7

Roles of Reactive Carbonyl Species (RCS) in Plant Response to Abiotic Stress

Mustafa Cemre Sonmez, Side Selin Su Yirmibesoglu, Rengin Ozgur, Baris Uzilday, and Ismail Turkan

Abstract

Abiotic and biotic stress conditions lead to production of reactive carbonyl species (RCS) which are lipid peroxide derivatives and have detrimental effects on plant cells especially at high concentrations. There are several molecules that can be classified in RCS; among them, 4-hydroxy-(E)-2-nonenal (HNE) and acrolein are widely recognized and studied because of their toxicity. The toxicity mechanisms of RCS are well known in animals but their roles in plant systems especially signaling aspects in metabolism need to be addressed. This chapter focuses on the production mechanisms of RCS in plants as well as how plants scavenge and modify them to prevent irreversible damage in the cell. We aimed to get a comprehensive look at the literature to summarize the signaling roles of RCS in plant metabolism and their interaction with other signaling mechanisms such as highly recognized reactive oxygen species (ROS) signaling. Changing climate promotes more severe abiotic stress effects on plants which also decrease yield on the field. The effects of abiotic stress conditions on RCS metabolism are also gathered in this chapter including their signaling roles during abiotic stresses. Different methods of measuring RCS in plants are also presented in this chapter to draw more attention to the study of RCS metabolism in plants.

Key words Environmental stress conditions, Reactive carbonyl species (RCS), Reactive oxygen species (ROS), Signaling mechanisms

1 Introduction

Environmental stress conditions significantly damage plants and impact agricultural and ecological vegetation in numerous ways. Abiotic stresses are increasing and continue to grow with worsening climate change, and biotic stresses are predicted to be affected [1]. A key result of these stresses in the plant cell is oxidative stress, which is caused by accumulation of reactive oxygen species (ROS), and has been studied extensively for the last few decades [2]. A related stress, however, caused by reactive carbonyl species (RCS), has been gaining attention of the scientific community recently. RCS, similar to ROS, participate in signaling at low concentrations,

Francisco J. Corpas and José M. Palma (eds.), *ROS Signaling in Plants: Methods and Protocols*, Methods in Molecular Biology, vol. 2798, https://doi.org/10.1007/978-1-0716-3826-2_7,

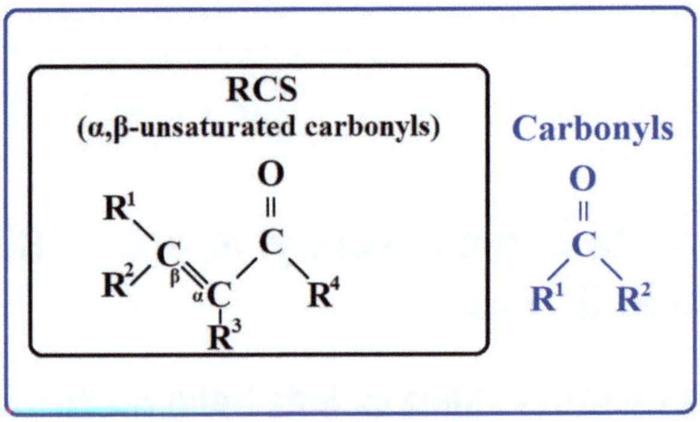

Fig. 1 Structures and relationship of carbonyls and RCS. *R* groups denote "rest of the molecule." (The diagram sizes are not-to-scale)

and their damaging side dominates at high concentrations. RCS are aldehydes and ketones, with α,β-unsaturated bond, meaning they have a C–C double bond conjugated to the carbonyl double bond (it should be noted that, although they have high reactivity and roles in signaling, this definition excludes the carbonyls methylglyoxal and glyoxal). RCS are part of a bigger group of molecules, called reactive electrophile species (RES), and RCS are RES oxylipins [3]. However, neither every RES, nor every RES oxylipin, nor every carbonyl is an RCS. RCS belong to all three groups of these molecules, but what distinguishes them is the α,β-unsaturated bond, on top of them being a carbonyl [4].

Figure 1 shows the relationship between RCS and carbonyls. As can be seen, RCS structure is part of the general structure given for carbonyls; however, the α,β-unsaturated bond and its position differentiate it from other carbonyls. The *R* groups can differ from a simple hydrogen molecule to complex structures, such as ring groups.

RCS, as highly reactive electrophiles, are deleterious to many cell components, since majority of them are nucleophiles. Sites such as thiol, imidazole, and hydroxyl groups are especially vulnerable [5]. They are more toxic than saturated carbonyls. RCS toxicity is mainly attributed to inhibition of target enzymes, and not only exogenous addition but also endogenous generation of RCS has been shown to cause toxicity. For this reason, RCS are scavenged both enzymatically and nonenzymatically. On the other hand, RCS also act downstream of ROS signaling, take part in stress signaling, and are part of programmed cell death and pathogenesis [6–8]. Signaling roles of RCS will be elaborated upon later in this chapter.

RCS, like ROS, are produced even under relaxed conditions, and their production increases under stress [9]. There are more than a dozen RCS found in plants. Acrolein and 4-hydroxy-(E)-2-nonenal (HNE) are the most toxic RCS that were found to increase early in reaction to ROS. Malondialdehyde (MDA), however, is another commonly found RCS that appears at later stages of oxidative stress, and it has lower reactivity [3]. It should also be mentioned that MDA has multiple forms, and while it is relatively stable in enolate form near neutral pH, it is electrophilic and reactive in protonated forms at lower pH [6]. But even considering this, it is around one magnitude lower in activity compared to HNE [10]. In theory, relatively less reactive RCS such as MDA should affect cells at higher concentrations, while acrolein and HNE should be at lower concentrations [4].

The qualities of RCS mentioned so far, signaling/damaging duality, increasing under stress, being downstream of ROS, and their variety, make them a noteworthy field of study for signaling under stress. In this chapter, we present findings for their production sites, scavenging and modification mechanisms, signaling, interaction with other reactive signaling mechanisms, and examine their signaling under different stresses. Finally, we are going to present different methods of measuring RCS in plants, and present a conclusion. It is our hope that by focusing on signaling and qualities that are important to it, and various measurement methods, we will be able to offer the reader an informative outlook of the spectacular work done on the rising role of RCS stress signaling, both in theory and practice.

2 Pathways and Subcellular Locations of RCS Production

2.1 Pathways of RCS Production in Plants

RCS are created from lipid hydroperoxides (LOOH), of which production has been associated with lipid membranes, especially ones rich with polyunsaturated fatty acids (PUFAs). LOOHs can be formed from membrane lipids (LH) either nonenzymatically, or in other words by autoxidation, or enzymatically. A summary of pathways of LOOH formation is given in Fig. 2, which will be elaborated upon step by step.

In the nonenzymatic formation, ROS react with fatty acid tails of membrane lipids, particularly of PUFAs, and they lead to LOOH formation. There are two scenarios through which PUFA and ROS interaction can lead to RCS formation, as described in [7] and [11]. In the first one, hydroxyl radical ($HO^{\bullet}$) and hydroperoxyl radical ($HO_2^{\bullet}$) initiate radical chain reactions, that end with an abstraction of a hydrogen atom from a PUFA. As a result, from this PUFA, a lipid radical ($L^{\bullet}$) is formed (Eq. 1).

$$LH + HO^{\bullet}/HO_2{}^{\bullet} \rightarrow L^{\bullet} \quad (1)$$

This $L^{\bullet}$ then, by addition of molecular oxygen (O_2), forms a lipid peroxyl radical ($LOO^{\bullet}$) (Eq. 2).

$$L^{\bullet} + O_2 \rightarrow LOO^{\bullet} \quad (2)$$

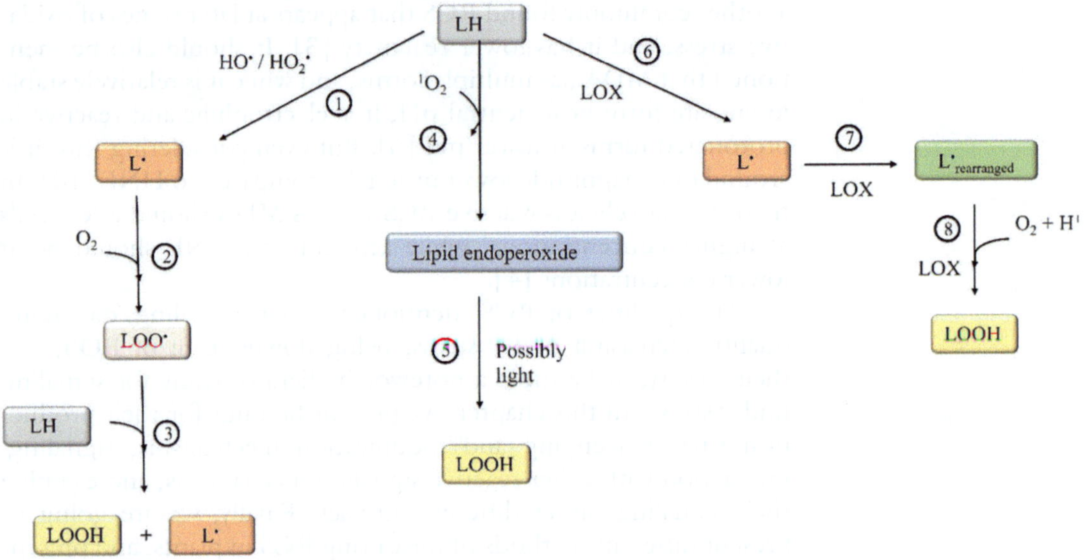

Fig. 2 Pathways of LOOH formation in the plant cell. Numbers in circles denote the equations provided in this section. *HO˙* hydroxyl radical, *HO_2˙* hydroperoxyl radical, *L˙* lipid radical, *LH* lipid, *LOO˙* lipid peroxyl radical, *LOOH* lipid hydroperoxide, *L˙$_{rearranged}$* rearrenged lipid radical, *LOX* lipoxygenase, *$^1O_2$* singlet oxygen

Finally, LOO˙ abstracts a hydrogen atom from a neighboring lipid molecule, forming LOOH, meanwhile forming a new L˙ (Eq. 3). This new L˙ can lead to a chain reaction of Eqs. 2 and 3, leading to formation of many more LOOHs formed. In other words, it propagates, and this is why this way of LOOH formation is called radical chain reactions. This scenario, of course, must have a termination step as well, like any other chain autoxidation reaction, where two radicals react to form a nonradical product [12].

$$LOO^{\bullet} + LH \rightarrow LOOH + L^{\bullet} \quad (3)$$

In the second scenario, singlet oxygen ($^1O_2$) interacts with a PUFA's double bond, and creates a lipid endoperoxide adduct (Eq. 4).

$$LH + {}^1O_2 \rightarrow \text{Lipid endoperoxide} \quad (4)$$

This lipid endoperoxide is then converted to LOOH by yet unknown mechanisms, however, light may play a role in this (Eq. 5).

$$\text{Lipid endoperoxide} \rightarrow LOOH \quad (5)$$

In the enzymatic formation, lipoxygenases (LOXs) (EC 1.13.11.-) oxygenate PUFAs containing one or several (1Z,4Z)-pentadiene moieties to form (1S,2E,4Z)-hydroperoxides ((1S,2E,4Z)-LOOHs). In plants, two major PUFAs containing this moiety are linoleic acid (18:2) ('carbon number':'double bond') and α-linolenic acid (18:3). When the reaction specifics are considered, LOXs can be classified based on the carbon atom they oxygenate on linoleic acid, which is either the ninth or 13th carbon, that correspond to 9-LOX and 13-LOX enzymes [13, 14]. For the 18-C lipids, the LOX reaction starts with hydrogen atom abstraction, almost exclusively at C-11 in plants (Eq. 6).

$$LH \rightarrow L^{\bullet} \tag{6}$$

A radical rearrangement follows (Eq. 7).

$$L^{\bullet} \rightarrow L^{\bullet}_{\text{rearranged}} \tag{7}$$

Finally, ends with oxygen insertion and proton addition at [+2] or [−2] of the initial hydrogen abstraction site, resulting in −OOH formation at C-9 or C-13 (Eq. 8) [14].

$$L^{\bullet}_{\text{rearranged}} + O_2 + H^+ \rightarrow LOOH \tag{8}$$

As of writing this chapter, there are 34 LOXs reviewed for *Arabidopsis thaliana* in UniProt database [15], suggesting the existence of various ways of accomplishing this reaction.

Once LOOHs are formed, they can be converted to RCS by nonenyzmatic or enzymatic mechanisms. In the nonenzymatic pathway, a redox catalyst, such as a transition metal, and a reductant, such as a lipid, are needed. In the presence of a redox catalyst, LOOH leads to the formation of a lipid alkoxyl radical ($LO^{\bullet}$) [7]. An example, using iron, is given in Eq. 9.

$$LOOH + Fe^{2+} \rightarrow LO^{\bullet} + Fe^{3+} + {}^{-}OH \tag{9}$$

This $LO^{\bullet}$ oxidizes the close by reductants, and forms further radicals. For example, by oxidizing the adjacent lipid molecules, $LO^{\bullet}$ leads to the formation of $L^{\bullet}$ and $LOO^{\bullet}$ molecules. The products of this reaction spontaneously decompose into carbonyls. Some of these carbonyls, since they carry unsaturated bond(s), are further reduced to produce more LOOH, which then lead to more carbonyl formation. Whether by this propagation or by the initial reaction, some of the carbonyls formed belong to the RCS subgroup [4]. A PUFA can lead to formation of different LOOHs, and a LOOH can lead to formation of different RCS; therefore, a single PUFA species can result in the formation of many different RCS [7].

2.2 Subcellular Locations of RCS Production in Plants

Considering the nature of RCS production, it would be expected that RCS are produced near lipid membranes. This seems to be the case, as there is evidence of RCS production in chloroplasts, mitochondria, peroxisomes, and cell membranes [3]. Following this thought, lipid membranes near high ROS production would have higher RCS production. This, again, makes chloroplasts, mitochondria, peroxisomes, and in some cases cell membranes, candidates for higher RCS production sites, as the first three are well-known sites of ROS production, and cell membrane and apoplast host ROS producers, such as NADPH oxidases (NOXs), class III peroxidases, lipoxygenases, quinine reductase, and amine and germin-like oxalate oxidases [16]. RCS production near lipid membranes containing higher amount of PUFAs would be expected to be higher than lipid membranes with lower PUFA content. Chloroplasts, especially their thylakoid membranes, could be a prime suspect in this regard, as these lipid membranes fill the innards of chloroplasts. For example, *Nerium oleander* has 77.1% PUFA in its thylakoids [17], which would make it fit well with this narrative. On the other hand, thylakoid membrane of cucumber has 78.4% unsaturated fatty acids [18], of which 49.95% are PUFA. Monounsaturated fatty acids (MUFAs) are mostly incapable of peroxidation [19] and therefore not good candidates for RCS production. Is this percentage of PUFA enough for high RCS production? It could be the case, especially considering the high amounts of thylakoid membranes in chloroplasts, but saturated/MUFA/PUFA composition of a species can differ based on the maturation stage [20]. This implies that RCS production could differ between stages of maturation. In addition to this, it is known that PUFA content changes based on the tissue. As a result, to test whether a specific amount of PUFA percentage is enough for high RCS production for an organelle, we would need studies that comparatively examine the lipid composition of each organelle and the amount of RCS production in a given tissue. Furthermore, studies would need to examine this relationship on a suborganeller level, too. For example, chloroplast needs to be examined by comparing the relationship of RCS production with PUFA levels of thylakoid membrane, inner membrane, and outer membrane. It should also be kept in mind that PUFA percentage by itself does not define the amount of RCS, and RCS scavenging and ROS production also play roles, but based on the literature so far, we had expected there to be relatively strong correlations between PUFA levels and membranes that are involved in RCS production, compared to membranes that are less involved.

As for the RCS production differentiation between organs, in leaf tissues of *A. thaliana* under optimal conditions, singlet oxygen was found to be responsible for over 80% of nonenzymatic lipid peroxidation. Meanwhile, in the nonphotosynthetic tissues, since singlet oxygen generation is low, free radicals are the main source [21]. From this information, we could extrapolate that, in photosynthesizing tissues, chloroplast is an extremely important source of RCS, maybe even the biggest one. Contrasting nonenzymatic production to enzymatic production for RCS or lipid peroxidation would allow researchers to discern the specific levels of production for these processes.

3 RCS Scavenging Mechanisms and Targets of RCS in the Cell

3.1 RCS Scavenging Mechanisms

As explained previously, RCS resembles ROS in many aspects, including the requirement of balancing between production and scavenging. Therefore, as in the case of ROS, cells need to remove RCS via detoxification mechanisms. Plant cells utilize two main types of RCS scavenging mechanisms to control endogenous RCS levels: enzymatic and nonenzymatic.

Although there are many studies that showed that polyphenols like phloretin, pelargonidin, and catechins have the ability to trap some RCS, such as acrolein and HNE in vitro, the main players of nonenzymatic RCS scavenging are the thiol compounds cysteine and the reduced form of glutathione (GSH), used as low-molecular-weight scavengers to detoxify RCS by forming conjugates [8, 22]. Among these, GSH is very crucial to the scavenging of certain RCS types due to its role in cellular redox balance [23].

The other class of RCS scavengers are those that function through enzymatic reactions. Enzymatic reactions for RCS scavenging are classified into five groups: (1) 2-alkenal reductases (AER) and alkenal/one oxidoreductases (AOR) reduce the α,β unsaturated bond of an RCS molecule; (2) aldo-keto reductases (AKR) catalyze the reduction of a carbonyl to alcohol; (3) aldehyde dehydrogenases (ALDH) catalyze the oxidation of a carbonyl to a carboxylic acid; (4) glutathione transferases (GST) catalyze the formation of the covalent bond between an electrophilic compound (such as an RCS) and GSH to detoxify the former; (5) aldehyde oxidases (AO) catalyze the oxidation of a carbonyl to carboxylate, using O_2 as the electron acceptor [8]. A visualization of the working of these enzymes is provided in Fig. 3.

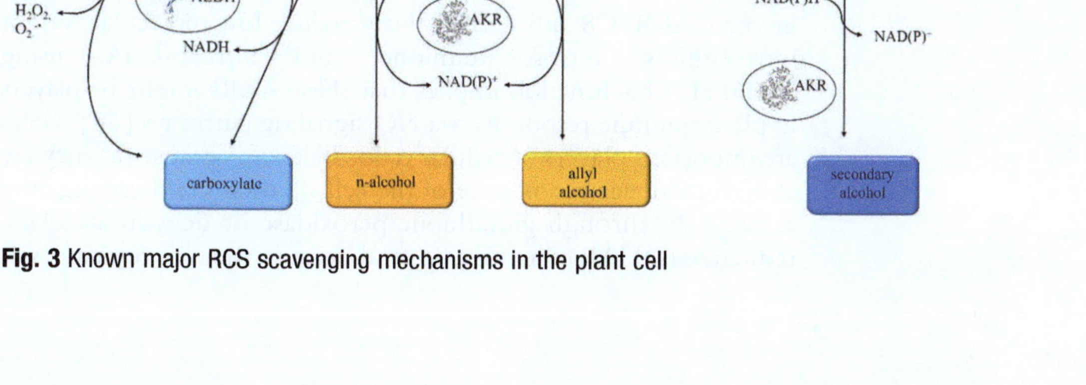

Fig. 3 Known major RCS scavenging mechanisms in the plant cell

Among these enzymes, AER and AOR are known for their ability to reduce RCS, by using NAD(P)H, at C–C bond that is conjugated to the carbonyl group, and therefore they do not react with the carbonyl group [24]. Moreover, cells use enzymes belonging to the AKR family or short-chain dehydrogenase/reductase family to reduce the aldehyde or ketone groups, by utilizing NAD (P)H [24]. Furthermore, ALDH is used in oxidization reactions of an aldehyde to a carboxylic acid with NAD^+, while GST catalyzes the RCS binding to GSH. As a result of these reactions, both these enzymes influence the cellular GSH pool [25]. Lastly, AOR can oxidize a carbonyl to carboxylate using O_2 as the electron acceptor [26]. Although these enzymes are classified as mentioned, they also have multiple isozymes, which can have different substrate specificities. For example, in *A. thaliana*, affinity of AER to smaller RCS, like acrolein, is lower than that of longer-chain RCS [8]. AKR isozymes have relatively broad substrate specificity for various carbonyls [27]. ALDH uses broad range of substrates, such as sinapaldehyde and coniferaldehyde for ALDH2C4, acetyl-CoA for ALDH3F1, and succinic semialdehyde for ALDH5 [28, 29]. In *A. thaliana*, genomic analysis showed that there are about 50 isogenes of GST in the genome, and less than half of the 28 total Tau-class GST isozymes use acrolein or HNE or both as the substrate [23, 30]. For AO, there are four different isoforms that were characterized (AO1-AO4), and it is shown that each AO has a strong preference for different aldehydes, such as indole-3-acetaldehyde, a precursor of IAA for AO1; 1-naphthaldehyde for AO2; abscisic acetaldehyde for AO3; and benzaldehyde for AO4 [31].

As can be seen from the aldehyde preferences of isozymes, besides their RCS scavenging ability, all these enzymes are important players in metabolic pathways. For example, ALDH2C4 has an important role in ferulic acid and sinapic acid biosynthesis pathway [32]. ALDH7B4 gene is expressed in all tissues under various stresses such as dehydration, salt, and wounding, because of its promoter's inducible nature by stress and aldehydes [33]. ALDH11A3 is also named as non-phosphorylating glyceraldehyde-3-phosphate dehydrogenase, and serves in export shuttle mechanism of NADPH from chloroplast to cytosol [34]. ALDH2B7 is predicted to be localized at mitochondrion, and uses acetaldehyde and glycolaldehyde as a substrate in acetic acid fermentation pathway [35, 36]. In [37] authors showed that, in vitro, AKR4C8 and AKR4C9 can reduce low molecular weight nitrosothiols, nitrosoglutathione, and nitroso-CoA, using NADPH. This function implies that these AKRs might be players in plant immune responses, via NO signaling pathways [38]. GSTs are important players of cellular redox balance, because (a) they are able to influence redox state of the cell via their GSH conjugating activity, (b) through glutathione peroxidase or dehydroascorbate reductase (DHAR) activities, (c) and because they are regulators of environmental stress adaptation and tolerance, pathogenic attack responses, and toxic material detoxification processes [39]. Moreover, GSTs have roles in stress signaling and xenobiotic detoxification pathways, such as GSTU3, GSTU4, GSTU5, and GSTU25 in H_2O_2 detoxification and GSTU24 in herbicide tolerance [40–42].

Due to their close relationship with the crucial metabolic pathways, all mentioned RCS scavenging enzymes are important elements of the cell, not only by maintaining the cellular redox balance but also by eliminating the deteriorating effect of RCS on other cellular compartments.

3.2 Targets of RCS in the Cell

Due to their reactive nature, RCS could target many cellular moieties, such as proteins and members of pathways like gene expression, photosynthesis, respiration, germination, and stress tolerance [3, 43]. Although it is known that ROS can directly modify proteins via nitrosylation or nitration of tyrosine residues, carbonylation of specific amino acids on their Cys residues by oxidation or creating carbonyl groups in the side chains of certain amino acids, disulfide crosslink formation, glycoxidation adducts, and glutathionylation, ROS also have an indirect protein modification ability to conjugate proteins with the breakdown products of lipid peroxidation [43]. RCS directly modify proteins at their Cys, Lys, His, and amino groups via forming Michael adducts or reaction of saturated carbonyls to create Schiff's bases [8]. Modification of proteins is related to the structure of RCS, which is determined by carbon chain lengths, number of unsaturated bonds, and the extent of oxygenation. These structural differences result in differences in polarity, hydrophobicity/hydrophilicity, solubility, and volatility of RCS, that are directly related to reactivity with proteins [4, 7, 44]. For instance, in [45] authors state that, when reacting with GSH, the second-order rate constant is over 100 $M^{-1}\ s^{-1}$ for HNE and 4-oxo-(E)-2-nonenal (ONE), which have the same structure except the degree of oxidation at fourth carbon. On the other hand, HNE has 1.3 $M^{-1}\ s^{-1}$ second-order rate constant. Moreover, authors of [46] showed that, by using anti-HNE antibodies, 34 proteins, which are mainly related to ROS generation and the redox network, were modified by HNE from salt-stressed *A. thaliana* samples [46]. Also, many proteins important for cellular redox balance, such as heat shock protein (HSP), thioredoxin (TRX), glutaredoxin (GRX), ascorbate peroxidase (APX), glyceraldehyde 3-phosphate dehydrogenase (GAPDH), dehydroascorbate reductase (DHAR), monodehydroascorbate reductase (MDHAR), superoxide dismutase (SOD), catalase (CAT), peroxidase (POX), and NADPH/respiratory burst oxidase proteins (NOX/RBOH) could be the target of RCS [25]. RCS have a broad range of target proteins, and modification of these proteins is heavily implied to alter their functions [8]. It is known that carbonylated proteins are degraded by 20S proteasome system, while 26S proteasome system is used for ubiquitinylated proteins [28]. When the level of carbonylation of proteins exceeds the capacity of proteasome system, they tend to accumulate as cytotoxic aggregates, and this accumulation results in many cellular pathways impairments [43]. There are many studies showing that this balance is regulated via autophagy, and RCS accumulation was found to induce programmed cell death (PCD) [8], which might also function to get rid of excess RCS in the organism [47, 48].

4 Foundations of RCS Signaling: Hormesis and Other Considerations

The RCS signaling in regard to specific stresses is discussed later, while the foundations of this signaling and its possible relationship to the hormesis paradigm are discussed as follows. Signaling potential of RCS and RES was first discovered during drug studies. In vertebrate cells, RCS or RES drugs were found to induce detoxification genes. Later, in plants, RCS that were produced as a result of *Pseudomonas syringae* infection were found to increase glutathione-S-transferase 1 (GST1) gene. Following studies suggested that endogenously produced RCS stimulated cell survival genes in plants [6]. In [7], the authors proposed five qualities of RCS that might make them viable signals: (1) longer lives of RCS, enabling them to diffuse across longer distances; (2) reversibility of RCS modification in certain cases; (3) ability of RCS to modify proteins by Michael addition and Schiff base formation; (4) diversity of RCS types, which might increase the specificity of signaling (e.g., for different organelles); (5) RCS levels acting as possible sensor targets for oxidative stress perception. Indeed, RCS are downstream of ROS [6–8], as they were found to increase after ROS stimulation, and scavenging RCS was found to inhibit the ROS response [3]. These findings support the argument that RCS could be oxidative stress sensor targets. On the other hand, longer reach of RCS would enable them to diffuse across membranes, a quality that is lacking or limited in ROS [2, 5]. The authors [5] note that the ability of RCS to diffuse through membranes and the tight enzymatic control on RCS production/degradation are strong arguments for its role as a signaling molecule, and further add that RCS activate specific receptors. However, this last part was confirmed in only non-plants so far. It is also mentioned that Keap1-Nrf2 pathway regulates RCS signaling in mammalians, and while *A. thaliana* does not have Nrf2 homologues, it has genes that might be homologous to Keap1. Furthermore, in our opinion, the diversity of RCS, combined with their downstream properties, could be a major factor in answering the question of how ROS achieve signal specificity with so few species.

A study that examined induced transcription under exogenous RCS treatments in *A. thaliana* whole plants found that 30% of the genes induced by 12-oxo-phytodienoic acid (OPDA) and 60% induced by A1-phytoprostanes (PPA_1) relied on TGA transcription factors, which activate salicylic acid inducible elements. More specifically, TGA2, TGA5, and TGA6 were involved. Of the genes that showed higher than twofold induction, 49–53% contained a TGA motif (TGACG). Furthermore, among the PPA_1 treated *A. thaliana* cell cultures, 24% of genes in both the upregulated and downregulated gene categories belonged to signaling. Noticeably, 17% of the upregulated genes were detoxification genes, while this was 1% in the downregulated genes [49]. In other studies, it was found that heat shock proteins (HSPs) were induced by RCS. It is probable that HSPs were produced as a result of a general stress response, acting as chaperones to protect proteins from RCS. RCS have also been shown to affect transcriptions of genes related to cell division, cell growth, auxin signaling, and programmed cell death (PCD) [6, 9]. For example, 10% of the downregulated genes in [49] belonged to cell wall metabolism category. In the same study, OPDA and PPA_1 treatments led to inhibition of cell division and root growth. Taken together, these findings paint a picture where RCS stimulate their detoxification process and induce protective chaperones, increasing the organism's tolerance against RCS damage. They also lead to stress-related changes in growth and related processes, such as cell division. This is probably, at least in part, related to the stress adaptation process, since the energy of the organism has to be allocated between different processes under stress [50].

To make better sense of the general picture of RCS signaling, perhaps we should look to the evolutionary roots of ROS signaling. The author [51] postulates that ROS metabolism originally evolved to sense harmful oxygen levels in the atmosphere, since in the ancient Earth, before life had adapted to high oxygen levels, molecular oxygen was extremely harmful to organisms. But after some evolutionary time passed, it follows, ROS were incorporated into all kinds of metabolism. However, despite the variety of ROS metabolism, we know that it is a key part of stress signaling [16]. This is most probably because stress creates disturbances in the oxygen metabolism, resulting in redox imbalances, which are conveyed via ROS. In this way, ROS signal their own threat to the organism. After all, no matter how incorporated ROS are into the signaling system, at high levels, they are a crucial threat to the organism, and stress leads to the production of higher levels of ROS [50]. Furthermore, dividing the effects of ROS on metabolism into categories like signaling or detoxification, even though contextually useful, does not reflect the complexity of metabolism. For instance, both ROS and RCS play signaling roles, but these signals also lead to an increase in ROS and RCS detoxification [16, 49]. Could we really say detoxification and signaling are entirely separate? We believe that this is not the case, for they are part of the same pathway. So, it can be said that ROS metabolism changed but also stayed the same, in the sense that ROS still signal disturbances in oxygen metabolism caused by stress. Considering that RCS are produced via ROS metabolism [8], there is sufficient reason to suggest that RCS function the same way. They might have been initially messengers of unsafe levels of oxygen, but in time they could have been incorporated into a more diverse metabolism. However, this incorporation into signaling, at least in a major way, still reflects disturbances in the metabolism under stress. As we will see in later sections, RCS are an important part of stress signaling. Thus, RCS, like ROS, are also messengers of their own vice; a threat that was assimilated into the signaling to warn the organism of incoming destruction. Furthermore, exogenous RCS exposure also leads to an increase in antioxidant activities [52], so it seems to be the case that RCS signal not only their own threat but also that of ROS.

4.1 RCS Signaling and Hormesis

Another quality of RCS that is of interest to signaling is related to hormesis paradigm. Hormesis is a dose-response paradigm that was originally founded in the field of toxicology. It describes a response pattern where a low dose of stress—eustress—is beneficial for different parameters, such as growth or antioxidant activity, whereas higher doses—distress—are inhibitory for these parameters. Thus, the parameters follow an inverted U-shaped dose-response. A molecule that creates this response in a parameter is named a hormetic molecule, while the affected parameter is called an endpoint. ROS are counted among hormetic molecules, and this includes those in plants, too. It is speculated that ROS participate in signaling at low doses, while at high doses they are harmful to the organism, and this quality is what makes ROS hormetic molecules [50]. Since, as previously explained, RCS also seem to exhibit this dichotomy [3], they are also potentially hormetic molecules. Indeed, RCS and other RES can improve cell survival, and it has been proposed that endogenous RES can be beneficial for cells [6]. So far, there is no study that explicitly examines endogenous or exogenous RCS exposure and hormesis induction. However, an RCS adjacent RES, methylglyoxal, has been shown to induce a hormetic response for tolerance in mammalian cells and yeast [53]. Furthermore, it has been demonstrated that exogenous HNE, 4-hydroxy-2-hexenal (HHE), and acrolein treatments led various enzymatic antioxidant activities to show or approximate this inverted U-shaped dose-response in the glycophytic *A. thaliana* and/or halophytic *Eutrema parvulum*. These antioxidants included SOD, CAT, POX, APX, and the water-water cycle enzyme glutathione reductase (GR). Furthermore, NOX, a superoxide-producing enzyme that is part of redox signaling and tolerance response, also seemed to exhibit this response in *A. thaliana* [52]. In addition, priming, that is, the increase in tolerance of an organism to a subsequent stress by exposure to a previous mild stress, is also included in the hormetic paradigm [50], and there are instances of this for RCS; RCS priming increased thermotolerance [54], tolerance against photooxidative stress [55], and tolerance against singlet oxygen [56]. Therefore, we propose that beneficial effects of RCS can be explained by the hormesis paradigm, and there is sufficient reason to think this is based on the signaling/damaging dichotomy.

5 RCS Signaling Cross-Talk with Other Reactive Species Signaling Pathways

Plants are exposed to numerous environmental alterations due to their sessile nature, and they must continually perceive these changes to optimize their performance under these conditions. They use this environmental monitoring information to create a response in a remarkably sophisticated and coordinated fashion, which requires interactions and cross-talk with several metabolic-molecular processes. In this manner, plants frequently use the formation and scavenging of "reactive species" as a quick cellular response to external stimuli [57]. Reactive chemical species, such as ROS, reactive nitrogen species (RNS), RCS, and reactive sulfur species (RSS) are among the most prevalent molecules. However, these species must be strictly controlled via balancing production and scavenging, because excess generation of these reactive species is known to disturb the redox homeostasis, leading to further changes in the plant, such as degeneration of biomolecules, disruption of electron transport chains (ETCs), oxidative damage, cell membrane fragmentation, and finally, PCD [11]. Furthermore, there is a complicated interplay between all these reactive species during stress conditions. They are also present under non-stress conditions, but their cellular quantities rise further under stress conditions. Under non-stress conditions, these radicals are scavenged and kept at non-toxic levels in the cells, and this redox balance is essential for optimal plant development and consequently yields [57]. These could be speculated as a cost for using these species in the organism.

Although plant researchers have been focused on the dual nature (destructive/beneficial) of ROS and RNS for a long time, several studies showed that relatively novel reactive members, RCS and RSS, appear to exhibit similar qualities [11]. For example, it is known that seed germination is controlled by ROS homeostasis via its interaction with gibberellic acid (GA) and abscisic acid (ABA), and this is mediated by NO—an RNS—by its interaction with ABA signaling [58, 59]. The authors [60] showed that seed germination was inhibited by the application of glucose, NaCl, and methyl viologen at varied doses and time periods to cucumber, tobacco, and rice. These caused an increase in oxidative stress, which led to higher levels of RCS, and plant development was enhanced when these chemicals were detoxified by aldo-keto reductase-1 (AKR1) under the same stress levels [60]. As can be seen from this example, RCS play crucial roles in seedling development and germination, albeit negatively under excess stress. Another conclusion is that RCS must be detoxified under stress, in order to enhance cellular activity and development in seedlings. Another example is related to the role of RCS and RSS as primary stress indicators, similar to ROS and RNS. It has also been observed that biotic or abiotic stress induces the ROS and RNS production, and they have role in influencing the expression levels of several regulatory genes; governing the interactions with receptors; altering the activity of transcription factors (TFs); transcribing stress-related genes, known as NO-responsive genes; and participating in plant defense, oxidative stress response, and hormone interactions in metabolism and development [11, 61]. According to the authors [62], the RSS, especially H_2S, play an important role in alleviating the negative effects of abiotic stresses on plants, by increasing antioxidant production, lowering the MDA and H_2O_2 contents in the leaves, and elevating ABA synthesis in the roots, in wheat. Similarly, the authors [52] showed that RCS imitate ROS signals and target proteins for redox processes, by oxidizing cysteine sulfhydryl groups in a reversible manner in order to control their activity. Also, it was shown that exogenous RCS application resulted in an elevation in the gene expressions of the membrane-bound SOS1 and tonoplast-localized NHX1 and NHX5 in *E. parvulum*, while the activities of various ROS scavenging enzymes, such as CAT and APX remained unchanged. The last examples are related to PCD, that is an especially crucial mechanism in hypersensitive response (HR), and various studies have revealed ROS, RNS, and RSS participation in PCD, under biotic and abiotic stress conditions. Accordingly, it is shown that, in soybean, pathogen-induced H_2O_2 burst causes HR, which could be inhibited by diphenyleneiodonium (DPI) supplementation, and this H_2O_2–induced response was time-dependent, as necrosis was observed in prolonged exposure, whereas PCD was induced by a single pulse [63–65]. Moreover, the authors [66] showed that the expressions of pathogenesis-related (PR) genes of *A. thaliana*, *AtPR1* and *AtPR2*, were elevated more significantly after virulent bacteria DC3000 inoculation in wild type Col-0 than the *atgsnor1–3* and NO-induced mutant line *atill6*. In the case of RSS, it is known that they react with other reactive species to initiate HR via specific processes, like S-nitrosation of Cys890 at D subunit of RBOH under pathogen attack, which hindered the binding of FAD cofactor and tempered the formation of ROS, resulting in inhibition of HR and cell death [67]. As can be seen from these examples and previous sections, all these reactive species work in a specific and most of the time synergistic way in cellular metabolism. Yet, this situation brings about a new question: What is at the heart of cross-talk between these reactive species?

The answer to this question lies in the balance between reactive species production and scavenging. Despite the fact that reactive species are categorized according to their fundamental elements (oxygen, nitrogen, sulfur, or carbon), there are specific metabolic checkpoints where they share the same targets or scavenging routes [16]. As a result, there is a strict regulatory mechanism in place that regulates the generation and detoxification of reactive species, for the plant to withstand environmental stresses [11]. In terms of cross-talk, it is known that there is a link between ROS and RNS, because the overproduction of ROS during abiotic stress leads to increased NO production, mainly via nitrate reductase (NR) and nitric oxide synthase-like activity of L-arginine; in addition, ROS scavenging and Ca^{2+} channel inhibitor applications inhibit H_2O_2-mediated NO production [16, 68]. Moreover, [69] showed that H_2O_2 is required for ABA-induced NO synthesis, however, the treatment of an NO scavenger, NOS inhibitor, or NO donor did not affect ABA-induced H_2O_2 production. Also, peroxynitrite ($ONOO^-$) causes tyrosine nitration of proteins as a post-translational modification (PTM), and S-nitrosation of some ROS detoxification enzymes, such as glycolate oxidase and CAT, decreases their activities [70, 71]. In a recent development, it has been shown that a nitro-fatty acid, nitro-linolenic acid (NO_2-Ln), is involved in response to oxidative stress, and that NO_2-Ln responsive genes include response to ROS and oxidative stress genes [72]. It is very probable that NO_2-Ln to be interacting with RCS, and a key point in illuminating this relationship lies in finding out whether it is upstream or downstream of RCS signaling. The possible role of other nitro-fatty acids in such a cross-talk also remains to be seen, but is well within the confines of possibility. Similar to RNS, RSS, which are also known as second-generation reactive species because they are formed with reactions of ROS and sulfur, have a role in some PTMs, such as *S*-cysteinylation, *S*-sulfenylation, *S*-nitrosation, *S*-persulfidation, *S*-cyanylation, *S*-acylation, and *S*-glutathionylation [11, 73]. They also regulate the stability and reactivity of various proteins, including plasma membrane antiporter (Na^+/H^+) system and antioxidant enzymes such as SOD, CAT, and APX [11]. ROS and RSS share intertwined mechanisms, including production, metabolism, and scavenging processes [74]. From a proteomic perspective, RCS are also able to reversibly modify the Cys, His, and Lys residues of proteins, and they mimic ROS in protein regulation [7]. Also, RCS could have a regulatory role for ROS detoxification enzymes, such as CuZnSOD, because of the RCS and Zn^{2+} interaction [75]. Moreover, studies show that RCS have a signal transducer role downstream of ROS signaling pathway in many physiological responses, such as inducing root injury, PCD, senescence of leaves and fruits, ABA-mediated stomatal closure, and root response to auxin [76]. These examples support the hypothesis that ROS-RCS interaction shapes the antioxidant system and signaling pathways in many aspects, including phytohormone biosynthesis and other signals.

In summary, since all the reactive species are connected and share many pathways, including productive and regulatory ones, they are required to cross-talk to each other. Although ROS and RNS garnered great attention for a long time, the relatively novel species RCS and RSS will be key players in the near future, due to their regulatory roles in many metabolic pathways. Therefore, the interaction between these reactive species must be extensively analyzed, in order to carve a path for future studies of genomic, transcriptomic, proteomic, and signal network analyses.

6 RCS Signaling Under Abiotic Stresses

An overview of RCS signaling studies under abiotic and biotic stresses is given in Table 1, and there are studies that show priming [54–56]. In the case of [55], this priming is achieved by the induction of singlet oxygen-responsive genes, resulting in higher tolerance against this molecule. β-cyclocitral and β-ionone are created by oxidation of β-carotene, a carotenoid that is responsible singlet oxygen scavenging. β-cyclocitral was found to have much stronger induction of singlet oxygen responsive genes than β-ionone. It seems while RCS work as a stress signal in singlet oxygen metabolism, the effect differs for specific RCS produced. In the case of [54], thermotolerance was increased by the induction of heat stress response to master regulator TFs and chaperones, although other abiotic stress TFs were induced too. The induction of chaperones might not be just limited to heat stress response, as chaperones are thought to be induced as a result of a general RCS stress response [6]. This is supported by the induction of other stress genes. For [56], in *Chlamydomonas reinhardtii*, glutathione pool and redox state, and GTS protein levels were induced by low dose acrolein, while they were inhibited by higher dose. Low-dose acrolein treatments also were associated with higher survival against singlet oxygen. The findings of glutathione metabolism and survival rate demonstrate that full hormetic dose-response is exhibited in this study. Another related study that shows the hormetic dose-response is [52]. In it, it was shown that SOD, CAT, POX, APX, GR, and NOX exhibited a hormetic dose-response under various RCS treatments, implying the existence of an RCS-to-ROS-metabolism signaling. This is also supported by a study that overexpressed an aldose reductase gene, an RCS scavenger, which resulted in decreased ROS levels [83]. Lastly, lateral root density was shown to follow a hormetic dose-response as a result of RCS treatments [76]. In summary, these studies point to the existence of hormetic response as a result of RCS. However, the induced signaling pathways are complex and need further elucidation. In addition, the

Table 1
Studies related to RCS signaling under abiotic and biotic stresses

RCS	Species	RCS dose	Exogenous application method	Endogenous generation method	Additional manipulation	Sampled part	RCS molecular effect	End result	Ref.
Exogenous application									
Acrolein, HNE	*Nicotiana tabacum*	0.1–0.2 mM	Cell culture incubation with acrolein	N/A	Scavenger carnosine addition	Cultured cells	Glutathione pool exhausted at both doses; C1LP increases at both doses; C3LP increases at both doses, especially at 0.2 mM; ascorbate decreases and ROS increases at 0.2 mM	PCD initiated at 0.2 mM but not at 0.1 mM	[77]
OPDA, phytoprostanes	*Arabidopsis thaliana*	75 μM	Unspecified for whole plants; incubation with RCS in cell cultures	N/A	tga2–5-6 mutant, opr3 mutant	Leaves, roots; cultured cells	Detoxification, stress response (including AOX, HSFs, HSPs), and secondary metabolism were induced; TGA2, TGA5, and TGA6 were induced	Cell division and root growth were inhibited with RCS	[49]
Acrolein, HHE, HNE	*Arabidopsis thaliana*, *Eutrema parvulum*	0.1 μM, 1 μM, 10 μM	Infiltration into leaves with insulin syringes on the abaxial side	N/A	N/A	Leaves, roots	Antioxidants and NOX were stimulated, sometimes hormetically; SOS1, NHX1, and NHX5 were stimulated	RCS used were able to stimulate antioxidant activities, partially hormetically, and ion exchange channel transcripts	[52]
Reactive short-chain leaf volatiles	*Arabidopsis thaliana*, *Solanum lycopersicum*, *Oryza sativa*	3 μL	Volatilization from a paper towel in a box	N/A	Aor mutant, HSFA1s mutant	Unspecified (assumed to be leaves)	Heat stress related transcript factors were induced (including HFSs); DREB2A and ZATs were induced; chaperon proteins were induced	Thermotolerance was increased by RCS priming	[54]

(continued)

Table 1 (continued)

RCS	Species	RCS dose	Exogenous application method	Endogenous generation method	Additional manipulation	Sampled part	RCS molecular effect	End result	Ref.
Endogenous generation									
Acrolein, HNE, (E)-2-heptenal, (E)-2-pentenal	*Nicotiana tabacum*	N/A	N/A	ABA treatment, hydrogen peroxide treatment	AER-over-expression	Leaves	AER-overexpression partially inhibited stomatal closure in H_2O_2- and ABA-treated plants	RCS have a role in stomatal closure signaling, downstream of H_2O_2	[78]
Acrolein, crotonal-dehyde, HHE, HNE, MDA, (E)-2-hexenal	*Arabidopsis thaliana*	N/A	N/A	Salinity	N/A	Leaves	RCS and RCS-modified proteins were increased as a result of salinity; various proteins, including those of stress response, were modified by HNE	HNE is increased and modifies various proteins under salinity	[46]
Exogenous application and endogenous generation									
Acrolein, HHE, HNE, MDA	*Arabidopsis thaliana*, *Nicotiana tabacum*	0.1, 0.2, 0.5, 3, 10, 50 mM	Cell culture incubation with RCS	Generated by hydrogen peroxide application or salinity treatment	AER-over-expression; carnosine and hydralazine treatments	Cell culture, roots	DNA was fragmented; *n*-terminal deoxynucl-eotidyl transferase dUTP nick end labeling was positive; cytoplasm was retracted	Cells underwent PCD as a result of RCS; AER-overexpression reduced PCD under salinity or H_2O_2	[79]
Acrolein, crotonal-dehyde, HNE	*Arabidopsis thaliana*	0.05–5 mM	Pouring on the root surface through a pore filter	Auxin addition	Auxin-deficient mutants; scavenger carnosine addition	Leaves, roots	RCS led to auxin/indole-3-acetic acid protein degradation	RCS treatments led to hormesis in LR density	[76]
β-Cyclocitral and β-ionone	*Arabidopsis thaliana*	5 μL, 15 μL, 50 μL, 500 μL, or 1 mL	Volatilization from a wick cotton in an airtight box	In vitro oxidation of β-carotene by rose Bengal generated singlet oxygen under light	N/A	Leaves	β-Cyclocitral led to the initiation of singlet oxygen responsive genes	Increased tolerance against photooxidative stress as a result of RCS priming	[55]

Acrolein	*Chlamydomonas reinhardtii*	120–1200 ppm	Volatilization from a wick paper in an airtight petri dish	High light, high oxygen	LHCSR3-deficient mutant	Cell culture	In the mutant, GPX5 and GST1 were elevated, proteins were carbonylated more, glutathione was more oxidized; in WT and low-dose acrolein treatment, glutathione level and redox state was elevated	Acrolein damage was associated with light stress, but glutathione amount and redox state were elevated as a response	[80]
Acrolein in exogenous, various RCS in other treatments	*Chlamydomonas reinhardtii*	0.01–0.15 μL	Volatilization from a paper wick in an airtight petri dish	Light stress	sor1 mutant	Cultured cells	Glutathione pool and glutathione redox state both followed a hormetic dose-response; lower acrolein concentrations stimulated GST protein levels; transcripts that overlap with singlet oxygen treatment were enriched	Acrolein resulted in hormetic response and priming for tolerance against singlet oxygen	[56]
Acrolein, benzaldehyde, citral, HNE, MDA, 1-naphthal-dehyde	*Arabidopsis thaliana*	0.25, 1, 2, 5 mM	Vacuum infiltration	Generated by dark, drought, and UV-C	Aldehyde oxidase 4 knockout and over-expression	Siliques, rosette leaves	Senescence markers were enhanced	Premature senescence of siliques was observed as a result of RCS	[26]
MDA	*Arabidopsis thaliana*	4 μmol/L air volume	Volatilization from a cotton wick in an airtight box	Anaerobiosis, diquat treatment, heat stress	Fatty acid desaturase mutant, JA-insensitive mutant, ethylene-insensitive mutant, SA deaccum-ulator transgenic; treatment of MDA mimic carbonyls	Leaves	Abiotic stress and oxidative stress genes were upregulated; mimics suggested cross-linking of MDA was associated with its activity	MDA is suggested to regulate abiotic stress response as a signal	[81]

(continued)

Table 1 (continued)

RCS	Species	RCS dose	Exogenous application method	Endogenous generation method	Additional manipulation	Sampled part	RCS molecular effect	End result	Ref.
MDA	*Arabidopsis thaliana*, *Cucumis sativus*, *Spinachia oleaceae*	100–400 μM	MDA incubation with proteins	Heat stress	Protein incubation with peroxidized fatty acid methyl esters; linolenic acid-deficient mutant	Leaves	OEC33 protein was modified by MDA in vitro, and in vivo under heat stress; LHC proteins were modified in vivo under heat stress; linolenic acid deficiency inhibited modification	Heat stress leads to modification of photosynthesis complex membrane proteins by MDA	[82]

ABA abscisic acid, *AER* 2-alkenal reductase, *AOX* alternative oxidase, *C1LP* caspase-1-like protease, *C3LP* caspase-3-like protease, *COI1* corona insensitive 1, *CPK6* calcium dependent protein kinase 6, *DREB2A* dehydration-responsive element binding protein 2, *EIN2* ethylene insensitive 2, *GPX5* glutathione peroxidase 5, *GST* glutathione-S-transferase, *HSF* heat shock factor, HSFA1 class A1 heat shock factor, *HSP* heat shock protein, *JA* jasmonic acid, *LHCSR3* light harvesting complex stress related 3, *LR* lateral root, *MeJa* methyl jasmonate, *MDA* malondialdehyde, *NHX* Na+/H+ antiporter, *NOX* NADPH oxidase, *OEC33* oxygen-evolving complex 33 kDa protein, *opr3* OPDA reductase 3, *PCD* programmed cell death, *RCS* reactive carbonyl species, *RBOH* respiratory burst oxidase homologue, *SA* salicylic acid, *SID2* salicylic acid induction deficient 2, *sor1* singlet oxygen resistant 1, *SOS1* salt overly sensitive 1, *tga* TGACG-binding, *ZAT* zinc finger of *Arabidopsis thaliana*

dose-dependence of pathways does not apply to just hormetic signaling. For example, in tobacco PCD was initiated at 0.2 mM but not at 0.1 mM [77]. These studies imply that RCS response is dose-dependent, and for this reason it should be investigated with dose-dependence curves.

RCS were shown to modify proteins under various stresses, including salinity and high temperature [46, 82]. Particularly, MDA was able to modify oxygen evolving complex 33 KDA (OEC33) by in vitro RCS treatment and under in vivo heat stress [82]. MDA was also able to modify, possibly by crosslinking, various proteins and alter their activities [81]. These findings give further credence to the idea that MDA is not only a stress marker but also a signaling molecule. Continuing on, RCS were able to induce detoxification genes [49], as would be expected from such reactive species. In the same vein, acrolein, after a transitory drop in the following parameters, led to the elevation of total glutathione content and redox state in *C. reinhardtii* [80]. It seems RCS can stimulate the RCS tolerance mechanisms, whether by directly signaling or by interacting with signaling mechanisms. RCS were also found to be involved in PCD signaling. More specifically, RCS scavenging by scavenger addition or overexpression of a scavenger inhibited PCD [79], and RCS activity induced C1LP and C3LP proteases, which preceded PCD [77]. While these studies implicate RCS in PCD signaling, further studies are needed to confirm if PCD is caused by signaling, not toxicity. In the same vein, RCS are implied to be involved in senescence [26], but further studies are needed. In addition to these signaling functions, RCS were shown to be downstream of H_2O_2 in stoma closure signaling [78, 84].

In summary, RCS play or potentially play a signaling role in various pathways, such as stoma closure, oxygen-evolving complex regulation, PCD, senescence, ROS scavenging, species-specific ROS signaling, abiotic stress responses, chaperone metabolism, TF induction, and RCS detoxification. They also, at least in certain cases, lead to the hormetic stimulation of endpoints by signaling. These endpoints include both tolerance mechanisms and physiological parameters.

7 Biotic Stress–Related RCS Signaling

Due to their close relationship with ROS, RCS probably have a role in many biotic stress signaling pathways, but the literature is still at an early phase. However, there are studies that hint at a role for RCS in biotic stress signaling (Table 1). For example, stoma is a weak point against pathogens, for it provides a way of entry for them. ABA is a well-known regulator of stoma closure, and it has been shown that it regulates stoma closure under pathogen-associated molecular pattern (PAMP)-triggered immunity (PTI), at least partially by cross-talking with SA signaling [85]. It is shown by [78] that AER-overexpression inhibited stomatal closure in H_2O_2- and ABA-treated plants. In a follow-up study, it is also been shown that RCS act as a signal mediator downstream of H_2O_2 in ABA signaling pathway for stomatal closure [84]. The experiment included exogenous applications of acrolein and HNE to leaves of wild-type, plasma membrane NOX mutant, *atrbohD atrbohF*, and the Ca^{2+}-dependent kinase mutant, *cpk6*, with or without two RCS scavengers, named carnosine and pyridoxamine. It also included acrolein and HNE applications to wild-type and GSH-deficient *cad2–1* mutant. ABA and H_2O_2 both stimulated the formation of acrolein and HNE. RCS scavengers decreased ABA- and H_2O_2-induced stomatal closure, but did not affect ABA-induced H_2O_2 generation. The NOX mutant and the wild type both promoted stomatal closure in response to acrolein and HNE, but not the Ca^{2+}-dependent kinase mutant. Acrolein increased cytosolic free Ca^{2+}, stomatal closure with glutathione depletion in the guard cells, and activation of plasma membrane Ca^{2+}-permeable cation channels. These studies suggest that RCS work downstream of ROS in ABA-mediated stomatal closure, but they are upstream of Ca^{2+} signaling. Therefore, by the way of ABA signaling, RCS might be partaking in pathogen related stoma closure.

On a related study, it has been shown that RCS are involved in methyl jasmonate (MeJa)-induced stomatal closure, and function as a signal mediator downstream of H_2O_2 in MeJA signaling. The authors found that exogenous RCS, compared to exogenous H_2O_2, more effectively initiated Ca^{2+} elevation, and induced stomatal closure. Additionally, RCS scavengers inhibited MeJa-induced stomatal closure in both wild-type tobacco and Arabidopsis. Therefore, RCS appear to play an important role in the MeJa signaling related to stomatal closure. This data suggests that downstream signal transduction activities, such as the increase of Ca^{2+} in the guard cell MeJA signal pathway, are mediated by RCS. Considering that exogenous MeJa depletes GSH [77], and GSH-deficient mutant has a more enhanced stomatal closure compared to wild type when it is exposed to RCS [84], it can be suggested that RCS production and scavenging pathways are both used to regulate stomatal closure under biotic and abiotic stresses.

A question regarding these stomatal closure studies is related to cross-talk of hormones. Biotic stress tolerance is associated with phytohormones, including SA, JA, ethylene, ABA, gibberellin, auxin, cytokinin, and brassinosteroid. Their signaling pathways control various stress reactions, and they are implicated in immunity [86]. The immune response to biotrophic and hemibiotrophic infections is generally positively regulated by SA, but the immune response to necrotrophic pathogens is positively regulated by JA and ethylene. These are the core immune phytohormones, and as mentioned in the previous part of this section, their signaling networks are modified by other phytohormones [86]. Immunity regulates the manufacture of phytohormones, which are then detected by specific receptors that send signals to transcriptional complexes. Additionally, hormonal cross-talk takes place at the level of transcription TFs, which is governed by the core TFs in each signaling cascade [87]. Considering that RCS play a role in both ABA- and MeJa-mediated stomatal closure, it is possible that RCS play a role in these phytohormone signaling cascades as well. This aspect needs further elucidation.

On another topic, HR is an important part of defense response. It strategically kills cells in a region to limit the scope of infection, protecting the whole plant by sacrificing regions of infection. HR achieves this by inducing PCD in cells [88]. The authors [77] showed that at 0.2 mM, acrolein produced PCD in 5 h, whereas at 0.1 mM, it had no effect, and was sublethal. The cellular glutathione pool was depleted by both lethal and sublethal doses of acrolein in 30 min, although only the lethal dose significantly increased ROS, and caused ascorbate levels to drop after 1–2 h. They also discovered that acrolein significantly increased the activity of caspase-1-like protease (C1LP) and caspase-3-like protease (C3LP), the proteases that induce PCD, prior to such redox alterations. Acrolein's lethal dose raised C3LP activity by two times more than its sublethal level. In contrast, there was no difference in the C1LP activity increases brought about by the two doses. This data suggests that RCS are involved in PCD signaling, but also that it is a gradual process. This is implied by the exhaustion of glutathione pool, and near identical and differential increases in C1LP and C3LP activities in response to low and high doses, respectively. Considering that RCS are downstream of ROS production, and HR is preceded by a ROS burst [88], it is highly probable that RCS work downstream of this burst.

Finally, the various RCS treatments have been shown to induce gene expressions similar to that of avirulent pathogen infection. The elevated expressions included PR genes, implicating RCS in pathogenesis signaling [89]. Induction of PR genes work together with localized PCD to stave off infection [90]. Taken together with the roles of RCS in PCD, it seems RCS are involved in many steps of pathogenesis defense.

8 Methods Used in RCS Stress Signaling Studies

In Table 1, *C. reinhardtii* studies were also included, because of their relevance to plant biology. We can see that studies in Table 1 can be split into three kinds, and these categories and the number of studies in each category are as follows: exogenous RCS application (five studies), endogenous RCS generation (three studies), and the ones that do both (nine studies). Somewhat surprisingly, the last category has the most studies. However, since both methods manipulate the conditions of the organism to in order to observe resulting effects, it follows that inclusion of both methods should increase the accuracy and detail of a study. Therefore, studies that use both exogenous application and endogenous generation are best suited to study RCS responses under stress.

Another quality of these studies is the inclusion of additional manipulation conditions. These generally mean mutants lacking a function in a related gene (10 out of 17 studies), overexpressing plants (4 out of 17 studies), or addition of an RCS scavenger (5 out of 17 studies). As can be seen, the use of mutants is a very popular choice for manipulation. The mutant choice is diverse, and generally includes manipulation of a scavenger enzyme, an RCS metabolism–related hormone, a lipid metabolism gene, or a transcription factor. These mutants manipulate RCS metabolism, making it easier to see the role of the metabolisms in various pathways. For example, the authors [82] included linolenic acid mutant in their study, which also included heat stress and MDA treatments. By including this mutant in their study, they showed that light-harvesting complex (LHC) proteins were modified by MDA under heat stress in wild type plants, but not in the mutants. Therefore, this suggested that MDA production by linolenic acids was a source of LHC modification under stress. The inclusion of mutants allowed them to pinpoint the source of MDA. Mutants are specific in their manipulation, unlike the addition of scavengers, so they allow the researchers to make more accurate manipulations. It should also be mentioned that the choice of mutant depends on the mechanism studied, and it can be more unorthodox, such as the manipulation of an LHC protein [80]. This is not surprising, since organisms are complex and there are many interactions between pathways.

Although this biology of interaction is rich, there are certain candidate mutants, of which inclusion in future studies should allow researchers to elucidate specific pathways. These candidates have more potential to affect RCS metabolism, and are as follows: (i) HPS and HSF mutants, for they are an important part of organism's RCS response; (ii) RCS scavenger mutants, for they regulate RCS levels; (iii) lipid metabolism mutants, for they are the source of RCS; (iv) hormone mutants, for they affect the RCS metabolism; (v) antioxidant mutants, for ROS lead to RCS production; and (vi) NOX mutants, for they produce ROS as part of stress signaling. In the case of last two categories, additional manipulations should be made to ensure RCS, not ROS, are what is affecting the pathway. For example, exogenous application of RCS scavengers could be used to see if RCS inhibition results in a similar phenotype to the NOX mutant, or it could be used to check whether it results in an opposite effect of antioxidant mutants; measurements of RCS-modified proteins could be used to pinpoint

RCS effects; stimulating RCS metabolism by exogenous addition or overexpression could work to isolate RCS and ROS metabolisms.

Studies of flowers and fruits are particularly absent from the RCS stress studies, making them a field of study ripe for research. Furthermore, some studies reveal priming mechanisms, whether they call them priming or not [54–56] and non-priming hormesis [52]. While ROS priming, especially H_2O_2 priming, is a field that has been garnering attention [50], RCS priming is yet to emerge. However, the current findings and the previously mentioned theoretical background imply that RCS are also priming molecules. It should also be remembered that RCS have longer half-lives than ROS [8], increasing their viability as priming molecules. In conclusion, RCS priming is a field that is waiting to be developed.

9 Conclusion and Future Prospects

RCS, as a reactive species, exhibit damage at high doses and signaling at low dose duality, and for this reason their production and scavenging are tightly controlled. Low-dose RCS signaling is implicated in both abiotic and biotic stresses. RCS cross-talks with other reactive species and hormones, hinting at an intricate and complex hub of signaling pathways in stress signaling, of which elucidation requires further attention. Relating to this, both abiotic and particularly biotic stress signaling studies regarding RCS are still in their infancy. Various manipulation techniques, including the use of mutants, overexpressing transgenics, and scavenger additions, are particularly useful in elucidation of RCS signaling pathways. On top of all this, RCS are implied to be hormetic molecules, stimulating both tolerance mechanisms and other physiological parameters. This is supported by both theory and research, and it is suggested that the mentioned duality is behind this quality. Studies examining the dose-response for RCS in plants would be highly illuminating. In conclusion, RCS stress signaling is a field that has produced highly fruitful research, and it is still emerging. There are still a lot of areas that need elucidation, making the field exciting and potentially highly rewarding. We predict that, in time, the field will only get more attention, and will evolve to include complex cross-talks with other signaling elements.

References

1. Leisner CP, Potnis N, Sanz-Saez A (2022) Crosstalk and trade-offs: plant responses to climate change-associated abiotic and biotic stresses. Plant Cell Environ. https://doi.org/10.1111/pce.14532
2. Soares C, Carvalho MEA, Azevedo RA, Fidalgo F (2019) Plants facing oxidative challenges—a little help from the antioxidant networks. Environ Exp Bot 161:4–25. https://doi.org/10.1016/j.envexpbot.2018.12.009

3. Mano J, Biswas MS, Sugimoto K (2019) Reactive carbonyl species: a missing link in ROS signaling. Plants (Basel, Switzerland) 8(10): 391. https://doi.org/10.3390/plants8100391
4. Mano J (2012) Reactive carbonyl species: their production from lipid peroxides, action in environmental stress, and the detoxification mechanism. Plant Physiol Biochem 59:90–97. https://doi.org/10.1016/j.plaphy.2012.03.010
5. Semchyshyn HM (2014) Reactive carbonyl species in vivo: generation and dual biological effects. TheScientificWorldJOURNAL 2014: 417842. https://doi.org/10.1155/2014/417842
6. Farmer EE, Mueller MJ (2013) ROS-mediated lipid peroxidation and RES-activated signaling. Annu Rev Plant Biol 64:429–450. https://doi.org/10.1146/annurev-arplant-050312-120132
7. Yalcinkaya T, Uzilday B, Ozgur R, Turkan I, Mano J (2019) Lipid peroxidation-derived reactive carbonyl species (RCS): their interaction with ROS and cellular redox during environmental stresses. Environ Exp Bot 165:139–149. https://doi.org/10.1016/j.envexpbot.2019.06.004
8. Biswas MS, Mano J (2021) Lipid peroxide-derived reactive carbonyl species as mediators of oxidative stress and signaling. Front Plant Sci 12:720867. https://doi.org/10.3389/fpls.2021.720867
9. Alché JD (2019) A concise appraisal of lipid oxidation and lipoxidation in higher plants. Redox Biol 23:101136. https://doi.org/10.1016/j.redox.2019.101136
10. Esterbauer H, Schaur RJ, Zollner H (1991) Chemistry and biochemistry of 4-hydroxynonenal, malonaldehyde and related aldehydes. Free Radic Biol Med 11(1):81–128. https://doi.org/10.1016/0891-5849(91)90192-6
11. Zhou X, Joshi S, Patil S, Khare T, Kumar V (2022) Reactive oxygen, nitrogen, carbonyl and sulfur species and their roles in plant abiotic stress responses and tolerance. J Plant Growth Regul 41(1):119–142. https://doi.org/10.1007/s00344-020-10294-y
12. Chan HW-S, Coxon DT, Peers KE, Price KR (1982) Oxidative reactions of unsaturated lipids. Food Chem 9(1–2):21–34. https://doi.org/10.1016/0308-8146(82)90066-8
13. Feussner I, Wasternack C (2002) The lipoxygenase pathway. Annu Rev Plant Biol 53:275–297. https://doi.org/10.1146/annurev.arplant.53.100301.135248
14. Andreou A, Feussner I (2009) Lipoxygenases – structure and reaction mechanism. Phytochemistry 70(13–14):1504–1510. https://doi.org/10.1016/j.phytochem.2009.05.008
15. The UniProt Consortium (2023) UniProt: the universal protein knowledgebase in 2023. Uniprot: the universal protein knowledgebase in 2023. Nucleic Acids Res, 51(D1). https://doi.org/10.1093/nar/gkac1052
16. Hasanuzzaman M, Bhuyan MHMB, Zulfiqar F, Raza A, Mohsin SM, Mahmud JA, Fujita M, Fotopoulos V (2020) Reactive oxygen species and antioxidant defense in plants under abiotic stress: revisiting the crucial role of a universal defense regulator. Antioxidants (Basel, Switzerland) 9(8):681. https://doi.org/10.3390/antiox9080681
17. Raison JK, Roberts JKM, Berry JA (1982) Correlations between the thermal stability of chloroplast (thylakoid) membranes and the composition and fluidity of their polar lipids upon acclimation of the higher plant, nerium oleander, to growth temperature. Biochimica et Biophysica Acta (BBA) – Biomembranes 688(1):218–228. https://doi.org/10.1016/0005-2736(82)90597-1
18. Shu S, Yuan Y, Chen J, Sun J, Zhang W, Tang Y, Zhong M, Guo S (2015) The role of putrescine in the regulation of proteins and fatty acids of thylakoid membranes under salt stress. Sci Rep 5:14390. https://doi.org/10.1038/srep14390
19. Arranz L, Naudí A, De la Fuente M, Pamplona R (2013) Exceptionally old mice are highly resistant to lipoxidation-derived molecular damage. Age (Dordr) 35(3):621–635. https://doi.org/10.1007/s11357-012-9391-0
20. Zhang MP, Zhang CJ, Yu GH, Jiang YZ, Strasser RJ, Yuan ZY, Yang XS, Chen GX (2010) Changes in chloroplast ultrastructure, fatty acid components of thylakoid membrane and chlorophyll a fluorescence transient in flag leaves of a super-high-yield hybrid rice and its parents during the reproductive stage. J Plant Physiol 167(4):277–285. https://doi.org/10.1016/j.jplph.2009.09.017
21. Triantaphylidès C, Krischke M, Hoeberichts FA, Ksas B, Gresser G, Havaux M, Van Breusegem F, Mueller MJ (2008) Singlet oxygen is the major reactive oxygen species involved in photooxidative damage to plants. Plant Physiol 148(2):960–968. https://doi.org/10.1104/pp.108.125690
22. Sugimoto K, Matsuoka Y, Sakai K, Fujiya N, Fujii H, Mano J (2021) Catechins in green tea powder (matcha) are heat-stable scavengers of acrolein, a lipid peroxide-derived reactive

carbonyl species. Food Chem 355:129403. https://doi.org/10.1016/j.foodchem.2021.129403

23. Mano J, Kanameda S, Kuramitsu R, Matsuura N, Yamauchi Y (2019) Detoxification of reactive carbonyl species by glutathione transferase tau isozymes. Front Plant Sci 10:487. https://doi.org/10.3389/fpls.2019.00487
24. Yamauchi Y, Hasegawa A, Taninaka A, Mizutani M, Sugimoto Y (2011) NADPH-dependent reductases involved in the detoxification of reactive carbonyls in plants. J Biol Chem 286(9):6999–7009. https://doi.org/10.1074/jbc.M110.202226
25. Knieper M, Viehhauser A, Dietz KJ (2023) Oxylipins and reactive carbonyls as regulators of the plant redox and reactive oxygen species network under stress. Antioxidants (Basel, Switzerland) 12(4):814. https://doi.org/10.3390/antiox12040814
26. Srivastava S, Brychkova G, Yarmolinsky D, Soltabayeva A, Samani T, Sagi M (2017) Aldehyde oxidase 4 plays a critical role in delaying silique senescence by catalyzing aldehyde detoxification. Plant Physiol 173(4):1977–1997. https://doi.org/10.1104/pp.16.01939
27. Simpson PJ, Tantitadapitak C, Reed AM, Mather OC, Bunce CM, White SA, Ride JP (2009) Characterization of two novel aldo-keto reductases from Arabidopsis: expression patterns, broad substrate specificity, and an open active-site structure suggest a role in toxicant metabolism following stress. J Mol Biol 392(2):465–480. https://doi.org/10.1016/j.jmb.2009.07.023
28. Ben-Nissan G, Sharon M (2014) Regulating the 20S proteasome ubiquitin-independent degradation pathway. Biomol Ther 4(3):862–884. https://doi.org/10.3390/biom4030862
29. Islam MS, Ghosh A (2022) Evolution, family expansion, and functional diversification of plant aldehyde dehydrogenases. Gene 829:146522. https://doi.org/10.1016/j.gene.2022.146522
30. Mano J, Ishibashi A, Muneuchi H, Morita C, Sakai H, Biswas MS, Koeduka T, Kitajima S (2017) Acrolein-detoxifying isozymes of glutathione transferase in plants. Planta 245(2):255–264. https://doi.org/10.1007/s00425-016-2604-5
31. Wu J, Kamanga BM, Zhang W, Xu Y, Xu L (2022) Research progress of aldehyde oxidases in plants. PeerJ 10:e13119. https://doi.org/10.7717/peerj.13119
32. Nair RB, Bastress KL, Ruegger MO, Denault JW, Chapple C (2004) The Arabidopsis thaliana REDUCED EPIDERMAL FLUORESCENCE1 gene encodes an aldehyde dehydrogenase involved in ferulic acid and sinapic acid biosynthesis. Plant Cell 16(2):544–554. https://doi.org/10.1105/tpc.017509
33. Stiti N, Missihoun TD, Kotchoni SO, Kirch HH, Bartels D (2011) Aldehyde dehydrogenases in Arabidopsis thaliana: biochemical requirements, metabolic pathways, and functional analysis. Front Plant Sci 2:65. https://doi.org/10.3389/fpls.2011.00065
34. Rius SP, Casati P, Iglesias AA, Gomez-Casati DF (2006) Characterization of an Arabidopsis thaliana mutant lacking a cytosolic non-phosphorylating glyceraldehyde-3-phosphate dehydrogenase. Plant Mol Biol 61(6):945–957. https://doi.org/10.1007/s11103-006-0060-5
35. Kirch HH, Bartels D, Wei Y, Schnable PS, Wood AJ (2004) The ALDH gene superfamily of Arabidopsis. Trends Plant Sci 9(8):371–377. https://doi.org/10.1016/j.tplants.2004.06.004
36. Rasheed S, Bashir K, Kim JM, Ando M, Tanaka M, Seki M (2018) The modulation of acetic acid pathway genes in Arabidopsis improves survival under drought stress. Sci Rep 8(1):7831. https://doi.org/10.1038/s41598-018-26103-2
37. Treffon P, Rossi J, Gabellini G, Trost P, Zaffagnini M, Vierling E (2021) Quantitative proteome profiling of a S-Nitrosoglutathione reductase (GSNOR) null mutant reveals a new class of enzymes involved in nitric oxide homeostasis in plants. Front Plant Sci 12:787435. https://doi.org/10.3389/fpls.2021.787435
38. Borrowman S, Kapuganti JG, Loake GJ (2023) Expanding roles for S-nitrosylation in the regulation of plant immunity. Free Radic Biol Med 194:357–368. https://doi.org/10.1016/j.freeradbiomed.2022.12.009
39. Horváth E, Bela K, Gallé Á, Riyazuddin R, Csomor G, Csenki D, Csiszár J (2020) Compensation of mutation in Arabidopsis glutathione transferase (AtGSTU) genes under control or salt stress conditions. Int J Mol Sci 21(7):2349. https://doi.org/10.3390/ijms21072349
40. Sappl PG, Carroll AJ, Clifton R, Lister R, Whelan J, Harvey Millar A, Singh KB (2009) The Arabidopsis glutathione transferase gene family displays complex stress regulation and co-silencing multiple genes results in altered metabolic sensitivity to oxidative stress. The

Plant Journal: For Cell and Molecular Biology 58(1):53–68. https://doi.org/10.1111/j.1365-313X.2008.03761.x

41. Behringer C, Bartsch K, Schaller A (2011) Safeners recruit multiple signalling pathways for the orchestrated induction of the cellular xenobiotic detoxification machinery in Arabidopsis. Plant Cell Environ 34(11):1970–1985. https://doi.org/10.1111/j.1365-3040.2011.02392.x
42. Skipsey M, Knight KM, Brazier-Hicks M, Dixon DP, Steel PG, Edwards R (2011) Xenobiotic responsiveness of Arabidopsis thaliana to a chemical series derived from a herbicide safener. J Biol Chem 286(37):32268–32276. https://doi.org/10.1074/jbc.M111.252726
43. Tola AJ, Jaballi A, Missihoun TD (2021) Protein carbonylation: emerging roles in plant redox biology and future prospects. Plants (Basel, Switzerland) 10(7):1451. https://doi.org/10.3390/plants10071451
44. LoPachin RM, Gavin T, Petersen DR, Barber DS (2009) Molecular mechanisms of 4-hydroxy-2-nonenal and acrolein toxicity: nucleophilic targets and adduct formation. Chem Res Toxicol 22(9):1499–1508. https://doi.org/10.1021/tx900147g
45. Rudolph TK, Freeman BA (2009) Transduction of redox signaling by electrophile-protein reactions. Sci Signal 2(90):re7. https://doi.org/10.1126/scisignal.290re7
46. Mano J, Nagata M, Okamura S, Shiraya T, Mitsui T (2014) Identification of oxidatively modified proteins in salt-stressed Arabidopsis: a carbonyl-targeted proteomics approach. Plant Cell Physiol 55(7):1233–1244. https://doi.org/10.1093/pcp/pcu072
47. Avin-Wittenberg T, Bajdzienko K, Wittenberg G, Alseekh S, Tohge T, Bock R, Giavalisco P, Fernie AR (2015) Global analysis of the role of autophagy in cellular metabolism and energy homeostasis in Arabidopsis seedlings under carbon starvation. Plant Cell 27(2):306–322. https://doi.org/10.1105/tpc.114.134205
48. Signorelli S, Tarkowski ŁP, Van den Ende W, Bassham DC (2019) Linking autophagy to abiotic and biotic stress responses. Trends Plant Sci 24(5):413–430. https://doi.org/10.1016/j.tplants.2019.02.001
49. Mueller S, Hilbert B, Dueckershoff K, Roitsch T, Krischke M, Mueller MJ, Berger S (2008) General detoxification and stress responses are mediated by oxidized lipids through TGA transcription factors in Arabidopsis. Plant Cell 20(3):768–785. https://doi.org/10.1105/tpc.107.054809
50. Sonmez MC, Ozgur R, Uzilday B (2023) Reactive oxygen species: connecting eustress, hormesis, and allostasis in plants. Plant Stress 8:100164. https://doi.org/10.1016/j.stress.2023.100164
51. Mittler R (2017) ROS are good. Trends Plant Sci 22(1):11–19. https://doi.org/10.1016/j.tplants.2016.08.002
52. Yalcinkaya T, Uzilday B, Ozgur R, Turkan I (2019) The roles of reactive carbonyl species in induction of antioxidant defence and Ros signalling in extreme halophytic model Eutrema parvulum and glycophytic model Arabidopsis thaliana. Environ Exp Bot 160:81–91. https://doi.org/10.1016/j.envexpbot.2019.01.009
53. Kold-Christensen R, Johannsen M (2020) Methylglyoxal metabolism and aging-related disease: moving from correlation toward causation. Trends Endocrinol Metab 31(2):81–92. https://doi.org/10.1016/j.tem.2019.10.003
54. Yamauchi Y, Kunishima M, Mizutani M, Sugimoto Y (2015) Reactive short-chain leaf volatiles act as powerful inducers of abiotic stress-related gene expression. Sci Rep 5:8030. https://doi.org/10.1038/srep08030
55. Ramel F, Birtic S, Ginies C, Soubigou-Taconnat L, Triantaphylidès C, Havaux M (2012) Carotenoid oxidation products are stress signals that mediate gene responses to singlet oxygen in plants. Proc Natl Acad Sci USA 109(14):5535–5540. https://doi.org/10.1073/pnas.1115982109
56. Roach T, Stöggl W, Baur T, Kranner I (2018) Distress and eustress of reactive electrophiles and relevance to light stress acclimation via stimulation of thiol/disulphide-based redox defences. Free Radic Biol Med 122:65–73. https://doi.org/10.1016/j.freeradbiomed.2018.03.030
57. Kaur C, Kushwaha HR, Mustafiz A, Pareek A, Sopory SK, Singla-Pareek SL (2015) Analysis of global gene expression profile of rice in response to methylglyoxal indicates its possible role as a stress signal molecule. Front Plant Sci 6:682. https://doi.org/10.3389/fpls.2015.00682
58. Signorelli S, Considine MJ (2018) Nitric oxide enables germination by a four-pronged attack on ABA-induced seed dormancy. Front Plant Sci 9:296. https://doi.org/10.3389/fpls.2018.00296
59. Huang H, Ullah F, Zhou DX, Yi M, Zhao Y (2019) Mechanisms of ROS regulation of plant development and stress responses. Front Plant Sci 10:800. https://doi.org/10.3389/fpls.2019.00800

60. Nareshkumar A, Subbarao S, Vennapusa AR, Ashwin V, Banarjee R, Kulkarni MJ, Ramu VS, Udayakumar M (2020) Enzymatic and non-enzymatic detoxification of reactive carbonyl compounds improves the oxidative stress tolerance in cucumber, tobacco and rice seedlings. J Plant Growth Regul 39(3):1359–1372. https://doi.org/10.1007/s00344-020-10072-w
61. Choudhury S, Panda P, Sahoo L, Panda SK (2013) Reactive oxygen species signaling in plants under abiotic stress. Plant Signal Behav 8(4):e23681. https://doi.org/10.4161/psb.23681
62. Ma D, Ding H, Wang C, Qin H, Han Q, Hou J, Lu H, Xie Y, Guo T (2016) Alleviation of drought stress by hydrogen sulfide is partially related to the Abscisic acid signaling pathway in wheat. PLoS One 11(9):e0163082. https://doi.org/10.1371/journal.pone.0163082
63. Levine A, Tenhaken R, Dixon R, Lamb C (1994) H2O2 from the oxidative burst orchestrates the plant hypersensitive disease resistance response. Cell 79(4):583–593. https://doi.org/10.1016/0092-8674(94)90544-4
64. Levine A, Pennell RI, Alvarez ME, Palmer R, Lamb C (1996) Calcium-mediated apoptosis in a plant hypersensitive disease resistance response. Curr Biol 6(4):427–437. https://doi.org/10.1016/s0960-9822(02)00510-9
65. De Pinto MC, Locato V, De Gara L (2012) Redox regulation in plant programmed cell death. Plant Cell Environ 35(2):234–244. https://doi.org/10.1111/j.1365-3040.2011.02387.x
66. Khan M, Al Azawi TNI, Pande A, Mun BG, Lee DS, Hussain A, Lee BH, Yun BW (2021) The role of nitric oxide-induced ATILL6 in growth and disease resistance in Arabidopsis thaliana. Front Plant Sci 12:685156. https://doi.org/10.3389/fpls.2021.685156
67. Yun BW, Feechan A, Yin M, Saidi NB, Le Bihan T, Yu M, Moore JW, Kang JG, Kwon E, Spoel SH, Pallas JA, Loake GJ (2011) S-nitrosylation of NADPH oxidase regulates cell death in plant immunity. Nature 478(7368):264–268. https://doi.org/10.1038/nature10427
68. Corpas FJ, González-Gordo S, Palma JM (2022) NO source in higher plants: present and future of an unresolved question. Trends Plant Sci 27(2):116–119. https://doi.org/10.1016/j.tplants.2021.11.016
69. Zhang A, Jiang M, Zhang J, Ding H, Xu S, Hu X, Tan M (2007) Nitric oxide induced by hydrogen peroxide mediates abscisic acid-induced activation of the mitogen-activated protein kinase cascade involved in antioxidant defense in maize leaves. New Phytol 175(1):36–50. https://doi.org/10.1111/j.1469-8137.2007.02071.x
70. Radi R (2004) Nitric oxide, oxidants, and protein tyrosine nitration. Proc Natl Acad Sci USA 101(12):4003–4008. https://doi.org/10.1073/pnas.0307446101
71. Ortega-Galisteo AP, Rodríguez-Serrano M, Pazmiño DM, Gupta DK, Sandalio LM, Romero-Puertas MC (2012) S-Nitrosylated proteins in pea (Pisum sativum L.) leaf peroxisomes: changes under abiotic stress. J Exp Bot 63(5):2089–2103. https://doi.org/10.1093/jxb/err414
72. Mata-Pérez C, Sánchez-Calvo B, Padilla MN, Begara-Morales JC, Luque F, Melguizo M, Jiménez-Ruiz J, Fierro-Risco J, Peñas-Sanjuán A, Valderrama R, Corpas FJ, Barroso JB (2016) Nitro-fatty acids in plant signaling: nitro-linolenic acid induces the molecular chaperone network in Arabidopsis. Plant Physiol 170(2):686–701. https://doi.org/10.1104/pp.15.01671
73. Corpas FJ, González-Gordo S, Rodríguez-Ruiz M, Muñoz-Vargas MA, Palma JM (2022) Thiol-based oxidative posttranslational modifications (OxiPTMs) of plant proteins. Plant Cell Physiol 63(7):889–900. https://doi.org/10.1093/pcp/pcac036
74. Olson KR, Straub KD (2016) The role of hydrogen sulfide in evolution and the evolution of hydrogen sulfide in metabolism and signaling. Physiology (Bethesda) 31(1):60–72. https://doi.org/10.1152/physiol.00024.2015
75. Hao Q, Maret W (2006) Aldehydes release zinc from proteins. A pathway from oxidative stress/lipid peroxidation to cellular functions of zinc. FEBS J 273(18):4300–4310. https://doi.org/10.1111/j.1742-4658.2006.05428.x
76. Biswas MS, Fukaki H, Mori IC, Nakahara K, Mano J (2019) Reactive oxygen species and reactive carbonyl species constitute a feed-forward loop in auxin signaling for lateral root formation. Plant J Cell Mol Biol 100(3):536–548. https://doi.org/10.1111/tpj.14456
77. Biswas MS, Mano J (2016) Reactive carbonyl species activate caspase-3-like protease to initiate programmed cell death in plants. Plant Cell Physiol 57(7):1432–1442. https://doi.org/10.1093/pcp/pcw053
78. Islam MM, Ye W, Matsushima D, Munemasa S, Okuma E, Nakamura Y, Biswas S, Mano J, Murata Y (2016) Reactive carbonyl species mediate ABA signaling in guard cells. Plant

Cell Physiol 57(12):2552–2563. https://doi.org/10.1093/pcp/pcw166

79. Biswas MS, Mano J (2015) Lipid peroxide-derived short-chain carbonyls mediate hydrogen peroxide-induced and salt-induced programmed cell death in plants. Plant Physiol 168(3):885–898. https://doi.org/10.1104/pp.115.256834
80. Roach T, Baur T, Stöggl W, Krieger-Liszkay A (2017) Chlamydomonas reinhardtii responding to high light: a role for 2-propenal (acrolein). Physiol Plant 161(1):75–87. https://doi.org/10.1111/ppl.12567
81. Weber H, Chételat A, Reymond P, Farmer EE (2004) Selective and powerful stress gene expression in Arabidopsis in response to malondialdehyde. Plant J Cell Mol Biol 37(6):877–888. https://doi.org/10.1111/j.1365-313x.2003.02013.x
82. Yamauchi Y, Furutera A, Seki K, Toyoda Y, Tanaka K, Sugimoto Y (2008) Malondialdehyde generated from peroxidized linolenic acid causes protein modification in heat-stressed plants. Plant Physiol Biochem 46(8–9):786–793. https://doi.org/10.1016/j.plaphy.2008.04.018
83. Singh P, Kumar D, Sarin NB (2016) Multiple abiotic stress tolerance in Vigna mungo is altered by overexpression of ALDRXV4 gene via reactive carbonyl detoxification. Plant Mol Biol 91(3):257–273. https://doi.org/10.1007/s11103-016-0464-9
84. Islam MM, Ye W, Matsushima D, Rhaman MS, Munemasa S, Okuma E, Nakamura Y, Biswas MS, Mano J, Murata Y (2019) Reactive carbonyl species function as signal mediators downstream of H2O2 production and regulate [Ca2+]cyt elevation in ABA signal pathway in Arabidopsis guard cells. Plant Cell Physiol 60(5):1146–1159. https://doi.org/10.1093/pcp/pcz031
85. Cao FY, Yoshioka K, Desveaux D (2011) The roles of ABA in plant-pathogen interactions. J Plant Res 124(4):489–499. https://doi.org/10.1007/s10265-011-0409-y
86. Tsuda K, Somssich IE (2015) Transcriptional networks in plant immunity. New Phytol 206(3):932–947. https://doi.org/10.1111/nph.13286
87. Pieterse CM, Van der Does D, Zamioudis C, Leon-Reyes A, Van Wees SC (2012) Hormonal modulation of plant immunity. Annu Rev Cell Dev Biol 28:489–521. https://doi.org/10.1146/annurev-cellbio-092910-154055
88. Locato V, De Gara L (2018) Programmed cell death in plants: an overview. In: Methods in molecular biology (Clifton, N.J.), vol 1743, pp 1–8. https://doi.org/10.1007/978-1-4939-7668-3_1
89. Alméras E, Stolz S, Vollenweider S, Reymond P, Mène-Saffrané L, Farmer EE (2003) Reactive electrophile species activate defense gene expression in Arabidopsis. Plant J Cell Mol Biol 34(2):205–216. https://doi.org/10.1046/j.1365-313x.2003.01718.x
90. Zurbriggen MD, Carrillo N, Hajirezaei MR (2010) ROS signaling in the hypersensitive response: when, where and what for? Plant Signal Behav 5(4):393–396. https://doi.org/10.4161/psb.5.4.10793

Chapter 8

Measurements of Ascorbate and Dehydroascorbate in Plants Using High-Performance Liquid Chromatography

Tamami Hamada and Takanori Maruta

Abstract

The current concepts emphasize the fundamental role of reactive oxygen species (ROS) as signaling molecules that coordinate defense mechanisms, cell death, and the growth and development processes in plants. However, due to the inherent reactivity of ROS, achieving precise control over their levels within plant cells, both spatially and temporally, becomes important to effectively harness the potential of ROS signaling while concurrently minimizing the risk of oxidative damage. Ascorbate is an exceptional antioxidant and contributes to the antioxidant defense system in plants. Its role is further reinforced by the presence of ascorbate peroxidases and enzymes responsible for recycling ascorbate from its oxidized forms. Ascorbate metabolism plays a pivotal role in averting oxidative damage and facilitates meticulous regulation of ROS signal availability. This chapter outlines the preferred protocol for the measurement of ascorbate.

Key words Abiotic stress, Ascorbate, Ascorbate–glutathione cycle, Ascorbate peroxidase, Dehydroascorbate

1 Introduction

Under abiotic stress conditions, such as high light, heat, and drought, plant cells exhibit an elevated production of reactive oxygen species (ROS), including superoxide, hydrogen peroxide (H_2O_2), singlet oxygen, and the hydroxyl radical [1]. Despite the inherent reactivity of ROS, which has the potential to induce cellular harm, the current concepts underscore the pivotal role of ROS as signaling molecules that orchestrate defense mechanisms, cell death, and growth and development in plants as well as in animals and bacteria [2]. Nevertheless, due to the significant production of oxygen and the reducing power through photosynthesis, the utilization of ROS signals in plants always carries the risk of oxidative damage. Adding complexity to this scenario, ROS production sites within plant cells exhibit diversity, and distinct signaling pathways come into play contingent upon the types of ROS and

Francisco J. Corpas and José M. Palma (eds.), *ROS Signaling in Plants: Methods and Protocols*,
Methods in Molecular Biology, vol. 2798, https://doi.org/10.1007/978-1-0716-3826-2_8,

the sites of their production [3]. These pathways operate concurrently and converge under abiotic stress conditions, thereby fine-tuning physiological responses. Achieving precise spatiotemporal control over ROS levels in plant cells is crucial to effectively harness the potential of ROS signaling while mitigating the risk of oxidative damage.

Ascorbate, an excellent antioxidant, holds a vital role in modulating the availability and signaling function of ROS in plant cells [4]. Plants accumulate high levels of ascorbate, particularly in their leaves, where photosynthesis and photorespiration can lead to ROS production [5]. Plants have acquired specific metabolic pathways to maximize the redox buffering capacity of ascorbate. For instance, plants employ the D-mannose/L-galactose pathway, also referred to as the Wheeler-Smirnoff pathway, for ascorbate biosynthesis [6]. This unique pathway allows plants to synthesize ascorbate without generating H_2O_2, which occurs in the final step of the animal pathway [7]. Moreover, plants possess ascorbate peroxidases (APXs) that effectively scavenge H_2O_2 using ascorbate as an electron donor [8]. While ascorbate readily reacts with superoxide, singlet oxygen, and the hydroxyl radical, its chemical reaction with H_2O_2 is slow. The presence of APXs empowers plants to employ ascorbate in scavenging all types of ROS [4, 9]. In higher plants, APX isoforms are distributed across various cellular compartments, including the cytosol, peroxisomal membrane, mitochondrial matrix, chloroplast stroma, and thylakoid membrane [9]. It is noteworthy that the APX reaction oxidizes ascorbate into monodehydroascorbate (MDHA), which can spontaneously convert into dehydroascorbate (DHA). Therefore, it is crucial to recycle the oxidized forms back into their reduced form (ASC). This critical process involves the contribution of monodehydroascorbate reductases (MDARs) and dehydroascorbate reductases (DHARs), which are distributed across all cellular compartments where APX is present [10, 11]. The coordination of ascorbate biosynthesis, utilization (the APX reaction), and recycling mechanisms mitigate the risk of oxidative damage and enable precise spatiotemporal control over the availability of ROS signals.

Typically, abiotic stresses, particularly high-light exposure, enhance the total ascorbate pool (the sum of ASC and DHA) by activating both ascorbate biosynthesis and recycling mechanisms [5]. However, the total ascorbate pool is intricately determined by the balance between the rates of biosynthesis and degradation of this compound. Since the degradation process generally commences with DHA, the rate of degradation is significantly influenced by the rate of ascorbate oxidation [4]. Consequently, in instances of severe oxidative stress that triggers rapid ascorbate oxidation, the total pool can be decreased. This is one of the characteristics often observed in *Arabidopsis* mutants that exhibit a significant deficiency in ROS scavenging capabilities in chloroplasts or peroxisomes [12, 13], for instance. While it is generally believed that the redox state of ascorbate (the proportion of ASC within the total pool) serves as a more robust marker of oxidative stress in comparison to the total pool size, we consider that there is merit in revisiting this notion. We have successfully generated *Arabidopsis* mutant lines that simultaneously lack multiple mechanisms for ascorbate recycling [11, 14]. Intriguingly, these mutants are unable to accumulate ascorbate at high levels under high-light stress due to the hastened degradation of ascorbate. However, they maintain high ascorbate redox states, suggesting that DHA does not accumulate and undergoes rapid degradation inside *Arabidopsis* cells [11, 14]. In contrast, due to its relatively high stability in acidic environments, DHA exhibits a comparatively prolonged half-life outside the cell (apoplast), where ascorbate oxidase (AO) catalyzes the oxidation of ascorbate [15]. This redox regulation of ascorbate by AO in apoplasts has been suggested to play a role in plant stress responses [15]. Thus, our comprehension of the physiological significance of regulating the ascorbate redox cycle both inside and outside plant cells remains limited. Nevertheless, considering that ascorbate represents the predominant component in the antioxidant pool of leaf cells, insights into cellular ascorbate profiles offer valuable clues regarding cellular redox status and the occurrence of oxidative stress.

While various methods exist for quantifying ascorbate levels, two representative approaches stand out: one employs high-performance liquid chromatography (HPLC) [16] and another is a spectrophotometric assay using AO [17]. In comparison to alternative colorimetric assays, such as the bipyridyl method [18], these two approaches show a higher degree of specificity toward ascorbate. We have employed both these approaches in previous studies [16, 19], and based on our own experience, we assert that the HPLC assay is superior in terms of accuracy and reliability. One plausible explanation for this superiority lies in the exceptional stability of ascorbate in metaphosphate buffer, attributable to its robust metal-chelating properties. Metaphosphate buffer serves as the extraction medium and mobile phase in the HPLC assays outlined in this assay. In contrast, AO-based assays necessitate the use of either HCl or $HClO_4$ for ascorbate extraction. This step is performed to deactivate enzymes that affect ascorbate levels or redox states, followed by the requirement to dilute the extract with a specific buffer and subsequently neutralize it for the ensuing AO reaction. These additional steps not only increase the complexity of the method but also hold the potential to impact ascorbate stability and turnover. It is important to note that AO-based assays are specific to ascorbate and offer enhanced accessibility in the absence of an HPLC system. Here, we provide the preferred and updated laboratory protocol of ascorbate measurement using an ultra-fast liquid chromatography (UFLC) system.

2 Materials

1. 2% metaphosphate in water (v/v) as a buffer solution.
2. 1% metaphosphate in water (v/v) as the mobile phase, filtered on two layers of filter paper with 4 μm pores (ADVANTEC).
3. 0.22 μm filter (Merck Millipore).
4. 350 mM Tris (2-carboxyethyl) phosphine hydrochloride (TCEP) in water.
5. Ascorbate standards (31.25–500 μmol L^{-1}) in 2% metaphosphate.
6. Ultra-fast liquid chromatography (UFLC) system (Prominence UFLC, Shimadzu) equipped with an SPD-20A UV-Vis detector (Shimadzu).
7. C-18 column (LUNA C18(2), column 150 × 4.6 nm, Shimadzu).

3 Methods

1. Excise plant tissues (such as leaves roots), measure their fresh weight, and promptly submerge them in the liquid nitrogen in a 15 mL Falcon tube. Store these samples at −80°C until use.
2. Immerse the frozen samples in liquid nitrogen within a mortar and proceed to pulverize them using a pestle to a fine powder.
3. Add 2 mL of a 2% metaphosphate buffer for every 100 mg of the sample to the frozen mortar containing the powdered tissue (*see* **Note 1**). Keep the mortar on ice and cover it with aluminum foil to shield it from light exposure, as ascorbate is light-sensitive. During the thawing process, gently agitate the tissue to ensure thorough mixing with the buffer.
4. Transfer 1 mL of the resulting sample to a 1.5 mL tube.
5. Centrifuge the tubes for 10 min at 15,300 × g (4 °C) and filter the supernatant through a 0.22 μm membrane filter. The filtered samples can then be utilized for the quantification of the reduced form of ascorbate (TCEP-untreated). If necessary, dilute the filtered samples in 2% metaphosphate buffer depending on the nature of the plant materials (*see* **Note 2**).
6. For measurement of the total ascorbate (sum of reduced and oxidized forms), transfer 360 μL of the diluted solution into a new 1.5 mL tube. Add 40 μL of 350 mM TCEP to reduce oxidized ascorbate (DHA) (*see* **Note 3**). Incubate the mixture on ice or at 4°C for a minimum of 2 h (*see* **Note 4**).

7. Prepare ascorbate standards, ranging from 31.25 to 500 μmol L^{-1}, in 2% metaphosphate buffer (*see* **Note 3**).
8. Inject 5 μL of the samples and standards into the UFLC system. The mobile phase comprises 1% metaphosphate, and the flow rate is set to 0.5 mL min^{-1}.
9. Measure the peak areas from the standards using the HPLC instrument's integrated software and plot these values against their corresponding concentrations to construct the calibration curve (*see* **Note 5**). In our setup, the retention time of ascorbate is approximately 8.5 min.
10. Calculate total and reduced ascorbate contents by utilizing the peak area from the TCEP-treated and -untreated extracts, respectively, and the calibration curve (*see* **Note 5**).
11. Calculate DHA content as the difference between total and reduced ascorbate contents.
12. Calculate ascorbate redox state (reduced ascorbate/total ascorbate × 100 [%]) (*see* **Note 6**).

4 Notes

1. A sample does not always require 100 mg for the ascorbate measurement. In our experiments, since 2 mL of buffer is used for 100 mg of sample, the amount of buffer used can be changed according to the fresh weight of the sample. For sample uniformity, we usually use two or three fully expanded leaves (50–100 mg of fresh weight) from plants grown under long-day (16-h day/8-h night) conditions for 3 weeks (e.g., [14]). When grown on 1/2 MS medium containing 1% sucrose in a petri dish under long-day conditions for 2 weeks, about 7 and 40 plants will have 50 mg of shoots and roots, respectively (this study).
2. In our experience, ascorbate levels in *Arabidopsis* leaves grown under standard growth conditions without stress are 3~5 μmol gFW^{-1}. In this case, no sample dilution is necessary (see below). Under long-term high-light stress conditions, ascorbate levels in leaves can exceed 12 μmol gFW^{-1} (e.g., [11]), so we dilute the sample by one-half or one-fifth. Because ascorbate levels in plant tissues vary widely among species as well as environmental conditions [20], preliminary experiments should be used to estimate approximate ascorbate levels in advance and to consider the need for sample dilution.
3. Prepare the solutions containing ascorbate or TCEP freshly each day.

4. While dithiothreitol (DTT) is commonly employed for the reduction of DHA, our protocol utilizes TCEP as the reductant of choice. It is noteworthy that reducing DHA with DTT necessitates the neutralization of the sample solution and an incubation period, typically around 30 min, at room temperature. Under these conditions, both ascorbate and DHA may exhibit instability. In contrast to DTT, TCEP displays a wide effective pH range (from pH 1.5 to 8.5). In the conditions outlined in this protocol, 35 mM TCEP can achieve a complete reduction of DHA when applied on ice within a 2-h timeframe. In our hands, this treatment can be extended for up to 12 h or longer. Furthermore, the incubation of samples with TCEP can be conveniently conducted within the sample drawer of the HPLC/UFLC machine at 4°C, during the measurement of reduced ascorbate and standards.

5. Figure 1 illustrates the calibration curve derived from a set of five ascorbate standards, spanning concentrations from 31.25 to 500 μmol L^{-1}. The peak area exhibits a linear increase proportionate to the concentration of ascorbate. Figure 2 presents the peak area of ascorbate (A) and the calculated ascorbate levels (μmol gFW^{-1}; B), obtained from both the shoots and roots of *Arabidopsis* wild-type plants, as well as ascorbate-deficient mutants (*vtc2-4*, [21]). The highest levels of ascorbate are found in the shoots of the wild type, while the lowest levels are observed in the shoots of *vtc2-4* mutants. Notably, the

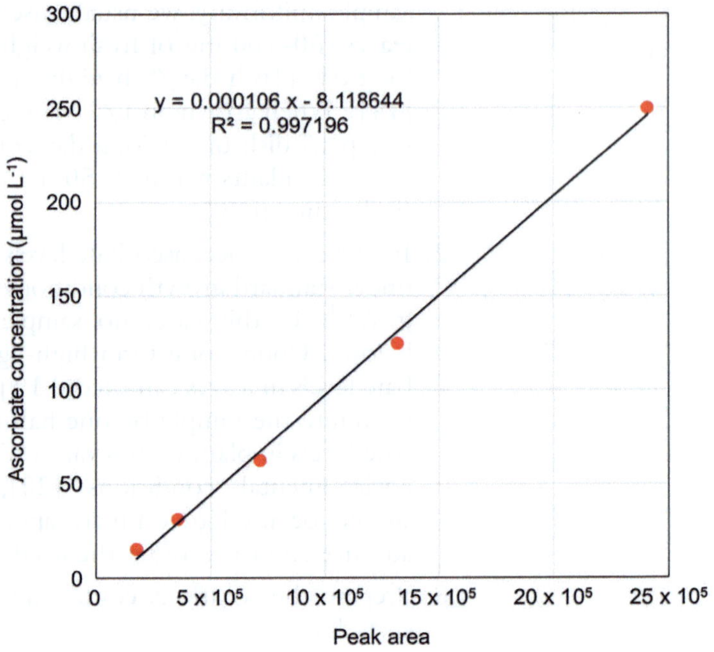

Fig. 1 The calibration curve derived from a set of five ascorbate standards

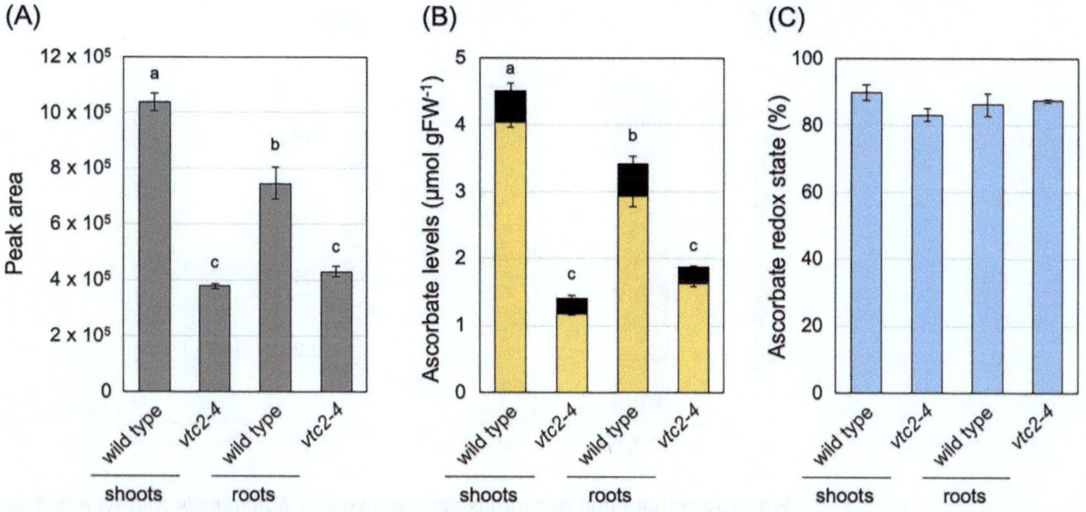

Fig. 2 Ascorbate levels in *Arabidopsis* wild type and *vtc2-4*. *Arabidopsis* wild-type plants (ecotype Col-0) and *vtc2-4* mutants were grown on half-strength MS medium containing 1% sucrose for 2 weeks according to [16]. (**a**) The peak area of total ascorbate (HPLC data) obtained from both the shoots and roots of wild type and *vtc2-4*. (**b**) The calculated ascorbate levels. Data are presented as the mean ± SE of three biological replicates. Different letters indicate significant differences ($P < 0.05$, Tukey–Kramer test). *ASC* reduced ascorbate, *DHA* oxidized ascorbate, *FW* fresh weight

mean peak area derived from undiluted wild-type shoot samples is approximately 50% of that obtained from the highest standard. Conversely, the mean peak area from undiluted *vtc2-4* shoot samples approximates fivefold the area obtained from the lowest standard. Thus, this calibration curve remains applicable for measuring ascorbate in mutants or species with even lower ascorbate levels than those observed in *vtc2* mutants.

6. While the ascorbate redox state has frequently served as an indicator of oxidative stress, our recent discoveries suggest the necessity for a more comprehensive examination of this concept [11, 14]. Figure 3 illustrates the levels of ASC (A) and DHA (B) in the shoots of *Arabidopsis* wild-type plants and *ao* mutants (SALK_108854), which lack the primary isoform of AO (At5g21100) [22]. There is no difference in ASC levels between the wild-type and *ao* mutants. In contrast, DHA levels in *ao* mutants are approximately half of those observed in the wild type. Considering that the *ao* mutants still retain around 15~20% of the wild-type AO activity [22], it is plausible that the majority of the detected DHA in *Arabidopsis* originates from the apoplast.

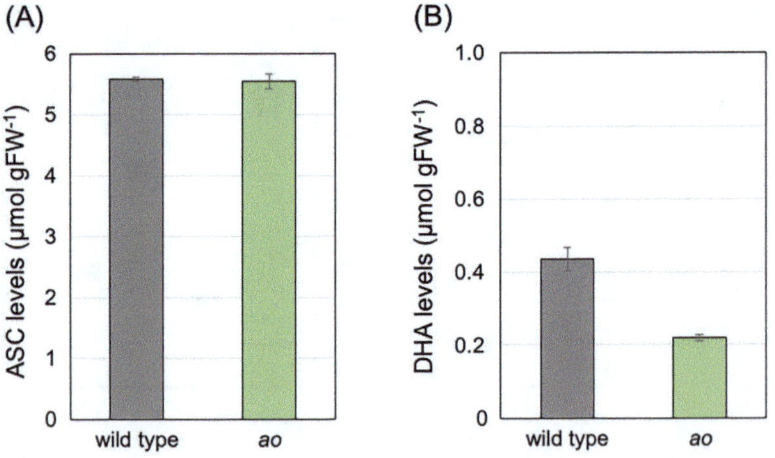

Fig. 3 Ascorbate and dehydroascorbate levels in *Arabidopsis* wild type and *ao*. *Arabidopsis* wild-type plants (ecotype Col-0) and *vtc2-4* mutants were grown on half-strength MS medium containing 0% sucrose for 2 weeks according to [16]. (**a**) ASC levels. (**b**) DHA levels. (**c**) Ascorbate redox states. Data are presented as the mean ± SE of three biological replicates. Significant differences (Student's t-test): $^{**}P < 0.01$ vs. values for wild type

Acknowledgments

This work was supported by JSPS KAKENHI Grant Numbers 18K19179 and 19K22284 (T.I. and T.M.).

References

1. Asada K (1999) The water-water cycle in chloroplasts: scavenging of active oxygens and dissipation of excess photons. Annu Rev Plant Physiol Plant Mol Biol 50:601–639. https://doi.org/10.1146/annurev.arplant.50.1.601
2. Mittler R, Zandalinas SI, Fichman Y, Van Breusegem F (2022) Reactive oxygen species signalling in plant stress responses. Nat Rev Mol Cell Biol 23(10):663–679. https://doi.org/10.1038/s41580-022-00499-2
3. Willems P, Mhamdi A, Stael S, Storme V, Kerchev P, Noctor G, Gevaert K, Van Breusegem F (2016) The ROS wheel: refining ROS transcriptional footprints. Plant Physiol 171(3):1720–1733. https://doi.org/10.1104/pp.16.00420
4. Smirnoff N (2018) Ascorbic acid metabolism and functions: a comparison of plants and mammals. Free Radic Biol Med 122:116–129. https://doi.org/10.1016/j.freeradbiomed.2018.03.033
5. Maruta T (2022) How does light facilitate vitamin C biosynthesis in leaves? Biosci Biotechnol Biochem 86(9):1173–1182. https://doi.org/10.1093/bbb/zbac096
6. Wheeler GL, Jones MA, Smirnoff N (1998) The biosynthetic pathway of vitamin C in higher plants. Nature 393(6683):365–369. https://doi.org/10.1038/30728
7. Wheeler G, Ishikawa T, Pornsaksit V, Smirnoff N (2015) Evolution of alternative biosynthetic pathways for vitamin C following plastid acquisition in photosynthetic eukaryotes. Elife 4. https://doi.org/10.7554/eLife.06369
8. Shigeoka S, Ishikawa T, Tamoi M, Miyagawa Y, Takeda T, Yabuta Y, Yoshimura K (2002) Regulation and function of ascorbate peroxidase isoenzymes. J Exp Bot 53(372):1305–1319
9. Maruta T, Sawa Y, Shigeoka S, Ishikawa T (2016) Diversity and evolution of ascorbate peroxidase functions in chloroplasts: more

than just a classical antioxidant enzyme? Plant Cell Physiol 57(7):1377–1386. https://doi.org/10.1093/pcp/pcv203

10. Tanaka M, Takahashi R, Hamada A, Terai Y, Ogawa T, Sawa Y, Ishikawa T, Maruta T (2021) Distribution and functions of monodehydroascorbate reductases in plants: comprehensive reverse genetic analysis of arabidopsis thaliana enzymes. Antioxidants 10(11). https://doi.org/10.3390/antiox10111726
11. Terai Y, Ueno H, Ogawa T, Sawa Y, Miyagi A, Kawai-Yamada M, Ishikawa T, Maruta T (2020) Dehydroascorbate reductases and glutathione set a threshold for high-light-induced ascorbate accumulation. Plant Physiol 183(1):112–122. https://doi.org/10.1104/pp.19.01556
12. Kerchev P, Waszczak C, Lewandowska A, Willems P, Shapiguzov A, Li Z, Alseekh S, Muhlenbock P, Hoeberichts FA, Huang J, Van Der Kelen K, Kangasjarvi J, Fernie AR, De Smet R, Van de Peer Y, Messens J, Van Breusegem F (2016) Lack of glycolate oxidase1, but not glycolate oxidase2, attenuates the photorespiratory phenotype of catalase2-deficient arabidopsis. Plant Physiol 171(3):1704–1719. https://doi.org/10.1104/pp.16.00359
13. Kameoka T, Okayasu T, Kikuraku K, Ogawa T, Sawa Y, Yamamoto H, Ishikawa T, Maruta T (2021) Cooperation of chloroplast ascorbate peroxidases and proton gradient regulation 5 is critical for protecting Arabidopsis plants from photo-oxidative stress. Plant J 107(3):876–892. https://doi.org/10.1111/tpj.15352
14. Hamada A, Tanaka Y, Ishikawa T, Maruta T (2023) Chloroplast dehydroascorbate reductase and glutathione cooperatively determine the capacity for ascorbate accumulation under photooxidative stress conditions. Plant J 114(1):68–82. https://doi.org/10.1111/tpj.16117
15. Foyer CH, Kyndt T, Hancock RD (2020) Vitamin C in plants: novel concepts, new perspectives, and outstanding issues. Antioxid Redox Signal 32(7):463–485. https://doi.org/10.1089/ars.2019.7819
16. Shiroma S, Tanaka M, Sasaki T, Ogawa T, Yoshimura K, Sawa Y, Maruta T, Ishikawa T (2019) Chloroplast development activates the expression of ascorbate biosynthesis-associated genes in Arabidopsis roots. Plant Sci 284:185–191. https://doi.org/10.1016/j.plantsci.2019.04.012
17. Noctor G, Mhamdi A, Foyer CH (2016) Oxidative stress and antioxidative systems: recipes for successful data collection and interpretation. Plant Cell Environ 39(5):1140–1160. https://doi.org/10.1111/pce.12726
18. Kampfenkel K, Van Montagu M, Inze D (1995) Extraction and determination of ascorbate and dehydroascorbate from plant tissue. Anal Biochem 225(1):165–167. https://doi.org/10.1006/abio.1995.1127
19. Maruta T, Yonemitsu M, Yabuta Y, Tamoi M, Ishikawa T, Shigeoka S (2008) Arabidopsis phosphomannose isomerase 1, but not phosphomannose isomerase 2, is essential for ascorbic acid biosynthesis. J Biol Chem 283(43):28842–28851. https://doi.org/10.1074/jbc.M805538200
20. Gest N, Gautier H, Stevens R (2013) Ascorbate as seen through plant evolution: the rise of a successful molecule? J Exp Bot 64(1):33–53. https://doi.org/10.1093/jxb/ers297
21. Lim B, Smirnoff N, Cobbett CS, Golz JF (2016) Ascorbate-deficient vtc2 mutants in Arabidopsis do not exhibit decreased growth. Front Plant Sci 7:1025. https://doi.org/10.3389/fpls.2016.01025
22. Yamamoto A, Bhuiyan MN, Waditee R, Tanaka Y, Esaka M, Oba K, Jagendorf AT, Takabe T (2005) Suppressed expression of the apoplastic ascorbate oxidase gene increases salt tolerance in tobacco and Arabidopsis plants. J Exp Bot 56(417):1785–1796. https://doi.org/10.1093/jxb/eri167

Chapter 9

Carotenoids and Tocopherol Profiling in Fleshy Fruits

Bruno Silvestre Lira, Aline Bertinatto Cruz, Magdalena Rossi, and Luciano Freschi

Abstract

Carotenoids and tocopherols are among the most powerful lipophilic antioxidants accumulated in fruit and vegetable crops. This chapter describes a method for the separation and quantification of carotenoids/chlorophylls and tocopherols based on microextraction followed by reverse- and normal-phase HPLC, respectively. Using this method, high-throughput, accurate analysis of these compounds can be performed in leaf and fruit samples.

Key words Antioxidants, Carotenoids, Flesh fruits, Ripening, Lycopene, Tocopherols, Tomato

1 Introduction

Carotenoids and tocopherols are two important classes of health-promoting compounds synthesized from a common precursor (GGDP, geranyl-geranyl pyrophosphate; Fig. 1). Four tocopherols, namely α-, β-, γ-, and δ-tocopherols, and numerous carotenoids, including carotenes and xanthophylls, are accumulated in a wide range of fruit and vegetable crops [1]. These compounds are paramount for plant development and yield, not only for being structural components of photosystems in the case of carotenoids [2, 3] but also for protecting membranes and photosynthetic structures from oxidative damage [4–6]. These antioxidant properties are also positive for human health. These compounds were reportedly associated with preventing and inhibiting the progression of several diseases, such as cancer and inflammatory syndromes, and for being associated with UV protection and ocular health [7–14].

Carotenoids and tocopherols can be simultaneously analyzed via liquid chromatographic (LC) separation followed by tandem mass spectrometry (MS) detection [15]. However, the much higher costs of LC-MS/MS apparatus, alongside their more demanding and expensive maintenance, make it prohibitive for

Francisco J. Corpas and José M. Palma (eds.), *ROS Signaling in Plants: Methods and Protocols*,
Methods in Molecular Biology, vol. 2798, https://doi.org/10.1007/978-1-0716-3826-2_9,

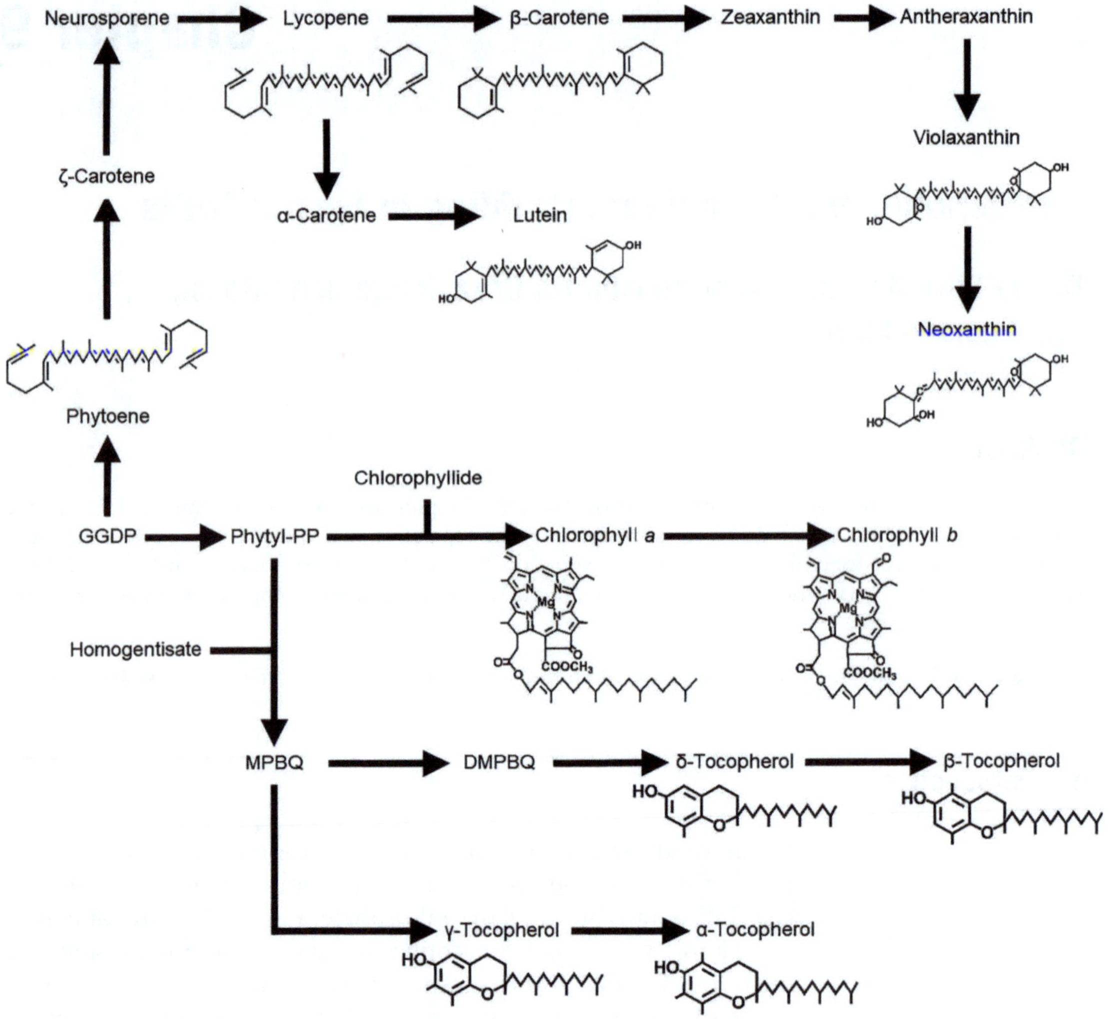

Fig. 1 Carotenoid, chlorophyll, and tocopherol biosynthetic pathway. Structures of detectable compounds are shown. GGDP, geranylgeranyl diphosphate; MPBQ, 2-methyl-6-phytyl-1,4-hydroquinone; DMPBQ, 2,3-dimethyl-6-phytyl-1,4-hydroquinone

high-throughput analysis of numerous samples for many laboratories. Therefore, high-performance liquid chromatography (HPLC) combined with diode array detector (HPLC-DAD) and HPLC with fluorescence detector (HPLC-FLD) remain the most used methods for profiling carotenoids, chlorophylls and tocopherols, respectively [16, 17].

This chapter describes the carotenoid and tocopherol separation and quantification via microextraction and reverse- and normal-phase HPLC. This method displays satisfactory accuracy and linearity to determine major carotenoids (lutein, neoxanthin, violaxanthin, β-carotene, lycopene, phytoene, and phytofluene),

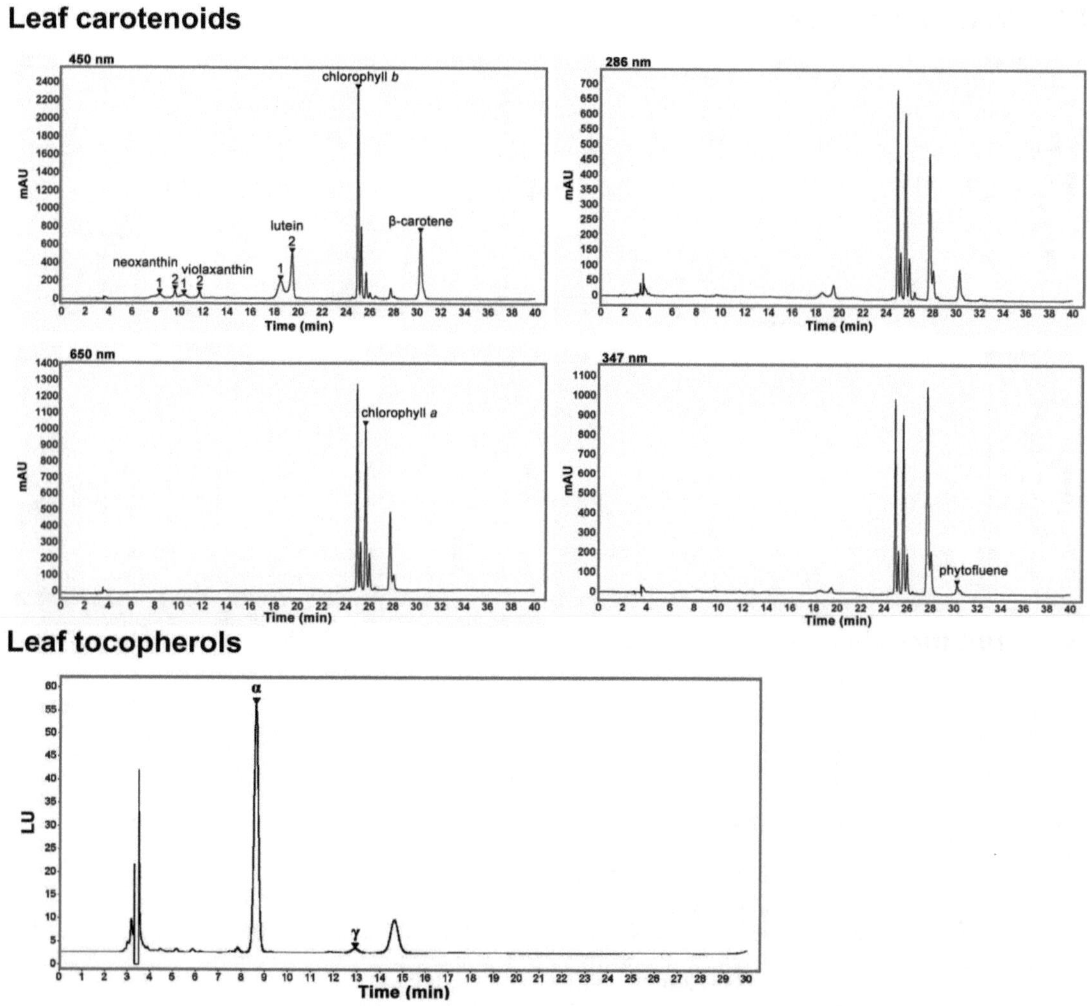

Fig. 2 HPLC profiles of the major carotenoids and tocopherols in tomato leaves. Peaks corresponding to neoxanthin, violaxanthin, lutein, chlorophyll *a*, chlorophyll *b*, β-carotene, phytofluene, α-tocopherol, and γ-tocopherol are indicated in the chromatograms

both chlorophylls (chlorophylls a and b), and all four tocopherol forms (α-, β-, γ-, and δ-tocopherols) in leaf and fruit tissues of tomato, a major fleshy fruit crop species [18–24] (Figs. 2 and 3).

2 Materials

2.1 Extraction Materials, Reagents, and Solutions

2.1.1 Carotenoid Extraction

1. 1.5 mL microtubes.
2. Saturated solution of sodium chloride. Weigh 3.7 g NaCl (ACS grade) in 10 mL ultrapure type I sterilized water. Store at room temperature.
3. Microtube vortex.

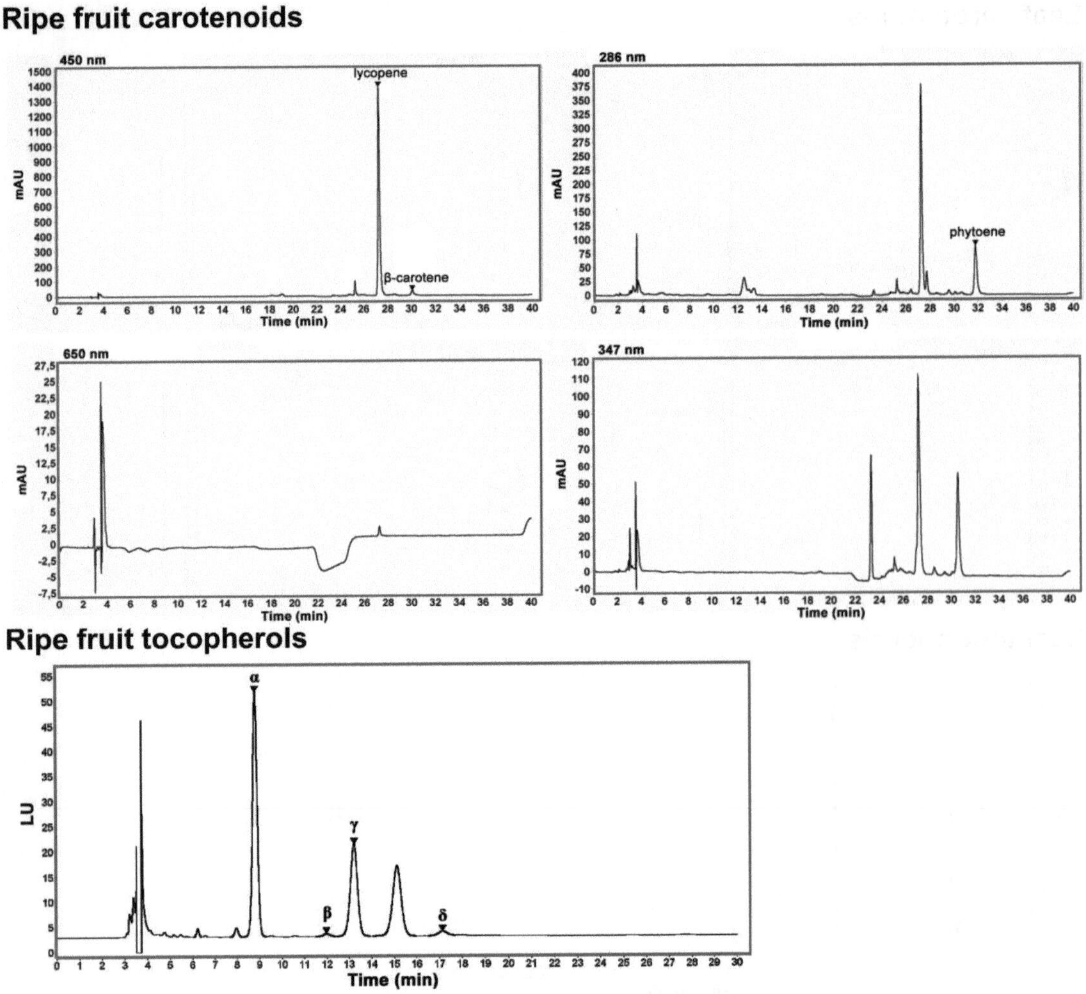

Fig. 3 HPLC profiles of the major carotenoids and tocopherols in tomato red ripe (15 d after breaker) fruits. Peaks corresponding to lycopene, β-carotene, phytoene, as well as α-, β-, δ-, and γ-tocopherol are indicated in the chromatograms

4. Dichloromethane (ACS grade). Store at room temperature.
5. 1:1 Hexane: ethyl ether solution (v:v; ACS grade). Store at room temperature.
6. Ultrasonic bath.
7. Refrigerated centrifuge with rotor for 1.5 mL microtubes.
8. Nitrogen gas, rotary evaporator, or vacuum concentrator compatible with 1.5 mL microtubes.
9. Ethyl acetate (ACS grade). Store at room temperature.
10. 0.45 μm polytetrafluoroethylene (PTFE) syringe filter.
11. 1 mL syringe.
12. Ambar HPLC vials.

13. 0.2–0.3 mL glass insert for HPLC vials.
14. HPLC coupled with a diode array detector (DAD, wavelength range: 200–700 nm) (*see* **Note 1**).
15. Reversed-phase column (Phenomenex Luna C18, 250 × 4.6 mm, 5 μm particle diameter, or similar).
16. Guard column (Phenomenex C18, 4 × 3.0 mm, or similar).

2.1.2 Tocopherol Extraction

1. 15 mL chloroform-resistant polypropylene tubes.
2. Methanol +0.1% BHT. Add 0.1 g BHT in 100 mL methanol (ACS grade). Store at 4 °C.
3. Chloroform +0.1% BHT. Add 0.1 g butylated hydroxytoluene (BHT) in 100 mL chloroform (ACS grade). Store at 4 °C.
4. Ultrasonic bath.
5. 20 mM Tris–HCl 1 M NaCl pH 7.5 solution. Add approximately 400 mL ultrapure type I sterilized water to a 500 mL graduated cylinder or a glass beaker. Weigh 1.21 g ultrapure Tris and 29.22 g NaCl and transfer to the cylinder. Adjust pH with HCl, then complete volume to 500 mL with ultrapure type I sterilized water. Store at 4 °C.
6. Refrigerated centrifuge with rotor for 15 mL tubes.
7. Nitrogen gas, rotary evaporator, or vacuum concentrator compatible with 15 mL tubes.
8. 9:1 Hexane: Methyl *tert*-butyl ether (MTBE) solution (v:v). Add hexane (HPLC grade) up to half the volume of a 1 L volumetric flask. With a volumetric pipette, add 100 mL of MTBE (HPLC grade) (*see* **Note 3**). Make up to 1 L with Hexane (HPCL grade).
9. 0.45 μm polytetrafluoroethylene (PTFE) syringe filter.
10. 1 mL syringe.
11. Ambar HPLC vials.
12. 0.2–0.3 mL glass insert for HPLC vials.
13. HPLC coupled with a fluorescence detector (wavelength range: 200–900 nm), seal pump, and inlet and outlet valves are suitable for normal phase application.

2.2 Standard Solution Preparation

2.2.1 Carotenoid Analysis

1. Weight 1 mg of each standard (lutein, neoxanthin, violaxanthin, β-carotene, lycopene, phytoene, phytofluene, chlorophyll *a* and chlorophyll *b*, HPLC suitable) in separate recipients (*see* **Note 2**).
2. Add 2 mL of ethyl acetate to dissolve each standard to obtain the standard stock solutions (500 μg/mL). Store at −80 °C.
3. Dilute with ethyl acetate and mix each standard separately to obtain a 10 μg/mL (per carotenoid) total carotenoid working standard solution. Store at −80 °C.

2.2.2 Tocopherol Analysis

1. Weight 5 mg of each standard (α-, β-, δ-, γ-tocopherol standards, HPLC suitable) in separate recipients (*see* **Note 2**).
2. Add 10 mL of hexane to dissolve the standards, and transfer 500 μL of each solution to a new tube.
3. Evaporate the hexane using nitrogen gas, rotary evaporator, or vacuum concentrator at a temperature below 30 °C.
4. Resuspend with 1 mL of methanol.
5. Measure the absorbance in the following wavelength to each form: α: 292 nm, β: 296 nm, γ: 298 nm, δ: 298 nm.
6. Divide the absorbance by the following coefficients to obtain the empirical concentration (μg/mL) of the respective stock solution: α: 0.0076; β: 0.0089, γ: 0.0091; δ: 0.0087. Store at −80 °C.
7. Dilute with hexane to get the working standard solution mix containing 5 μg/mL of each form (*see* **Note 2**). Store at −80 °C.

2.3 HPLC Solutions

1. Ethyl acetate (HPLC grade).
2. 9:1 acetonitrile:water. Measure 900 mL of acetonitrile (HPLC grade) in a 1 L volumetric flask. Separately, measure 100 mL of ultrapure type I sterilized water in a 100 mL volumetric flask. Mix both solutions and leave to reach room temperature before use.
3. 9:1 hexane: Methyl *tert*-butyl ether (MTBE) solution (v:v). Add hexane (HPLC grade) up to half the volume of a 1 L volumetric flask. With a volumetric pipette, add 100 mL of MTBE (HPLC grade) (*see* **Note 3**). Make up to 1 L with hexane (HPCL grade).

3 Methods

3.1 Carotenoid Analysis

1. Grind frozen or lyophilized samples in liquid N_2 to obtain a fine powder (*see* **Note 2**).
2. Weigh 200 mg frozen material (or 20 mg lyophilized material) of the previously powered sample in a 1.5 mL microtube.
3. Add 100 μL of saturated NaCl solution to each microtube and vortex for 30 s.
4. Add 200 μL of dichloromethane and vortex 30 s.
5. Add 500 μL of 1:1 hexane: ethyl ether (v/v).
6. Vortex for 1 min at maximum speed, and sonicate in an ultrasonic bath for 5 min (*see* **Note 4**).
7. Centrifuge at 13.000 *g* for 5 min at 4 °C.

8. Transfer the superior organic phase to a new 2 mL microtube.
9. Repeat **steps 5** to **8** three more times, adding the upper organic phase to the same collection microtube.
10. Dry the samples with nitrogen gas, rotary evaporator, or vacuum concentrator at a temperature below 30 °C.
11. Resuspend with 300 μL ethyl acetate (*see* **Note 5**).
12. Filter the samples with a 0.45 μm polytetrafluoroethylene (PTFE) syringe filter using 1 mL syringe.
13. Transfer the extracts to inserts in HPLC vials. Samples can be immediately analyzed or stored at −80 °C for up to a week.
14. The mobile phase consists of solution A (acetonitrile: ultrapure water 9:1) and B (ethyl acetate) and a gradient of %B: 0% to 5% from 0 to 4 min; 5% to 10% from 4 to 12 min; 10% to 20% from 12 to 17 min; 20% to 65% from 17 to 20 min; 65% from 20 to 35 min; 65% to 0% from 35 to 40 min. The solvent flow rate is 0.8 mL min^{-1} in a column at 25 °C (*see* **Note 6**).
15. Monitor using four wavelengths: 286 nm for phytoene, 347 nm for phytofluene, 650 nm for chlorophyll *a*, and 450 nm for lycopene, lutein, β-carotene, chlorophyll *b*, neoxanthin, and violaxanthin (*see* **Note** 7).
16. Use 20 μL of injection volume for samples. Obtain a standard curve by injecting 1.0, 2.5, 5, 10, 20, 30, 40, and 50 μL of the total carotenoid working standard solution (Table 1). Determine the endogenous metabolite concentration by comparing the peak areas of the chromatograms with the commercial standards (*see* **Note 8**).

3.2 Tocopherol Analysis

1. Grind frozen or lyophilized samples in liquid N_2 to obtain a fine powder (*see* **Note 2**).
2. Weight 25–30 mg of lyophilized sample in a 15 mL tube (*see* **Note 9**).
3. Add 1.5 mL of the methanol +0.1% BHT solution, and vortex for 1 min.
4. Add 1 mL of chloroform +0.1%BHT, and vortex for 1 min.
5. Sonicate in an ultrasonic bath for 5 min (*see* **Note 4**).
6. Add 2.5 mL of Tris–HCl 20 mM NaCl 1 M pH 7.5 to each tube, and vortex for 1 min.
7. Centrifuge at 3.000 *g* for 5 min at 4 °C.
8. Transfer the organic phase (bottom phase) to a new 15 mL tube.
9. Re-extract by adding 2 mL of chloroform +0.1% BHT to the initial tube.
10. Vortex for 1 min, and sonicate for 5 min.

Table 1
Retention time of detectable carotenoids and tocopherols

Compound	Wavelength	Retention time window
FLD		
α-Tocopherol	Excitation: 295 nm emission: 330 nm	8–9 min
β-Tocopherol		11–12 min
γ-Tocopherol		13–14 min
δ-Tocopherol		17–18 min
DAD		
Neoxanthin 1	450 nm	8–9 min
Neoxanthin 2		9.5–10 min
Violaxanthin 1		10–10.5 min
Violaxanthin 2		11–12 min
Lutein 1		18–19 min
Lutein 2		19–20 min
Chlorophyll *b*		24.5–25.5 min
Lycopene		27–28 min
β-Carotene		30–30.5 min
Chlorophyll *a*	650 nm	25.5–26 min
Phytoene	286 nm	31–33 min
Phytofluene	347 nm	30–31 min

11. Centrifuge at 3.000 g for 5 min at 4 °C.
12. Unite the organic phase with the one previously collected in **step 8**.
13. Adjust the volume to 4 mL with chloroform +0.1%BHT.
14. Remove any residual aqueous phase from the top of the collected organic phase and adjust the volume to 3 mL (*see* **Note 9**).
15. Dry the samples with nitrogen gas, rotary evaporator, or vacuum concentrator at a temperature below 30 °C.
16. Resuspend the dried pellets with 200 μL of 9:1 hexane: MTBE (v:v).
17. Filter the samples with a 0.45 μm polytetrafluoroethylene (PTFE) syringe filter using 1 mL syringe (*see* **Note 10**).
18. Transfer the extracts to HPLC vials. Samples can be immediately analyzed or stored at −80 °C for up to a week.
19. The mobile phase is an isocratic flow of 9:1 hexane: MTBE (v:v) for 30 min per sample in a column at 30 °C. No post-time is needed after each injection.
20. Set the fluorescence detector excitation signal at 296 nm and emission detection at 340 nm. Integration can be turned off for the first 5 min.

21. Use 20 μL of injection volume for samples (*see* **Note 11**). Obtain a standard curve by injecting 1.25, 2.5, 5, 10, and 20 μL of the total carotenoid working standard solution (Table 1). Determine the endogenous metabolite concentration by comparing the peak areas of the chromatograms with the commercial standards (*see* **Note 8**).

4 Notes

1. For higher stability of carotenoids and tocopherols, samples must be kept at 4 °C. If the HPLC system does not contain a thermostat coupled to the sample tray, samples can be maintained at 4 °C and four at a time loaded in the injector tray.
2. Standard preparation and sample extraction must be done in dim light. During the procedures, samples and reagents must be kept always on ice. After the extraction, drying and resuspension, samples must be stored at −80 °C.
3. Proper MTBE concentration is paramount to separate β- and γ-tocopherol.
4. Add ice to the ultrasonic bath prior to sonicating the extracts.
5. If any compound exceeds the detection capacity of the equipment, a dilution or smaller volumes may be injected to detect the abundant forms.
6. The HPLC analysis time can be shortened from 40 to 25 min when detecting only lutein, β-carotene, lycopene, phytoene, and phytofluene. In this case, the mobile phase consists of solution A (acetonitrile: ultrapure water 9:1) and B (ethyl acetate) and a gradient of %B: 20% to 65% from 0 to 10 min; 65% from 10 to 20 min; 65% to 0% from 20 to 25 min. The solvent flow rate is 1.0 mL min^{-1} in a column at 25 °C.
7. Neoxanthin, violaxanthin, and lutein are split into two peaks. For calculations, use the sum of both areas of these compounds.
8. As slight variations in the retention time of the compounds may occur due to equipment and mobile phase conditions, a sample of 10 μL of the standard curve may be injected to identify the peak of each compound whenever necessary.
9. The aqueous phase can be more easily visualized by keeping the samples for at least 1 h at −80 °C.
10. Do not let the rubber plunger directly contact the samples at the filtration step.
11. After each injection, autosampler rinsing with 9:1 hexane: MTBE (v:v) is strongly suggested to clear the needle surface.

Acknowledgments

This work was supported by the São Paulo Research Foundation (FAPESP) (grants 16/02033-1, 16/01128-9, 17/14953-0, 17/17935-3, 18/16389-8, 18/06436-9, 20/03720-8, 21/05714-8), Conselho Nacional de Desenvolvimento Científico e Tecnológico (CNPq) (grants 422287/2018-0, 305012/2018-5, 302565/2021-3), and Coordenação de Aperfeiçoamento de Pessoal de Nível Superior—Brazil (CAPES)—Finance Code 001. BSL was a recipient of a fellowship granted by GALY Company.

References

1. Corpas FJ, Freschi L, Palma JM (2023) ROS metabolism and ripening of fleshy fruits. In: Oxidative stress response in plants, Advances in botanical research, pp 205–238. https://doi.org/10.1016/bs.abr.2022.08.024
2. Tracewell CA, Vrettos JS, Bautista JA et al (2001) Carotenoid photooxidation in photosystem II. Arch Biochem Biophys 385(1): 61–69. https://doi.org/10.1006/abbi.2000.2150
3. Jensen PE, Bassi R, Boekema EJ et al (2007) Structure, function and regulation of plant photosystem I. Biochimica et Biophysica Acta (BBA)-Bioenergetics 1767(5):335–352. https://doi.org/10.1016/j.bbabio.2007.03.004
4. Foyer CH, Noctor G (2005) Redox homeostasis and antioxidant signaling: a metabolic interface between stress perception and physiological responses. Plant Cell 17(7): 1866–1875. https://doi.org/10.1105/tpc.105.033589
5. Kruk J, Holländer-Czytko H, Oettmeier W et al (2005) Tocopherol as singlet oxygen scavenger in photosystem II. J Plant Physiol 162(7):749–757. https://doi.org/10.1016/j.jplph.2005.04.020
6. Telfer A (2014) Singlet oxygen production by PSII under light stress: mechanism, detection and the protective role of β-carotene. Plant Cell Physiol 55(7):1216–1223. https://doi.org/10.1093/pcp/pcu040
7. Ju J, Picinich SC, Yang Z et al (2010) Cancer-preventive activities of tocopherols and tocotrienols. Carcinogenesis 31(4):533–542. https://doi.org/10.1093/carcin/bgp205
8. Thompson MD, Cooney RV (2019) The potential physiological role of γ-tocopherol in human health: a qualitative review. Nutr Cancer 72(5):808–825. https://doi.org/10.1080/01635581.2019.1653472
9. Lee ARYB, Tariq A, Lau G et al (2022) Vitamin E, alpha-tocopherol, and its effects on depression and anxiety: a systematic review and meta-analysis. Nutrients 14(3):656. https://doi.org/10.3390/nu14030656
10. Di Vincenzo A, Tana C, El Hadi H et al (2019) Antioxidant, anti-inflammatory, and metabolic properties of tocopherols and tocotrienols: clinical implications for vitamin E supplementation in diabetic kidney disease. Int J Mol Sci 20(20):5101. https://doi.org/10.3390/ijms20205101
11. Amengual J (2019) Bioactive properties of carotenoids in human health. Nutrients 11(10): 2388. https://doi.org/10.3390/nu11102388
12. Eggersdorfer M, Wyss A (2018) Carotenoids in human nutrition and health. Arch Biochem Biophys 652:18–26. https://doi.org/10.1016/j.abb.2018.06.001
13. Caseiro M, Ascenso A, Costa A et al (2020) Lycopene in human health. LWT 127:109323. https://doi.org/10.1016/j.lwt.2020.109323
14. Rodriguez-Concepcion M, Avalos J, Bonet ML et al (2018) A global perspective on carotenoids: metabolism, biotechnology, and benefits for nutrition and health. Prog Lipid Res 70:62–93. https://doi.org/10.1016/j.plipres.2018.04.004
15. Lee HW, Zhang H, Liang X et al (2020) Simultaneous determination of carotenoids, tocopherols and phylloquinone in 12 Brassicaceae vegetables. LWT 130:109649. https://doi.org/10.1016/j.lwt.2020.109649
16. Amorim-Carrilho KT, Cepeda A, Fente C et al (2014) Review of methods for analysis of carotenoids. TrAC Trends Anal Chem 56:49–73. https://doi.org/10.1016/j.trac.2013.12.011
17. Bao Y, Magallenes-Lundback M, Deason N et al (2020) High throughput profiling of tocochromanols in leaves and seeds of

Arabidopsis and Maize. Plant Methods 16(1): 126. https://doi.org/10.1186/s13007-020-00671-9

18. Lira BS, Gramegna G, Amaral P et al (2023) Phytol recycling: essential, yet not limiting for tomato fruit tocopherol accumulation under normal growing conditions. Plant Mol Biol. https://doi.org/10.1007/s11103-022-01331-3
19. Zuccarelli R, Rodríguez-Ruiz M, Lopes PJ et al (2021) Multifaceted roles of nitric oxide in tomato fruit ripening: NO-induced metabolic rewiring and consequences for fruit quality traits. J Exp Bot 72(3):941–958. https://doi.org/10.1093/jxb/eraa526
20. Alves FRR, Lira BS, Pikart FC et al (2020) Beyond the limits of photoperception: constitutively active PHYTOCHROME B2 overexpression as a means of improving fruit nutritional quality in tomato. Plant Biotechnol J 18(10):2027–2041. https://doi.org/10.1111/pbi.13362
21. Rosado D, Trench B, Bianchetti R et al (2019) Downregulation of PHYTOCHROME-INTERACTING FACTOR 4 influences plant development and fruit production. Plant Physiol 181(3):1360–1370. https://doi.org/10.1104/pp.19.00833
22. Cruz AB, Bianchetti RE, Alves FRR et al (2018) Light, ethylene and auxin Signaling interaction regulates carotenoid biosynthesis during tomato fruit ripening. In: Frontiers in Plant Science 9:1370. https://doi.org/10.3389/fpls.2018.01370
23. Bianchetti RE, Lira BS, Monteiro SS et al (2018) Fruit-localized phytochromes regulate plastid biogenesis, starch synthesis, and carotenoid metabolism in tomato. J Exp Bot 69(15):3573–3586. https://doi.org/10.1093/jxb/ery145
24. Lira BS, Rosado D, Almeida J et al (2016) Pheophytinase knockdown impacts carbon metabolism and nutraceutical content under normal growth conditions in tomato. Plant Cell Physiol 57(3):642–653. https://doi.org/10.1093/pcp/pcw021

Chapter 10

Anthocyanins: Quantification by Tandem Mass Spectrometry

Cindy Dias, Manuela Pintado, and Clara Sousa

Abstract

Mass spectrometry is a high throughput technique widely used for metabolic fingerprinting of plant material. Among the diverse plant metabolites, pigments such as anthocyanins play a determinant role in plant defence mechanisms, protecting them from biotic and abiotic stresses. Anthocyanins are phenolic water-soluble glycosides or acyl-glycosides of anthocyanidins which could be accurately detected and quantified through mass spectrometry. This chapter describes how to extract anthocyanins from higher plant materials and quantify them through a liquid chromatography–mass spectrometry (LC–MS) based method.

Key words Anthocyanins, Pigments, Metabolites, Plant, HPLC, Solid phase extraction (SPE), Mass spectrometry (MS)

1 Introduction

Anthocyanins belong to the large natural group of phenolic compounds, particularly the flavonoid group widely found in fruits and vegetables. An increasing interest in these water-soluble organic pigments appears due to their demonstrated numerous biological activities [1]. Besides protecting plants from abiotic and biotic stresses, anthocyanins exhibit potential health benefits due to their antioxidant [2], antibacterial [3], and anti-inflammatory [4] properties, among others. Additionally, these molecules can be used as natural coloring agents [5]. Due to their versatility, properties, and being safe molecules, anthocyanins became the target of numerous studies and are of large interest to the food, feed, medicine, and cosmetic industry.

In this context, an efficient extraction, purification, and quantification of anthocyanins is of utmost relevance. Here, we demonstrate a low-cost ultrasound-assisted method for extraction (UAE) of anthocyanins and its purification using solid-phase extraction

Francisco J. Corpas and José M. Palma (eds.), *ROS Signaling in Plants: Methods and Protocols*,
Methods in Molecular Biology, vol. 2798, https://doi.org/10.1007/978-1-0716-3826-2_10,

Table 1
Structure of the main anthocyanidin compounds

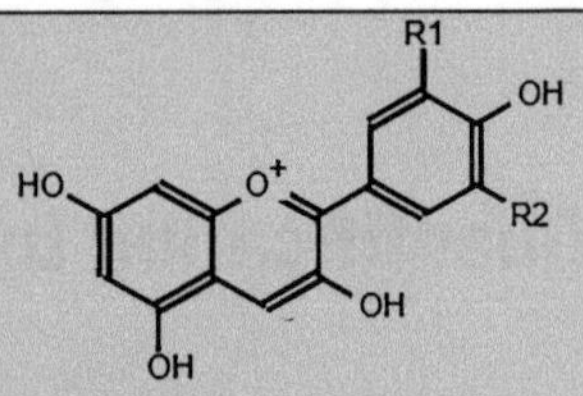

Anthocyanidin	Structure R1	R2	Main colour
Cyanidin	–OH	–H	Orange/Red
Delphinidin	–OH	–OH	Bluish/Red
Malvidin	$-OCH_3$	$-OCH_3$	Bluish/Red
Pelargonidin	–H	–H	Orange
Peonidin	$-OCH_3$	–H	Orange/Red
Petunidin	$-OCH_3$	–OH	Bluish/Red

(SPE) for subsequent quantification through tandem mass spectrometry. Briefly, a sample pre-treatment through lyophilization of the plant material is required. The subsequent extraction step, due to the polarity of anthocyanins, is undertaken with a polar organic solvent slightly acidic to improve anthocyanin extractability and stability. A mixture of methanol:water:formic acid is herein used as the extraction solvent. Since during the extraction many impurities (sugars, proteins, lipids, amino acids) are dragged from the plant material, a further purification step is essential to obtain anthocyanins with high stability and purity for identification and quantification. Here, we use an SPE C18 cartridge for the proper purification and recovery of anthocyanins. Finally, identification and quantification are performed using tandem mass spectrometry through a high-performance liquid chromatography system coupled to a quadrupole time-of-flight mass spectrometer (HPLC-QTOF-MS). The mass spectrometer is operated in positive electrospray ionization mode. Although more than 700 anthocyanins have been described in literature, they are derived from six main structures: cyanidin, delphinidin, malvidin, pelargonidin, peonidin, and petunidin (Table 1) founded in natural sources bound to sugars [6]. The most common sugars bound to anthocyanidins are glucose, galactose, arabinose, rhamnose, and xylose [7]. Usually, anthocyanin quantifications are related to a single selected anthocyanin compound due to their countless specimens. In this protocol, we select malvidin-3-glucose.

2 Materials

2.1 Anthocyanin Ultrasound-Assisted Extraction (UAE)

1. Lyophilized plant material.
2. Anthocyanins extraction solution: methanol:H_2O:formic acid (75:24:1 v/v) solution. Add 75 mL of methanol, 24 mL of ultra-pure water, and 1 mL of formic acid to a beaker (*see* **Note 1**). Store at 4 °C until use (*see* **Note 2**).
3. Ultrasonic bath.
4. Centrifuge.

2.2 Solid-Phase Extraction (SPE)

1. SPE C18 cartridge (3 mL, 500 mg) (*see* **Note 3**).
2. SPE vacuum manifold system.
3. Vacuum pump.
4. Cartridge activation solvent: methanol (*see* **Note 2**).
5. Cartridge conditioning and washing solvent (extraction solution): methanol:H_2O:formic acid (75:24:1 v/v) (*see* **Notes 2** and **4**).
6. Cartridge elution solvent: methanol acidified with 1% (v/v) formic acid. For a final volume of 100 mL, mix 99 mL of methanol with 1 mL of formic acid (*see* **Note 2**).
7. Collection tubes.
8. Syringe filters of 0.22 μm pore size.
9. Syringe of 1 mL.
10. Vials and lids for the HPLC-QTOF-MS equipment.

2.3 High-Performance Liquid Chromatography – Mobile Phases

1. Mobile phase A: H_2O:formic acid (99.9:0.1 v/v). Add 0.1 mL of formic acid to 99.9 mL of ultra-pure water and filtrate (*see* **Note 2**). Recommended are membrane-type filters with pore sizes of 0.22 μm (*see* **Note 5**). Then place the mobile phase in an ultrasonic bath for 15 min for degassing (*see* **Note 6**).
2. Mobile phase B: acetonitrile:formic acid (99.9:0.1). Add 0.1 mL of formic acid to 99.9 mL of acetonitrile (*see* **Note 2**) and follow the same procedure as described for mobile phase A.
3. Glass vacuum filtration system.

2.4 Anthocyanins Standard Solutions

1. Internal standard: daidzin.
2. Standard for quantification: malvidin-3-glucose.

3 Methods

3.1 Anthocyanin Ultrasound-Assisted Extraction (UAE)

1. Lyophilized plant material: plant material used for anthocyanins extraction should be lyophilized immediately after harvesting or, if not possible, stored at −80 °C for no more than 2 weeks (*see* **Note 7**). The anthocyanin extraction should be performed also immediately after the lyophilization step (*see* **Note 8**).
2. Weight 25 mg of lyophilized plant material to a 2 mL Eppendorf. Triplicates of the extraction are advised for proper statistical data analysis.
3. To each Eppendorf, add 1 mL of the anthocyanin extraction solution and gently shake the Eppendorf till the solution reach all the sample powder (*see* **Note 9**).
4. Place the Eppendorfs in an ultrasonic bath for 15 min.
5. Centrifuge the Eppendorf tubes at 12,000 rpm for 10 min and recover the supernatant. The supernatant will be further used for the SPE step (*see* **Note 10**).

3.2 Solid-Phase Extraction (SPE)

1. SPE cartridges need to be activated and conditioned before the supernatant loading. Activate and conditionate the C18 column with 3 mL methanol followed by 3 mL of the extraction solution, respectively (*see* **Note 11**). Collect the different fractions and discard them.
2. Load 1 mL of sample in the column which should be passed through the column at a rate of about 1–2 drops/s (*see* **Note 12**). Discard the flow-through.
3. Wash the column with 2 × 3 mL of the extraction solution at the same flow rate as the sample loading to remove impurities will have less affinity to the column. Discard the flow-through.
4. Elute the column with 1 mL of acidified methanol with 1% (v/v) formic acid at a rate of 1–2 drop/s (*see* **Note 13**).
5. Filter the elute using a 0.22 μm filter and fill the HPLC-QTOF-MS vials. Place the lids, and the sample is now ready for injection.

3.3 HPLC-QTOF-MS Conditions

All the prepared samples are subjected to reversed-phase chromatography using the Thermo system, Ultimate 3000. Chromatographic separation is performed at a flow rate of 0.250 mL/min using a column (Thermo, Acclaim RSLC 120-C18, 2.2 μm, 2.1 × 100 mm) which is maintained at 35 °C. The solvent gradient should be settled as follows: 0–10 min (10% B), 10–14 min (21% B), 14–18 min (27%), 18–20 min (58% B), 20–24 (100% B), and 24–26 min (10% B). The injection volume is at 5 μL, and the autosampler should be kept at 4 °C.

Mass spectra are acquired in positive mode from 50 to 1000 m/z. The parameters for mass scan are as follows: capillary voltage, 4.0 kV; drying gas temperature, 200 °C; drying gas flow, 8.0 L min^{-1}; nebulizing gas pressure, 1.5 bar; collision RF, 750 Vpp; transfer time, 60 μs; and prepulse storage, 5 μs. Internal mass calibration uses sodium formate clusters with the sodium formate delivered by a syringe pump at the start of each chromatographic analysis. The LC-ESI-QTOF data is acquired with the Bruker Compass HyStar 6.0 software1.

3.4 Anthocyanins Quantification

1. Recovery ratio determination: to determine the percentage of anthocyanins losses during the extraction and purification steps, an internal standard must be used (daidzin) (*see* **Note 14**). Three samples (lyophilized plant material + extraction solvent) should be prepared twice ("twin 1" and "twin 2") for the recovery ratio determination. The internal standard need to be added to "twin 1" sample prior extraction and to "twin 2" sample after extraction and immediately before the HPLC-QTOF-MS analysis. The concentration of internal standard added to both samples has to be the same to produce a chromatographic peak of a similar magnitude than the remaining analytes of the sample. The intensity ratio of the internal standard peaks (prior to HPLC-QTOF-MS analysis and extraction) is used to estimate the losses during extraction and further correct the anthocyanins concentration in the samples. The final recovery ratio is the mean value obtained from the three samples.
2. Calibration curve for quantification: Anthocyanins are usually quantified using a common standard as malvidin-3-glucose (*see* **Note 15**). Around 10 standard solutions need to be prepared in a concentration range which includes the anthocyanin concentration of the samples. This step needs a prior sample injection to evaluate the magnitude of the chromatographic signals in the samples. After the preparation of the malvidin-3-glucose standard solutions, the internal standard (daidzin) needs to be added to all the solutions in the same concentration. The calibration curve is obtained using the intensity of the signal of malvidin-3-glucose divided by the intensity of the internal standard peak.
3. Anthocyanins quantification: Prior to HPLC-QTOF-MS measurements, the internal standard needs to be added to all the samples in the same concentration as used in the standard solutions used to produce the calibration curve. The intensity of the signal obtained for each anthocyanin is further divided by the peak intensity of the internal standard prior the concentration determination through the calibration curve.

4 Notes

1. When alcohol dissolves in water, these molecules fit themselves in between water molecules and the volume decreases. So, be aware that the final volume will not be 100 mL.
2. All the reagents should be of LC–MS grade.
3. The SPE cartridge size/volume should take into consideration the volume of the sample. In this case, as the final volume after extraction is 1 mL, a 3 mL column is sufficient. While the SPE sorbent weight should consider the concentration/amount of the analyte. If it is expected to have a high concentration of the analyte, the use of small sorbent weight will lead to saturation faster not allowing the maximum retention of the analyte.
4. Generally, the sample solvent loaded into the column is the same used for conditioning the column and the washing solvent is the same used for conditioning the column.
5. Filtration removes particles that can clog the HPLC system and column.
6. Degassing will remove the dissolved air responsible for the formation of bubbles that can cause problems in the HPLC system. For that reason, connect the HPLC mobile phase line directly to the bottle after ultra-sonification.
7. Plant material should not be stored for a long time at –80 °C because it promotes cell wall disruption and consequently degradation of the bioactive compounds.
8. The pre-treatment reduces the particle size and improves extraction efficiency.
9. When mixing the extraction solution with the lyophilized samples, care should be taken to ensure that (i) all the powder is in contact with the methanol solution including the particles at the bottom of the Eppendorf, and (ii) that no powder is attached to the lid or the side wall of the Eppendorf. Both issues can compromise the accurate anthocyanin quantification.
10. Centrifugation will remove any unwanted particulates which will facilitate SPE purification.
11. This is a very important step that will prepare the sorbent for the interaction with anthocyanins. After initiating this step, it is very important to prevent the sorbent from drying during conditioning (dry sorbent will affect the capacity of the analytes to interact with the sorbent). Thus, allow about 1 mm of the conditioning solvent to remain above the top tube frit and turn off the pump until sample application. We recommend using the same flow rate for activation and conditioning. We find that it is easier if practice before.

12. To obtain better retention results, allow the sample to enter the packed sorbent completely before turning on the vacuum pump. Use moderate pressure. It is very important that the sample is loaded at an appropriate flow rate. A high flow rate can lead to inconsistent retention.
13. A smaller elution volume allows a more concentrated elute. The elute solvent should have a higher polarity than the sample solvent to remove anthocyanins trapped in the column.
14. The selected internal standard should, ideally, not be present in the analysed samples. A first screening is advised to see if daidzin is present in the analysed samples. The screening could easily be performed through an MRM experiment of the internal standard and also of the samples. If daidzin is present in the samples, phloridzin could be an alternative to the internal standard.
15. If different standards are available, the corresponding calibration curve could be obtained. The concentrations of the corresponding anthocyanins are, in such scenario, directly determined.

Acknowledgements

This work was supported by National Funds from FCT—Fundação para a Ciência e a Tecnologia through project UIDB/50016/2020.

References

1. Tan J, Han Y, Han B, Qi X, Cai X, Ge S, Xue H (2022) Extraction and purification of anthocyanins: a review. J Agric Food Res 8:100306
2. Jiao X, Li B, Zhang Q, Gao N, Zhang X, Meng X (2018) Effect of in vitro-simulated gastrointestinal digestion on the stability and antioxidant activity of blueberry polyphenols and their cellular antioxidant activity towards HepG2 cells. Int J Food Sci Technol 53:61–71
3. Pertuzatti PB, Barcia MT, Rebello LPG, Gómez-Alonso S, Duarte RMT, Duarte MCT, Godoy HT, Hermosín-Gutiérrez I (2016) Antimicrobial activity and differentiation of anthocyanin profiles of rabbiteye and highbush blueberries using HPLC–DAD–ESI-MS n and multivariate analysis. J Funct Foods 26:506–516
4. Eroglu Ozkan E, Seyhan MF, Kurt Sirin O, Yilmaz-Ozden T, Ersoy E, Hatipoglu Cakmar SD, Goren AC, Yilmaz Aydogan H, Ozturk O (2021) Antiproliferative effects of Turkish pomegranate (*Punica granatum* L.) extracts on MCF-7 human breast cancer cell lines with focus on antioxidant potential and bioactive compounds analyzed by LC-MS/MS. J Food Biochem 45:e13904
5. Tena N, Asuero AG (2022) Up-to-date analysis of the extraction methods for anthocyanins: principles of the techniques, optimization, technical progress, and industrial application. Antioxidants 11:286. https://doi.org/10.3390/antiox11102028
6. Zhao Y-W, Wang C-K, Huang X-Y, Hu D-G (2021) Anthocyanin stability and degradation in plants. Plant Signal Behav 16:1987767
7. Alappat B, Alappat J (2020) Anthocyanin pigments: beyond aesthetics. Molecules 25:5500

Chapter 11

Common Methods of Extraction and Determination of Phytomelatonin in Plants

Antonio Cano, Josefa Hernández-Ruiz, and Marino B. Arnao

Abstract

The presence of melatonin in plants, called phytomelatonin, has gained great interest in recent years. The determination of phytomelatonin levels in plant extracts for both physiological and plant foodstuff studies requires sophisticated techniques due to the low endogenous levels of this indolic compound with hormonal nature. This chapter presents the most common and advanced techniques in the determination of phytomelatonin, with special emphasis on the techniques of extraction, cleaning, separation, detection, identification, and quantification. Multiple examples and recommendations are presented for a clear overview of the pros and cons of phytomelatonin determinations in plant tissues, seeds, and fruits, mainly.

Key words Phytomelatonin, Plant Melatonin, Extraction Methods, Analytical Methods, Melatonin

1 Introduction

Melatonin (*N*-acetyl-5-methoxytryptamine) is an indoleamine derived from tryptophan. Discovered in 1958 in the pineal gland of cow [1], and in 1995 measured in plants, the so-called phytomelatonin [2–4] is considered a universal molecule present in all organisms, from bacteria, algae and fungi to arthropods, reptiles, fish, birds, and mammals [5, 6]. Melatonin is a small molecule (Fig. 1) with a molecular weight of 232.28 Da; it is an amphipathic molecule with a solubility of 0.1 mg/mL in water and 8 mg/mL in ethanol and also is oxidizable and light sensitive [7]. Melatonin presents an indole ring in its structure, similar to the plant hormone auxin (indole-3-acetic acid, IAA) [8, 9].

In mammals, melatonin acts as a hormone regulating multiple rhythms such as sleep-wake cycles, temperature, endocrine hormones, sexual behavior, and mood, among others. Also, it regulates some immune responses, and sets the central and peripheral biological clocks on time [10]. In plants, phytomelatonin acts as a master regulator. It plays an important role in the plant resistance

Francisco J. Corpas and José M. Palma (eds.), *ROS Signaling in Plants: Methods and Protocols*,
Methods in Molecular Biology, vol. 2798, https://doi.org/10.1007/978-1-0716-3826-2_11,

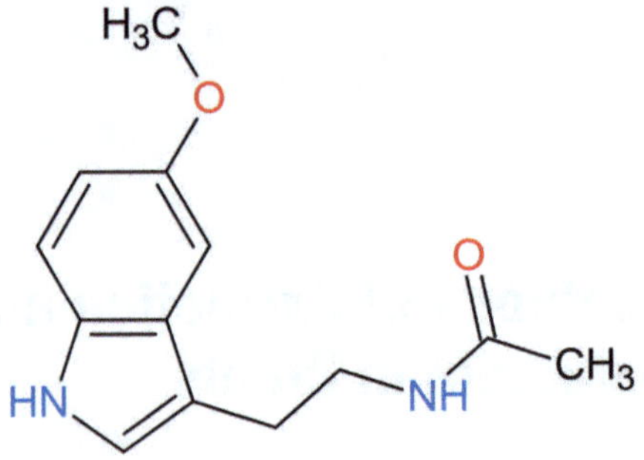

Fig. 1 Melatonin structure

responses to biotic or abiotic stresses, including extreme temperature, salinity, drought, UV radiation, and soil and air pollutant presence, among others [9, 11–13]. Also, phytomelatonin regulates seed germination, rooting, plant growth, leaf senescence, flowering, and fruit ripening, obtaining a great relevance in the last years in plant science, especially in horticultural crops [14–16].

In both animals and plants, melatonin is an excellent antioxidant in vivo, being a powerful free radical scavenger, and more specifically scavenging reactive oxygen and nitrogen species (ROS/RNS). In plants it has been shown that phytomelatonin behaves as a regulator of homeostatic balance between ROS/RNS levels and antioxidative enzyme defenses such as peroxidases, catalases, superoxide dismutases, respiratory burst oxidases (RBOHs) and those from the ascorbate-glutathione cycle [14, 17, 18]. Also, phytomelatonin regulates plant hormone levels and the expression of several hormone-responsive genes under different conditions [13, 19–21].

2 Phytomelatonin Plant Content

The phytomelatonin content in plants is highly variable, depends on the species, and is affected by genetic factors and external environment. Generally, it occurs at very low endogenous levels, being higher in leaves and roots, and lower in fruits, seeds, and flowers [22, 23]. With respect to phylogenetic classification, there is no clear predominance of some family plants over others, in terms of their phytomelatonin endogenous content. However, medicinal/aromatic plants (MAPs) are considered to contain the highest levels of phytomelatonin, possibly due to the extreme conditions in which many of them grow up, but also with great variability in the data [24–27].

One of the common problems in phytomelatonin studies is the large dispersion in data on endogenous levels in plant tissues. This is usually due to the imprecision of some extraction protocols and the use of not very adequate determination techniques, with quantification by excess due to overcrossing in immunological techniques with other indolic compounds (very common and diverse in plants), and/or by false positives in chromatographic techniques, as we will see later [28].

3 Phytomelatonin Biosynthesis Pathway

Melatonin biosynthesis in plants has been studied extensively (Fig. 2) [29]. The precursor of phytomelatonin is tryptophan, as in animal biosynthesis, and practically the same enzymes are involved. The first step is the conversion of tryptophan into tryptamine, catalyzed by the tryptophan decarboxylase (TDC); then, the action of tryptamine 5-hydroxylase (T5H) converts tryptamine in 5-hydroxytryptamine, also known as serotonin. The next step is the acetylation of serotonin by the serotonin N-acetyltransferase (SNAT), which is subsequently methylated by the action of acetyl serotonin methyl transferase (ASMT); in plants, this step can also be done by the enzyme caffeic *O*-methyltransferase (COMT). These are the main steps in the melatonin biosynthesis in plants, but this path could be modified in senescence or stress situations. So, serotonin could be transformed into 5-methoxytryptamine by the action of ASMT/COMT and generate melatonin by the further action of SNAT. In addition, melatonin may be synthetized through the formation of *N*-acetyltryptamine by the enzyme SNAT, and converted to phytomelatonin by T5H, although this route has not been demonstrated. An interesting fact is the finding in rice of four genes encoding histone deacetylase (DAC) that could reverse the acetylation steps from serotonin to *N*-acetylserotonin and 5-methoxytryptamine to phytomelatonin [14, 30].

4 Methods of Extraction of Phytomelatonin

Phytomelatonin concentration in plant tissues is low, and therefore extraction and concentration steps are required for its determination. The complexity of plant material is a disadvantage, and further purification steps are recommended. Homogenization with subsequent liquid–liquid extraction (LLE) or solid-phase extraction (SPE) is among the most used classical extraction procedures. The extraction solvent is another important aspect to maximize the isolation of phytomelatonin; being an amphipathic molecule, either lipophilic or hydrophilic solvent could be used for the extraction. The typical protocol for phytomelatonin analysis consists of an extraction step carried out in aqueous or organic solvent, clean-up step for removing undesirable compounds, and an analysis step

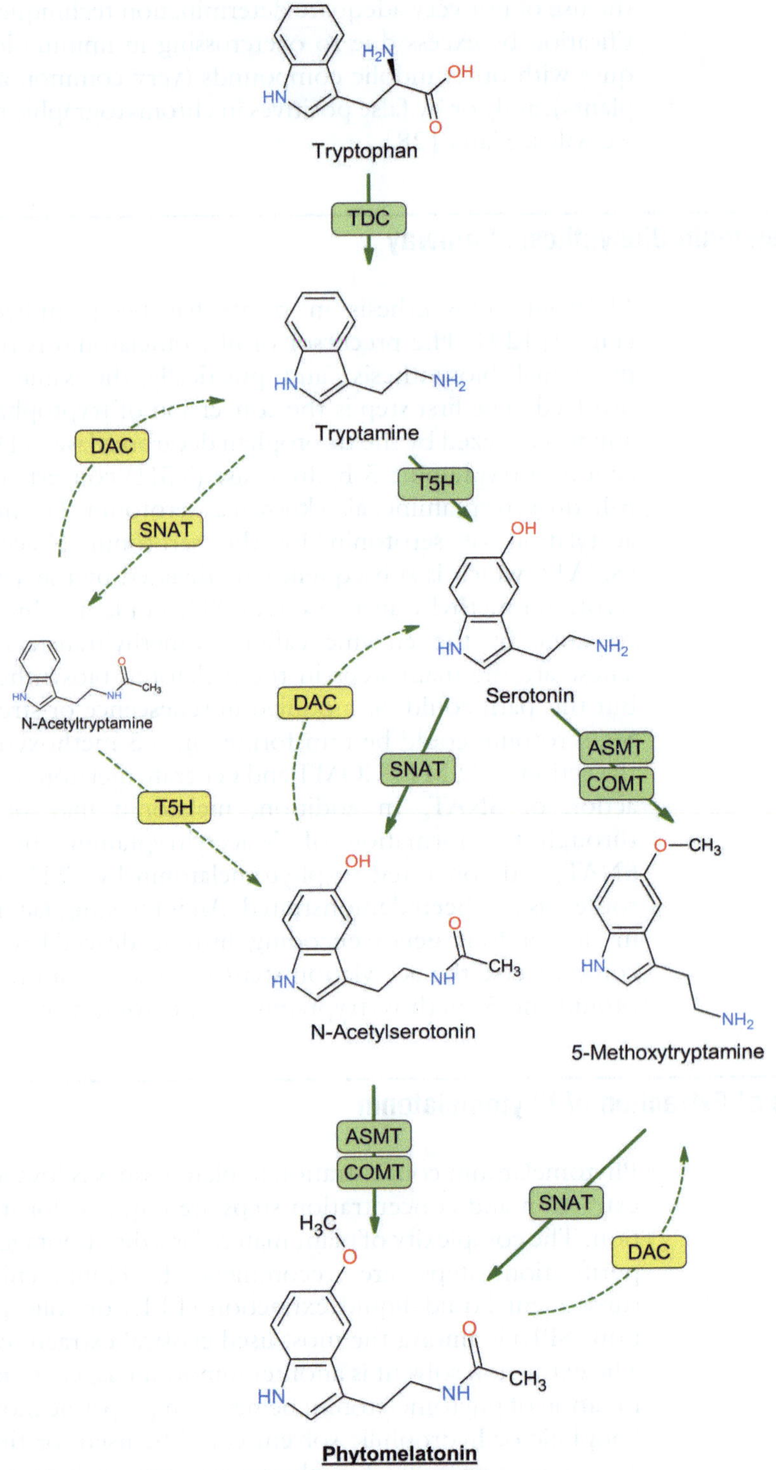

Fig. 2 *Phytomelatonin biosynthesis pathway* (*TDC* tryptophan decarboxylase, *T5H* tryptamine 5-hydroxylase, *SNAT* serotonin *N*-acetyltransferase, *ASMT* acetylserotonin methyltransferase, *COMT* caffeic *O*-methyltransferase, *DAC* deacetylase

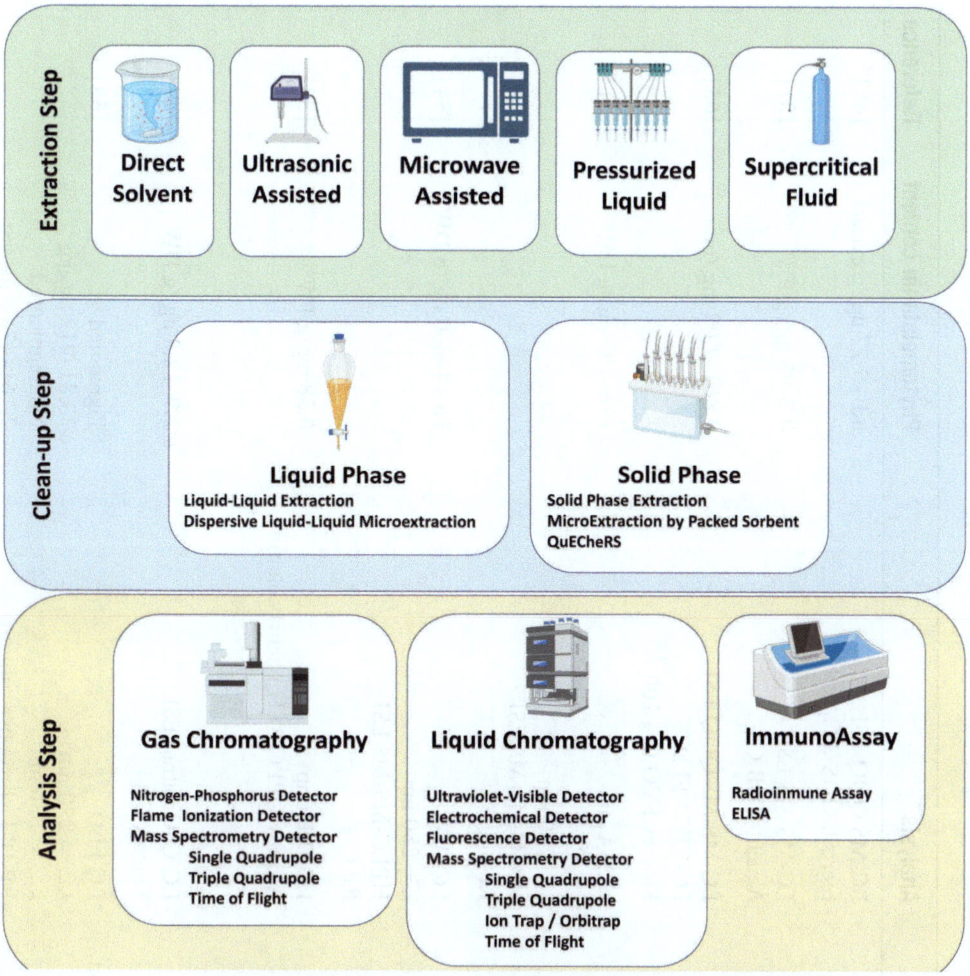

Fig. 3 Extraction, clean-up, and analysis: steps most used in phytomelatonin determination

for identification and quantitation of phytomelatonin; different protocols for each step are available (Fig. 3). Different methods are described for the extraction, purification, and analysis of phytomelatonin in different plant samples (Table 1).

4.1 Liquid–Liquid Extraction (LLE)

Homogenization followed by liquid–liquid extraction (LLE) is widely used in phytomelatonin extraction due to the different solubility of the molecule between aqueous and organic solvents. The first use of LLE for phytomelatonin extraction was made by Dubbels and coworkers in 1995; the homogenization was carried out with sodium carbonate, centrifuged, and then extracted with diethyl ether, giving phytomelatonin concentration up to 2000 pg/mg of protein [3]. Other homogenization buffers and organic solvents were used such as ethyl acetate and sodium phosphate buffer pH 7.5 [28]; or extracted with methanol, evaporated, and

Table 1
Different extraction and analytical methods used in the determination of phytomelatonin in some plant species and products

Species/product	Extraction	Analysis	Phytomelatonin content	Reference
Morus alba leaves	UAE 1:2 MeOH 20 min	LC-MS QQQ ESI Inertsill ODS-3 C_{18}	nd—9.77 μg/g extract	[48]
Glycine max L.	LLE Ethyl acetate	LC-MS-MS ESI Agilent-XDB C_{18}	0.2–95 ng/g FW	[31]
Food supplements (tabs)	–	LC-MS-MS QQQ ESI LC-VWD-FLD Kinetex EVO C_{18} 100 Å	0.861–2.09 mg/unit	[65]
Brassica ssp. seeds	UAE-SPE UAE-LLE	ELISA kit LC-PDA $\lambda = 280$ nm HPLC-MS-MS ESI RP C_{18}	176–197 ng/g DW	[41]
Brassica campestris seeds	SC-CO_2	LC-PDA $\lambda = 280$ nm HPLC-MS-MS ESI RP C_{18}	56.44–660 ng/g DW	[54]
Oryza sativa ssp.	LLE ACN or MeOH 1 M HCl Ether	LC-Orbitrap Acquity UPLC BEH phenyl column	0.58 ng/g FW	[33]
Prunus cerasus fruits	SPE C_{18}	LC-Q-Orbitrap ESI Kinetex C_{18}	2.9–12.3 ng/g DW	[44]
Oryza sativa grain	Pressurized liquid extraction 80–100% ethyl acetate in MeOH	LC-FLD $\lambda_{exc} = 290$ nm $\lambda_{emi} = 330$ nm RP 18 Lichrospher column	Pigmented 73.81–212 μg/kg Nonpigmented 27.61–47.83 μg/kg	[53]

Allium sativum L. bulbs	DSE MeOH	Thin-layer chromatography (TLC) methods coupled with densitometric detection	4.2–199.6 mg/kg DW	[39]
Oryza sativa grain	Microwave-assisted extraction (MAE)	LC-FLD $\lambda_{exc} = 290$ nm $\lambda_{emi} = 330$ nm RP-18 Lichrospher column	16.0–264.0 μg/kg Avg.: 54.17 ± 13.48 μg/kg	[52]
Various fruits or vegetables	LLE Sodium carbonate Diethylether	RIA GC-MS Derivatized pentafluorpropionic anhydride	0.1–47 ng/100 g Nd in potato	[3]
Oryza sativa grain	UAE 200 watts and 24 kHz	LC-FLD RP 18 Acquity UPLC® BEH column $\lambda_{exc} = 290$ nm $\lambda_{emi} = 334.9$ nm	7.13–216 μg/kg	[49]
Lupinus albus	DSE 15 h Overnight dark >90% LLE <55% Na-phosphate buffer pH 7.5 UAE	LC-FLD Spherisorb-S5 ODS2 $\lambda_{exc} = 280$ nm $\lambda_{emi} = 350$ nm Isocratic water:Acetonitrile (50:50) Flow rate of 0.2 L/min	1.8–20 ng/g FW	[28]
Hordeum vulgare	DSE 15 h Overnight dark >90% LLE Na-phosphate buffer pH 7.5 UAE	LC-FLD Spherisorb-S5 ODS2 $\lambda_{exc} = 280$ nm $\lambda_{emi} = 350$ nm	0.58–24.1 ng/g FW	[28]
Walnuts *Juglans regia* L.	LLE 40% recovery	LC-ECD 200–900 mV	3.5 ± 1.0 ng/g	[32]
Wines	DLLME Acetonitrile (disperser solvent) Chloroform (extracting solvent) NaCl Recovery: 92–107%	LC-FLD YMC-pack ODS-A $\lambda_{exc} = 286$ nm $\lambda_{emi} = 298$ nm	0.63–7.44 ng/mL	[36]

(continued)

Table 1
(continued)

Species/product	Extraction	Analysis	Phytomelatonin content	Reference
Grape-related foodstuffs	MEPS C_8 sorbent Eluted with methanol	LC-FLD Microsorb MV-Rainin C_8 RP $\lambda_{exc} = 298$ nm $\lambda_{emi} = 386$ nm	Grape 1.2–1.5 ng/g Must/wine 0.3–1.1 ng/mL	[46]
Chinese medicinal herbs	SPE C_{18} SPE cartridge	LC-FLD RP 5 μm Hypersil ODS (4.6 × 250 mm) $\lambda_{exc} = 280$ nm $\lambda_{emi} = 348$ nm	12–190 ng/g	[40]
Wines	QuEChERS SPE DLLME	LC-FLD	0.74–1.36 ng/mL	[37]
Sesame	QuEChERS Methanol C18 dispersant	LC-QQQ C_{18} column MRM: m/z 233 → 174, 159 and 130	Sesame 0.41–25.18 ng/g	[56]
Different plant/food samples	UAE SPE DSE	LC-MS-MS APCI$^+$ and ESI$^+$	4.9–440 ng/g	[63]
Different melatonin products		LC-Orbitrap DSI m/z 233.2 to 174.2 Collision energy 20 eV LOD: 0.04 μM LOQ: 0.17 μM	1.16–16.08 mM	[69]
Plant and cell culture	UAE 50% MeOH	LC-MS Q ESI$^+$ Waters Acquity BEH column	nd—40.05 ng/g FW	[58]
Wines	SPE Strata X-polymeric reversed phase	LC-MS Q ESI and APCI LC-FLD $\lambda_{exc} = 280$ nm $\lambda_{emi} = 350$ nm	0.038–0.063 μg/L	[42]

Coffee tissues	DSE buffer	LC-Q-ToF ESI$^+$ ToF (kV): 9.10 LC-FLD $\lambda_{exc} = 280$ nm $\lambda_{emi} = 350$ nm	0.407–96.54 μg/g FW 115.25 μg/g DW	[71]
Must fermentation process	SPE C_{18}	LC-Orbitrap HESI$^+$ ZORBAX RRHD SB-C_{18} LOD: 0.0047 ng/mL LOQ: 0.0144 ng/mL	nd—0.14 ng/mL	[72]
Lupinus albus	DSE Ethyl acetate BHT 15 h dark and shaking	LC-ECD RP-C_{18} column (ODS2) +850 mV LC-MSD QQQ ESI$^+$ SIM 233–174	28.1–93 ng/g FW	[8]
Monocot species	DSE Ethyl acetate BHT 15 h dark and shaking	LC-ECD RP-C_{18} column (ODS2) +850 mV LC-MSD/QQQ ESI$^+$-SIM 233–174	26.7–124.7 ng/g FW	[38]
Morning glory plants Tomatoes	SPE C_{18}	RIA Radioactive melatonin (1 mCi/mL [O-methyl-3H] melatonin GC-MS Derivatized with pentafluoro propionic anhydride	Morning glory 20–30 pg/g RIA 5–10 pg/g GC-MS Tomatoes 3–14 pg/g RIA 0.5–3 pg/g GC-MS	[75]
Olive oil	LLE	ELISA	50–119 pg/mL	[77]
Grape skins	UAE SPE	LC-FLD $\lambda_{exc} = 280$ nm $\lambda_{emi} = 348$ nm ELISA	0.005–0.965 ng/g	[78]

redissolved in sodium phosphate before LLE against ethyl acetate two times, first at pH 8.0 and then at pH 3.1 [31]; or extracted with methanol, dried, and redissolved in phosphate-buffered saline (pH 7.4) and LLE with chloroform and a recovery of 40% [32]; extracted with acetonitrile or methanol (12 h at 4°C, in the dark), solvent evaporation, and redissolved in 1 M HCl and LLE with ether, the recoveries were > 80%, but methanol extraction was more affected by plant matrix [33].

4.2 Dispersive Liquid–Liquid Microextraction (DLLME)

Another type of LLE is the dispersive liquid–liquid microextraction (DLLME) [34, 35] where a dispersant soluble in both aqueous and organic phase are added to the extraction media, generating a third phase that allows a higher mass transfer of the target to the extraction solvent. DLLME was applied to determine phytomelatonin in wines with acetonitrile as the disperser solvent, chloroform as the extracting solvent and NaCl, with a high efficiency [36, 37].

4.3 Direct Solvent Extraction (DSE)

A simpler procedure is the direct solvent extraction (DSE) where samples are extracted with a suitable organic solvent for a determined time; after that, the solvent is evaporated and sample is redissolved before analysis. Phytomelatonin extraction by DSE in lupin tissues [8] and monocot species [38] was accomplished using ethyl acetate for 15 h in the dark with continuous shaking. Comparing DSE with LLE in samples of *Lupinus albus* L. and *Hordeum vulgare* L., Arnao's group found that direct solvent extraction (DSE) using ethyl acetate or chloroform (overnight in dark) was more effective in recovering phytomelatonin (>90%) than LLE after homogenization in phosphate buffer with the same solvents (<55%). This recovery increased after successive LLE steps (~80%). The use of sonication after extraction and before solvent elimination improved the phytomelatonin recovery from the samples. Therefore, DSE is an easy and affordable procedure with high recovery rates for melatonin extraction in plants [28]. DSE was also used for the extraction of phytomelatonin from garlic bulbs [39].

4.4 Solid-Phase Extraction (SPE)

Other of the most used extraction and purification technique is solid-phase extraction (SPE). In this case, the phytomelatonin is retained in a column with sorbent, and subsequently eluted with a suitable solvent [40–42]. Three main sorbents used in SPE are inorganic materials, silica-bonded C_8/C_{18} sorbents, and polymer-based materials [43]. The elution of phytomelatonin from the cartridge is usually carried out using 100% methanol as solvent. SPE has some advantages against LLE, requires fewer amounts of solvent, is a more environment-friendly procedure, and presents higher accuracy and reliability [41]. SPE was applied in cherry fruits using a C_{18} cartridge; after extraction with ethyl acetate and redissolved in water before SPE step, phytomelatonin was recovered by

0.1% formic acid in methanol with a recovery of about 100% [44], while in wines, C_{18} cartridge and methanol were used as elution solvent [37]. As an SPE micro-procedure, microextraction by packed sorbent (MEPS) [45] was also used in wine and grape analysis, using C_8 sorbent with a recovery of >90% [46]. The application of heat and pressure can speed up the segregation between phytomelatonin and the matrix in samples [47], so techniques assisted by ultrasonic or microwave have been used for phytomelatonin extraction.

4.5 Ultrasonic-Assisted Extraction (UAE)

During ultrasonic-assisted extraction (UAE) bubbles are formed due to cavitation in the solvents, so changes in pressure and temperature are produced, and the mass transfer to solvent is enhanced. UAE also needs a further step for cleaning and removing interfering compounds prior to analysis. Chakraborty and Bhattacharjee [41] carried out an exhaustive study on UAE of phytomelatonin from mustard seeds. Homogenization with ethanol:water (1:1) and subsequent ultrasonication in an ice bath were insufficient to detect phytomelatonin by ELISA or HPLC-MS due to co-extractants. Further purification step was made using LLE or SPE in order to detect phytomelatonin, showing a higher correlation between ELISA and HPLC-MS values for the SPE than for LLE, indicating that SPE is an adequate step for phytomelatonin extraction and purification [41]. A study on Chinese medicinal herbs subjected the samples to UAE for 30 min in a 1:10 (w/v) ratio in methanol at room temperature; an SPE procedure was needed before analytical determination by liquid chromatography [40]. For mulberry leaves and teas, UAE was done with methanol 1:2 (w/v) for 20 min, and no subsequent steps were carried out before phytomelatonin determination by LC/MS-MS [48]. A study to extract phytomelatonin from rice grain showed that the optimal solvent was 50% methanol in water at 18.5°C for 10 min, with a sample-to-solvent ratio of 1: 2.5 (pH 3.5), an amplitude of 30%, and cycles of 0.2 s^{-1} [49].

4.6 Microwave-Assisted Extraction (MAE)

Microwave-assisted extraction (MAE) uses heat and pressure to separate the compound in matrix based on their different ability to absorb radiation [50] and break weak hydrogen bounds [51]. The use of MAE in rice grain showed that the optimal conditions were temperature between 125 and 175°C—as greater temperature reduced yield due to phytomelatonin instability—microwave power between 500 and 1000 W, time (5–15 min), solvent (10–90% ethyl acetate in methanol), and ratio of solvent to sample (10:1–20:1) [52].

4.7 Pressurized Liquid Extraction (PLE)

Pressurized liquid extraction (PLE) offers some advantages over traditional extraction techniques, a lower solvent consumption, higher extraction yields, and reproducibility [50, 51]. The use of organic solvent at elevated temperature and pressure increases the efficiency of the extraction process. Phytomelatonin extraction from rice grains was carried out with a Dionex ASE 200 extractor (Dionex, Sunnyvale, CA, USA), the optimized conditions were: extraction solvent 80–100% ethyl acetate in methanol, pressurized at 100–200 atm, and heated between 100 and 200 °C. The recovery was 98.9%, and the whole process took 12 min [53].

4.8 Supercritical Fluid Extraction (SFE)

The optimized condition for phytomelatonin extraction from yellow mustard seed using supercritical carbon dioxide was studied by Chakraborty and Bhattacharjee, finding that the optimal extraction conditions were 300 bar, 50 °C for 120 min, obtaining 660.72 ± 41.05 ng/g of dry mustard powder [54].

4.9 QuEChERS (Quick, Easy, Cheap, Effective, Rugged, and Safe)

The quick, easy, cheap, effective, rugged, and safe (QuEChERS) method [55], a more effective and ecology-friendly method, has been used for phytomelatonin extraction. This green analytical chemistry procedure requires smaller amount of non-halogenated and low-toxicity solvents, reagents, and lab equipment. It has a low cost and short time. QuEChERS extraction methods have been applied to extract phytomelatonin from wines with acetonitrile; $MgSO_4$ and NaCl were added after extraction, centrifuged and treated with primary secondary amine (PSA), Bulk Carbograph, and $MgSO_4$ in a dispersive SPE tube [37]. In a phytomelatonin biosynthesis study in sesame plants, C_{18} dispersant was used after a methanolic extraction in dark at 4 °C overnight [56].

5 Analytical Methods

Many analytical techniques have been used for determining phytomelatonin content after extraction. Liquid chromatography is among the most used methodologies, although immunoassay is becoming increasingly important when quantifying phytomelatonin. The first study for determining phytomelatonin used gas chromatography with mass spectrometry detection (GC-MS) technique [57]; also, phytomelatonin was detected in 1995 using radioimmune assay (RIA) and liquid chromatography coupled with mass spectrometry (LC-MS) detector [2, 3].

5.1 Liquid Chromatography (LC)

Liquid chromatography is the most used technique for analyzing melatonin in animal and plant samples. There is a great variety of coupled detectors (photometric and diode array, electrochemical, fluorimetric, and mass spectrometry), column types, and conditions to perform the correct analysis [43]. Regarding the chromatographic columns, the most commonly used include C_{18}, C_8, phenyl [33, 58], and C_4 column [59]. Comparing the different columns, the best results were obtained with a C_{18} column [60]. In order to overcome poor retention of polar compounds such as phytomelatonin by the C_{18} columns, the use of hydrophilic interaction liquid chromatography (HILIC) was proposed; this type of column is more effective, lowers the analytical time, and requires a less organic solvent [61]. For the mobile phase, typical mixtures are MeOH/H_2O or ACN/H_2O. The use of modifiers, such as formic acid and acetic acid, is also recommended. Ionic liquids have also been used as modifiers, improving the peak height of phytomelatonin [62]. A higher separation, resolution and less analytical time can be achieved using ultra-high-pressure liquid chromatography (UHPLC) with columns of lower particle size [63, 64].

5.2 UV-Vis, Photometric Diode Array (PDA)

In the case of UV-Vis detector, phytomelatonin is usually monitored at 274 nm [65] or 280 nm [41, 54]. UV-Vis detector was used in the quantification of phytomelatonin in mustard seed [41, 54], and for food supplements [65].

5.3 Fluorescence Detector (FLD)

Fluorescence detector is more sensitive than UV-Vis to determine and quantify phytomelatonin [65]. For FLD, the emission wavelength is usually set between 224 and 298 nm with the emission wavelength in the range of 330 to 392 nm [28, 43, 47, 65]. Some studies in rice [49, 52, 53], *Lupinus albus*, *Hordeum vulgare* [28], wine [36, 42], and food supplements [65] quantified phytomelatonin with a fluorescence detector with a high sensitivity and accuracy.

5.4 Electrochemical Detector (ECD)

Another detector used in the analysis of phytomelatonin is the electrochemical detector (ECD), which is used in the analysis of health products and supplements. In this case, the potential selections were 50 mV for screening and 850 mV for analytical, and 1 μA of current intensity [66]. Walnut phytomelatonin content was quantified by HPLC-ECD with an eight-channel coulometry array with 100 mV for the first channel, while increasing the potential by 100 mV for the remaining channels [32]. Phytomelatonin in lupin tissues and monocot species were also determined with HPLC-ECD with +850 mV [8, 38].

5.5 Mass Detector (MSD)

Mass detectors (MSD) are one of the most used in determining phytomelatonin, providing the high sensitivity of MS with the selectivity of LC. Two main mass detector types could be used: low mass resolution (low resolution mass spectrometry, LRMS) and high mass resolution (high resolution mass spectrometry, HRMS). Mass spectrometry can be used to screen unknown compounds though the full scan mode, or to identify and quantify a determined compound, then the selected reaction monitoring (SRM) or multiple reaction monitoring (MRM) modes can be used. The ion mode for phytomelatonin usually is positive mode, and the transition from 233 to 216/174 was selected. For ionization, different sources could be used, electrospray ionization (ESI) is the most used, but atmospheric pressure chemical ionization (APCI) [61], nanospray ionization (nESI) [67], matrix-assisted laser desorption ionization (MALDI) [68], and droplet spray ionization (DSI) [69] are other ionization methods used in phytomelatonin detection by MS.

Mass detector is mainly to confirm the mass of the analyte and identify it by its molecular fragments. It is a reliable technique that requires a clean extract to avoid interferences or matrix effects. Many types of mass detector have been developed, from single quadrupole (Q), triple quadrupole (QQQ), time of flight (Q-ToF), ion-trap (IT-MS), or Orbitrap-MS, increasing the resolution of the mass obtained for the analyte. In the determination of phytomelatonin, all these described MS detectors have been used [70].

5.6 Single Quadrupole (Q)

Single quadrupole was used to determine phytomelatonin in various plants and cell cultures [58] with a limit of detection of 95.3 pg/mL. In wines, using the same mass detector the LOD was 0.0023 μg/L [42].

5.6.1 Triple Quadrupole (QQQ)

The use of QQQ detector allows further fragmentation of molecular ion and determines that the transition for phytomelatonin molecule is from m/z 233 to m/z 174 [48], so a better identification of phytomelatonin is obtained, and interference signals are avoided. Phytomelatonin in lupin tissues and monocot species were identified by SRM (233 → 174) by LC coupled to a QQQ detector [8, 38]. Phytomelatonin and isomers in Italian grape products were analyzed by UPLC with triple quadrupole detector using MRM transitions 233 → 174, 233 → 159, 233 → 143 and 233 → 131 [64]. Triple quadrupole was also used in ESI^+ mode in mulberry leaves [48], soybean with ESI^+ mode [31], and food supplement with ESI^+ and SRM 233.1 → 130.1 and 233.1 → 174.0 with 40 eV and 15 eV of collision energy, respectively [65].

5.6.2 Time of Flight (Q-ToF)

Time-of-flight mass detector is used for determining phytomelatonin in green and roasted beans [71] and operates in a positive mode with a Q-ToF (kV) of 9.10; the phytomelatonin so determined ranged from 7.1 to 8.9 μg/g in *Coffea canephora* and from 6.8 to 7.8 in *C. arabica*.

5.6.3 Ion-Trap (IT-MS) and Orbitrap MS

This detector provides a high-resolution m/z, so lower matrix interferences are obtained, and identification is very accurate. Phytomelatonin in rice leaves and roots was determined by a UPLC-Orbitrap with an Acquity UPLC BEH phenyl column with a gradient mobile phase using water and acetonitrile with ESI^+ and full scan mode. The linear range was 0.01–20 ng/mL, with a LOQ and LOD of 0.03 and 0.11 pg, respectively [33]. Some pharmaceutical products were analyzed for phytomelatonin content using an

Orbitrap operating in collision-induced dissociation with a collision energy of 20 eV, in ESI^+, and 233.2 to 174.2 transition [69]; the limit of detection and quantification were 0.04 μM and 0.17 μM, respectively. From cherries and tomatoes, phytomelatonin and d_4-melatonin, (linear ion trap)-LTQ-Orbitrap in ESI^+ with MRM 233.2 to 174.0915 and 159.0677 for phytomelatonin, and 237 to 178.1164 for d_4-melatonin was used, the method presented a LOD of 10 pg/g, and no phytomelatonin was detected for cherries cultivars, whereas tomato phytomelatonin ranged from lower than LOD to 149 pg/g [44]. Must and wine samples were analyzed using LTQ-Orbitrap with a heated electrospray ionization source (HESI) in positive mode, with a normalized collision energy of 20 eV, and the limit of detection and quantification were 0.0047 and 0.0144 ng/mL, respectively [72].

5.7 Gas Chromatography (GC)

GC is useful in the separation and identification of molecules with a low mass or being volatiles. It does not have matrix effects or ion suppression and presents a great chromatographic resolution. GC is generally coupled with mass detector. Phytomelatonin analysis is carried out after a derivative reaction with pentafluorpropionic anhydride [3]. A study using a Q-TOF, a higher resolution mass detector, coupled with GC was made for separation and elucidation of phytomelatonin in the stem bark of *Crataeva nurvala* Buch-Ham [73].

5.8 Immunoassay

Immunoassay are based on the recognition of part of molecule by an antibody obtained against that molecule. Among the different strategies of an immunoassay determination, the competitive reactions between a tracer molecule and the molecule of interest are among the most used strategies. In this case, the large number of molecules in the sample, the lower response of the assay. Two different immunoassays have been used for melatonin determination, radioimmunoassay (RIA) using ^3H- or ^{125}I-labeled tracers, and enzyme-linked immuno sorbent Assay (ELISA) using biotinylated melatonin. These techniques are widely used in human and animal sample determinations [74].

5.8.1 RIA

The first identification of melatonin in plants was made by radioimmunoassay and confirmed by HPLC-MS [3]. van Tassel determined phytomelatonin in different organs of morning glory plants (*Pharbitis nil* Choisy), and tomatoes using RIA [75]. Despite the high sensitivity of the RIA, this type of test is not widely used in the determination of phytomelatonin due, possibly, to its high cost and the need for radioactivity detectors [74].

5.8.2 ELISA Kits

Many commercial kits for phytomelatonin quantification are available for plant samples. Different antigens can be used to produce phytomelatonin antibodies [74, 76], and depending on the antigen used, the results of phytomelatonin quantitation could vary depending on the cross-reaction with other similar compounds. Nowadays, there are a lot of different ELISA kits available in the market, each with different antigen again melatonin, so papers are being published with widely different results. A comprehensive review of the immunoassay determination of melatonin can be consulted [74]. A competitive enzyme immunoassay kit (Melatonin ELISA, IBL-Hamburg, Germany) was used to assay phytomelatonin in olive oil obtaining values from 50 to 119 pg/mL [77], in grape skin with values between 0.005 and 0.965 ng/g [78]. In another study, ultrasonicated in vitro potato plants were analyzed for phytomelatonin using melatonin kit (ab213978; Abcam Plc, Cambridge, UK), values from 3 to 9 ng/g were obtained [79].

6 Conclusion

- The low content of phytomelatonin present in plant tissues and organs makes the use of concentration steps very advisable; additional purification steps are recommended; and better results seem to be obtained with SPE than with LLE procedure.
- Liquid–liquid clean-up is simple, cheap, and presents an appropriate recovery rate, but it is a time-consuming procedure, with a large solvent consumption and waste generation. Errors in measured volume, emulsion formation, and poor phase separation could lead to a lower phytomelatonin extraction, and therefore to an inaccurate determination [43].
- The use of pressurized liquid or dispersive liquids increases the yield, requires lower solvents, and is more environment-friendly; but a lot of parameters must be optimized, and expensive equipment are required.
- Solid-phase extraction and clean-up are easy, have high selectivity, with many different sorbents to choose from, and are susceptible to automatization and minimization, reducing the solvent and sorbent consumption. The main problem is the cartridge clogging, and that sometimes is a time-consuming process.
- Ultrasonic- and microwave-assisted extraction are easy to perform and do not require expensive equipment. They improve the phytomelatonin yield when applied after direct solvent extraction, but the drawback of this assisted extraction is the high temperatures that could appear during the process.
- QuEChERS is simple and rapid, with high recovery and yield, but automatization is not always possible.
- Immunoassay analysis relies on the antibodies against the exposed part of the molecule, and the specificity and accuracy

could be affected by the inevitable cross-reactivity of antibodies [74, 80]. In RIA, ^{125}I-2-iodomelatonin use has serious problems, such as being a radioactive compound and requiring extra care, making its use not recommended. Another aspect of ^{125}I-2-iodomelatonin in RIA kits is that it has a molecular weight of 356 Da compared with melatonin (232 Da), while biotinylated or conjugated to BSA melatonin in ELISA kits has a molecular weight of more than 500 Da, and these different sizes can have significant impact on the kinetics of the binding and specificity [74]. In addition, plants have multiple melatonin-like compounds that could provide a poor phytomelatonin quantification. RIA showed higher values of phytomelatonin compared to GC-MS in morning glory plants [75].

- Liquid chromatography coupled with some of the multiple available detectors is the most used technique for phytomelatonin analysis, detection, and quantification. UV-Vis and FLD analysis are cheap and easy to use. But there is a great number of compounds with fluorescence and absorbance ability in plants, so co-elution and quenchers would be a major problem in the determination of phytomelatonin by these detectors. However, MS-MS detectors are necessary for identification and validation of the quantification.
- For MS detectors, ionization methods or RP column could be different depending on the matrix or sample type, HILIC–APCI$^+$ ionization was better for melatonin standards, while C_8 – APCI$^+$ was the best choice for grape skin and C_{18} – ESI$^+$ was suitable for wine for phytomelatonin. On the other hand, C_8 – ESI$^+$ was the most appropriate for plant tissues of *Arabidopsis thaliana* [61]. Currently, triple quadrupole detection and Orbitrap techniques are the most reliable in the identification and quantification of phytomelatonin.

Acknowledgments

The research was funded by the Ministerio de Ciencia e Innovación (Spain), "Proyectos de I+D+," Programas Estatales de Generación de Conocimiento y Fortalecimiento Científico y Tecnológico del Sistema de I+D+i y de I+D+i Orientada a los Retos de la Sociedad, del Plan Estatal de Investigación Científica y Técnica y de Innovación 2017–2020, Grant PID2020-113029RB-I00, funded by MCIN/AEI/ 10.13039/501100011033. More information can be found on Phytohormones and Plant Development Research Group web page at https://www.um.es/en/web/phytohormones (accessed October 22, 2023)*a*.

References

1. Lerner AB, Case JD, Takahashi Y, Lee TH, Mori W (1958) Isolation of melatonin, the pineal gland factor that lightens melanocytes [1]. J Am Chem Soc 80:2587–2587. https://doi.org/10.1021/ja01543a060
2. Kolar J, Machackova I, Illnerova H, Prinsen E, van Dongen W, van Onckelen H (1995) Melatonin in higher plant determined by radioimmunoassay and liquid chromatography-mass spectrometry. Biol Rhythm Res 26:406–409
3. Dubbels R, Reiter RJ, Klenke E, Goebel A, Schnakenberg E, Ehlers C, Schiwara HW, Schloot W (1995) Melatonin in edible plants identified by radioimmunoassay and by HPLC-MS. J Pineal Res 18:28–31
4. Hattori A, Migitaka H, Iigo M, Itoh M, Yamamoto K, Ohtani-Kaneko R, Hara M, Suzuki T, Reiter RJ (1995) Identification of melatonin in plants and its effects on plasma melatonin levels and binding to melatonin receptors in vertebrates. Biochem Mol Biol Int 35:627–634
5. Hardeland R (2019) Melatonin in the evolution of plants and other phototrophs. Melatonin Res 2:10–36. https://doi.org/10.32794/mr11250029
6. Hardeland R, Poeggeler B (2003) Non-vertebrate melatonin. J Pineal Res 34: 233–241. https://doi.org/10.1034/j.1600-079X.2003.00040.x
7. Garcia-Parrilla MC, Cantos E, Troncoso AM (2009) Analysis of melatonin in foods. J Food Compost Anal 22:177–183. https://doi.org/10.1016/j.jfca.2008.09.009
8. Hernández-Ruiz J, Cano A, Arnao MB (2004) Melatonin: a growth-stimulating compound present in lupin tissues. Planta 220:140–144. https://doi.org/10.1007/s00425-004-1317-3
9. Arnao MB, Hernández-Ruiz J (2014) Melatonin: plant growth regulator and/or biostimulator during stress? Trends Plant Sci 19:789–797. https://doi.org/10.1016/j.tplants.2014.07.006
10. Socaciu AI, Ionut R, Socaciu MA, Ungur AP, Bârsan M, Chiorean A, Socaciu C, Râjnoveanu AG (2020) Melatonin, an ubiquitous metabolic regulator: functions, mechanisms and effects on circadian disruption and degenerative diseases. Rev Endocr Metab Disord 21: 465–478. https://doi.org/10.1007/s11154-020-09570-9
11. Arnao MB, Hernández-Ruiz J (2015) Functions of melatonin in plants: a review. J Pineal Res 59:133–150. https://doi.org/10.1111/jpi.12253
12. Arnao MB, Hernández-Ruiz J (2015) Melatonin: synthesis from tryptophan and its role in higher plant. Amino Acids Higher Plants 390–435. https://doi.org/10.1079/9781780642635.0390
13. Arnao MB, Hernández-Ruiz J (2021) Regulatory role of melatonin in the redox network of plants and plant hormone relationship in stress. In: Gupta DK, Corpas FJ (eds) Hormones and plant response. Springer, Cham, pp 235–272
14. Arnao MB, Cano A, Hernández-Ruiz J (2022) Phytomelatonin: an unexpected molecule with amazing performances in plants. J Exp Bot. https://doi.org/10.1093/jxb/erac009
15. Arnao MB, Cano A, Hernández-Ruiz J (2023) Research in plant melatonin: original and current studies. Melatonin Res 6:224–228. https://doi.org/10.32794/mr112500151
16. Aghdam MS, Mukherjee S, Flores FB, Arnao MB, Luo Z, Corpas FJ (2021) Functions of melatonin during postharvest of horticultural crops. Plant Cell Physiol pcab175. https://doi.org/10.1093/pcp/pcab175
17. Arnao MB, Hernández-Ruiz J (2019) Melatonin and reactive oxygen and nitrogen species: a model for the plant redox network. Melatonin Res 2:152–168. https://doi.org/10.32794/11250036
18. Arnao MB, Hernández-Ruiz J (2019) Melatonin: a new plant hormone and/or a plant master regulator? Trends Plant Sci 24:38–48. https://doi.org/10.1016/j.tplants.2018.10.010
19. Arnao MB, Hernández-Ruiz J (2021) Melatonin as a regulatory hub of plant hormone levels and action in stress situations. Plant Biol 23:7–19. https://doi.org/10.1111/plb.13202
20. Arnao MB, Hernández-Ruiz J (2018) Melatonin and its relationship to plant hormones. Ann Bot 121:195–207
21. Arnao MB, Hernández-Ruiz J (2022) Melatonin against environmental plant stressors: a review. Curr Protein Pept Sci 22:413–429. https://doi.org/10.2174/1389203721999210101235422
22. Cheng G, Ma T, Deng Z, Gutiérrez Gamboa G, Ge Q, Xu P, Zhang Q, Zhang J, Meng J-F, Reiter R, Fang Y, Sun X (2021) Plant-derived melatonin from food: a gift of nature. Food Funct 12:2829–2849. https://doi.org/10.1039/D0FO03213A

23. Arnao MB (2014) Phytomelatonin: discovery, content, and role in plants. Adv Bot 2014: e815769. https://doi.org/10.1155/2014/815769
24. Losada M, Cano A, Hernández-Ruiz J, Arnao MB (2022) Phytomelatonin content in Valeriana officinalis L. and some related phytotherapeutic supplements. Int J Plant Based Pharm 2: 176–181. https://doi.org/10.55484/ijpbp.1079005
25. Arnao MB, Giraldo-Acosta M, Castejón-Castillejo A, Losada-Lorán M, Sánchez-Herrerías P, El Mihyaoui A, Cano A, Hernández-Ruiz J (2023) Melatonin from microorganisms, algae, and plants as possible alternatives to synthetic melatonin. Meta 13: 72. https://doi.org/10.3390/metabo13010072
26. Arnao MB, Hernández-Ruiz J (2018) The potential of Phytomelatonin as a nutraceutical. Molecules 23:238. https://doi.org/10.3390/molecules23010238
27. Arnao MB, Hernández-Ruiz J (2018) Phytomelatonin, natural melatonin from plants as a novel dietary supplement: sources, activities and world market. J Funct Foods 48:37–42. https://doi.org/10.1016/j.jff.2018.06.023
28. Arnao MB, Hernández-Ruiz J (2009) Assessment of different sample processing procedures applied to the determination of melatonin in plants. Phytochem Anal 20:14–18
29. Back K, Tan D-X, Reiter RJ (2016) Melatonin biosynthesis in plants: multiple pathways catalyze tryptophan to melatonin in the cytoplasm or chloroplasts. J Pineal Res 61:426–437. https://doi.org/10.1111/jpi.12364
30. Lee K, Lee HY, Back K (2018) Rice histone deacetylase 10 and Arabidopsis histone deacetylase 14 genes encode *N*-acetylserotonin deacetylase, which catalyzes conversion of *N*-acetylserotonin into serotonin, a reverse reaction for melatonin biosynthesis in plants. J Pineal Res 64:e12460. https://doi.org/10.1111/jpi.12460
31. Zou J, Yu H, Yu Q, Jin X, Cao L, Wang M, Wang M, Ren C, Zhang Y (2021) Physiological and UPLC-MS/MS widely targeted metabolites mechanisms of alleviation of drought stress-induced soybean growth inhibition by melatonin. Ind Crop Prod 163:113323
32. Reiter RJ, Manchester LC, Tan D (2005) Melatonin in walnuts: influence on levels of melatonin and total antioxidant capacity of blood. Nutrition 21:920–924. https://doi.org/10.1016/j.nut.2005.02.005
33. Ye T, Hao YH, Yu L, Shi H, Reiter RJ, Feng YQ (2017) A simple, rapid method for determination of melatonin in plant tissues by UPLC coupled with high resolution orbitrap mass spectrometry. Front Plant Sci 8:64. https://doi.org/10.3389/fpls.2017.00064
34. Pourhossein M, Shahtaheri SJ, Mazloumi A, Rahimi-Foroushani A, Helmi-Kohneshahri M, Khani HM (2018) Dispersive liquid–liquid microextraction for the determination of salivary melatonin as a biomarker of circadian rhythm. J Anal Chem 73:966–972. https://doi.org/10.1134/S106193481810009X
35. Talebianpoor MS, Khodadoust S, Rozbehi A, Akbartabar Toori M, Zoladl M, Ghaedi M, Mohammadi R, Hosseinzadeh AS (2014) Application of optimized dispersive liquid-liquid microextraction for determination of melatonin by HPLC-UV in plasma samples. J Chromatogr B 960:1–7
36. Viegas O, Esteves C, Rocha J, Melo A, Ferreira IMPLVO (2021) Simultaneous determination of melatonin and trans-resveratrol in wine by dispersive liquid–liquid microextraction followed by HPLC-FLD. Food Chem 339: 128091. https://doi.org/10.1016/j.foodchem.2020.128091
37. Eremia SAV, Albu C, Radu GL, Ion M (2023) Different extraction approaches for the analysis of melatonin from cabernet sauvignon and Feteasca Neagra wines using a validated HPLC-FL method. Molecules 28:2768. https://doi.org/10.3390/molecules28062768
38. Hernández-Ruiz J, Cano A, Arnao MB (2005) Melatonin acts as a growth-stimulating compound in some monocot species: melatonin as growth promoter in plants. J Pineal Res 39: 137–142. https://doi.org/10.1111/j.1600-079X.2005.00226.x
39. Muszynska B, Maslanka A, Sulkowska-Ziaja K, Opoka W, Szopa A (2014) Analysis of 5-methyltryptamine, L-tryptophan, 5-hydroxy-L-tryptophan, and melatonin in the bulbs of garlic by thin-layer chromatographic method coupled with densitometric detection. J Planar Chrom 27:210–216. https://doi.org/10.1556/jpc.27.2014.3.11
40. Chen G, Huo Y, Tan D-X, Liang Z, Zhang W, Zhang Y (2003) Melatonin in Chinese medicinal herbs. Life Sci 73:19–26. https://doi.org/10.1016/S0024-3205(03)00252-2
41. Chakraborty S, Bhattacharjee P (2020) Ultrasonication-assisted extraction of a phytomelatonin-rich, erucic acid-lean nutraceutical supplement from mustard seeds: an antioxidant synergy in the extract by reductionism. J Food Sci Technol 57:1278–1289. https://doi.org/10.1007/s13197-019-04161-2
42. Fracassetti D, Vigentini I, Lo Faro AF, De Nisi P, Foschino R, Tirelli A, Orioli M, Iriti M

(2019) Assessment of tryptophan, tryptophan Ethylester, and melatonin derivatives in red wine by SPE-HPLC-FL and SPE-HPLC-MS methods. Food Secur 8:99. https://doi.org/10.3390/foods8030099

43. Zhang X, Zhang Y, Zhou Y, Liu Z, Wei B, Feng X (2023) Melatonin in different food samples: recent update on distribution, bioactivities, pretreatment and analysis techniques. Food Res Int 163:112272. https://doi.org/10.1016/j.foodres.2022.112272
44. Reinholds I, Pugajeva I, Radenkovs V, Rjabova J, Bartkevics V (2016) Development and validation of new UPLC-HQOMS method for determination of melatonin in fruits. J Chromatogr Sci 54:977–984. https://doi.org/10.1093/chromsci/bmw030
45. Abdel-Rehim M (2011) Microextraction by packed sorbent (MEPS): a tutorial. Anal Chim Acta 701:119–128. https://doi.org/10.1016/j.aca.2011.05.037
46. Mercolini L, Mandrioli R, Raba J, Raggi MA (2012) Content of melatonin and other antioxidants in grape-related foodstuffs: measurement using a MEPS-HPLC-F method. J Pineal Res 53:21–28
47. Alessa H, Saber AL, Althakafy JT (2022) Up-to-date studies regarding the determination of melatonin by chromatographic methods. Arch Pharm 355:2100378. https://doi.org/10.1002/ardp.202100378
48. Panyatip P, Padumanonda T, Yongram C, Kasikorn T, Sungthong B, Puthongking P (2022) Impact of tea processing on tryptophan, melatonin, phenolic and flavonoid contents in mulberry (Morus alba L.) leaves: quantitative analysis by LC-MS/MS. Molecules 27:4979. https://doi.org/10.3390/molecules27154979
49. Setyaningsih W, Duros E, Palma M, Barroso CG (2016) Optimization of the ultrasound-assisted extraction of melatonin from red rice (Oryza sativa) grains through a response surface methodology. Appl Acoust 103:129–135
50. Camel V (2001) Recent extraction techniques for solid matrices—supercritical fluid extraction, pressurized fluid extraction and microwave-assisted extraction: their potential and pitfalls. Analyst 126:1182–1193. https://doi.org/10.1039/b008243k
51. Kaufmann B, Christen P (2002) Recent extraction techniques for natural products: microwave-assisted extraction and pressurised solvent extraction. Phytochem Anal 13:105–113. https://doi.org/10.1002/pca.631
52. Setyaningsih W, Palma M, Barroso CG (2012) A new microwave-assisted extraction method for melatonin determination in rice grains. J Cereal Sci 56:340–346
53. Setyaningsih W, Saputro IE, Barbero GF, Palma M, García Barroso C (2015) Determination of melatonin in Rice (Oryza sativa) grains by pressurized liquid extraction. J Agric Food Chem 63:1107–1115. https://doi.org/10.1021/jf505106m
54. Chakraborty S, Bhattacharjee P (2017) Supercritical carbon dioxide extraction of melatonin from Brassica campestris: in vitro antioxidant, hypocholesterolemic and hypoglycaemic activities of the extracts. Int J Pharm Sci Res 8:2486–2495
55. Varela-Martínez DA, González-Sálamo J, González-Curbelo MÁ, Hernández-Borges J (2020) Quick, easy, cheap, effective, rugged, and safe (QuEChERS) extraction. In: Poole CF (ed) Liquid-phase extraction. Elsevier, pp 399–437
56. Wang X, You J, Liu A, Qi X, Li D, Zhao Y, Zhang Y, Zhang L, Zhang X, Li P (2022) Variation in melatonin contents and genetic dissection of melatonin biosynthesis in sesame. Plan Theory 11:2005. https://doi.org/10.3390/plants11152005
57. Greer M, Williams CM (1967) Gas-chromatographic determination of melatonin and 6-hydroxymelatonin. Clin Chim Acta 15:165–168. https://doi.org/10.1016/0009-8981(67)90342-7
58. Erland LAE, Chattopadhyay A, Jones AMP, Saxena PK (2016) Melatonin in plants and plant culture systems: variability, stability and efficient quantification. Front Plant Sci 7. https://doi.org/10.3389/fpls.2016.01721
59. Oyama T, Nagai R, Fujimoto M, Konishi R, Mita M, Uezono K, Zaitsu K, Hamase K (2015) Development of a fully-automated on-line oxidation column-switching HPLC system for the determination of endogenous melatonin in human clinical samples. Anal Sci 31:1129–1135. https://doi.org/10.2116/analsci.31.1129
60. Wang Z, Jiang H, Chen X, Song X, Xu F, Chen F, Mao Z, Gao S, Chen W (2020) A rapid and sensitive method for simultaneous determination of eight protein-bound uremic toxins in human serum by UHPLC-MS/MS: application in assessing peritoneal dialysis. J Pharm Biomed Anal 186:113312. https://doi.org/10.1016/j.jpba.2020.113312
61. Gomez FJV, Hernández IG, Martinez LD, Silva MF, Cerutti S (2013) Analytical tools for elucidating the biological role of melatonin in plants by LC-MS/MS: liquid phase separations. Electrophoresis 34:1749–1756. https://doi.org/10.1002/elps.201200569

62. Danielson ND, Mansour FR, Zhou L, Connell CV, Dotlich EM, Gibler JN, Norman BE, Grossman S, Wei W, Zhang Y (2018) Liquid chromatography with alkylammonium formate ionic liquid mobile phases and fluorescence detection. J Chromatogr A 1559:128–135. https://doi.org/10.1016/j.chroma.2018.03.020
63. Gomez FJV, Raba J, Cerutti S, Silva MF (2012) Monitoring melatonin and its isomer in Vitis vinifera cv. Malbec by UHPLC-MS/MS from grape to bottle. J Pineal Res 52:349–355
64. Vitalini S, Gardana C, Simonetti P, Fico G, Iriti M (2013) Melatonin, melatonin isomers and stilbenes in Italian traditional grape products and their antiradical capacity. J Pineal Res 54: 322–333. https://doi.org/10.1111/jpi.12028
65. Barreiro Astray S, Barbosa-Pereira L, Lage-Yusty MA, López-Hernández J (2021) Comparison of analytical methods for the rapid determination of melatonin in food supplements. Food Anal Methods 14:734–741. https://doi.org/10.1007/s12161-020-01912-2
66. Erland LAE, Saxena PK (2017) Melatonin natural health products and supplements: presence of serotonin and significant variability of melatonin content. J Clin Sleep Med 13:275–281. https://doi.org/10.5664/jcsm.6462
67. Carter MD, Wade Calcutt M, Malow BA, Rose KL, Hachey DL (2012) Quantitation of melatonin and n-acetylserotonin in human plasma by nanoflow LC-MS/MS and electrospray LC-MS/MS: melatonin and N-acetylserotonin quantitation. J Mass Spectrom 47:277–285. https://doi.org/10.1002/jms.2051
68. Ferreira MS, De Oliveira DN, Mesquita CC, Barbosa APDL, Anhê GF, Catharino RR (2016) MALDI-MSI: a fast and reliable method for direct melatonin quantification in biological fluids. J Anal Sci Technol 7:28. https://doi.org/10.1186/s40543-016-0106-5
69. Jiang J, Zhang D, Zhang H, Yu K, He J, Zhang X, Li N (2019) Rapid determination of melatonin by droplet spray ionization mass spectrometry. Int J Mass Spectrom 444: 116191. https://doi.org/10.1016/j.ijms.2019.116191
70. Giraldo-Acosta M, Martínez-Andújar C, Martínez-Melgarejo PA, Cano A, Hernández-Ruiz J, Arnao MB (2022) Protective effect (Safener) of melatonin on Vigna Radiata L. seedlings in the presence of the fungicide copper oxychloride. J Plant Growth Regul. https://doi.org/10.1007/s00344-022-10886-w
71. Ramakrishna A, Giridhar P, Sankar KU, Ravishankar GA (2012) Endogenous profiles of indoleamines: serotonin and melatonin in different tissues of Coffea canephora P ex Fr. As analyzed by HPLC and LC-MS-ESI. Acta Physiol Plant 34:393–396. https://doi.org/10.1007/s11738-011-0829-2
72. Fernández-Cruz E, Álvarez-Fernández MA, Valero E, Troncoso AM, García-Parrilla MC (2016) Validation of an analytical method to determine melatonin and compounds related to l-tryptophan metabolism using UHPLC/HRMS. Food Anal Methods 9:3327–3336. https://doi.org/10.1007/s12161-016-0529-z
73. Bhattacharjee A, Shashidhara SC, Saha S (2014) Isolation, purification and structural elucidation of N-Acetyl-5-Methoxytryptamine (melatonin) from Crataeva nurvala Buch-ham stem bark. Am J Phytomed Clin Ther 2:301–309
74. Kennaway DJ (2020) Measuring melatonin by immunoassay. J Pineal Res 69:e12657. https://doi.org/10.1111/jpi.12657
75. Van Tassel DL, Roberts N, Lewy A, O'Neill SD (2001) Melatonin in plant organs: Melatonin in plant organs. J Pineal Res 31:8–15. https://doi.org/10.1034/j.1600-079X.2001.310102.x
76. Kennaway DJ (2019) A critical review of melatonin assays: past and present. J Pineal Res e12572. https://doi.org/10.1111/jpi.12572
77. De La Puerta C, Carrascosa-Salmoral MP, García-Luna PP, Lardone PJ, Herrera JL, Fernández-Montesinos R, Guerrero JM, Pozo D (2007) Melatonin is a phytochemical in olive oil. Food Chem 104:609–612. https://doi.org/10.1016/j.foodchem.2006.12.010
78. Iriti M, Rossoni M, Faoro F (2006) Melatonin content in grape: myth or panacea? J Sci Food Agric 86:1432–1438. https://doi.org/10.1002/jsfa.2537
79. Pesti-Asbóth G, Molnár-Bíróné P, Forgács I, Remenyik J, Dobránszki J (2022) Ultrasonication affects the melatonin and auxin levels and the antioxidant system in potato in vitro. Front Plant Sci 13:979141. https://doi.org/10.3389/fpls.2022.979141
80. Ye T, Yin X, Yu L, Zheng S-J, Cai W-J, Wu Y, Feng Y-Q (2019) Metabolic analysis of the melatonin biosynthesis pathway using chemical labeling coupled with liquid chromatography-mass spectrometry. J Pineal Res 66:e12531. https://doi.org/10.1111/jpi.12531

Chapter 12

Determination of Free Proline in Plants

Florencia Sena, Jorge Monza, and Santiago Signorelli

Abstract

Proline metabolism has been associated with the induction of reactive oxygen species (ROS), antioxidant enzymes, and the control of cellular redox status. Moreover, proline accumulation is a highly evolutionarily conserved response to diverse abiotic stresses in plants. Thus, proline quantification has been helpful in abiotic stress research as a stress marker. The need for a reliable, fast, and simple method to detect proline in plant tissues is a powerful resource to imply the physiological status of plants under abiotic stress. This chapter summarizes the main strategies for proline extraction and quantification, highlighting their limitations and advantages, and recommends and details a specific protocol for proline extraction and quantification. The chapter provides a friendly version of this protocol with notes useful for researchers to perform the protocol.

Key words Proline, Extraction, Detection, Abiotic stress marker, ROS, Spectrophotometric methods, Plants

1 Introduction

1.1 Proline Accumulation in Plants: A Good Stress Marker

Under abiotic stress, higher plants induce proline biosynthesis. This response has been reported as an early response to multiple abiotic stress conditions, which has been observed in the model plant *Arabidopsis thaliana* [1, 2], as well as other species of productive importance such as pastures and crops [3–5]. Even though this response is robust, where proline concentration can represent up to 80% of the total amino acid pool under stress conditions [6], its role under stress is not yet known with certainty [4]. Different hypotheses have been handled for the cellular function of proline accumulation under stress, such as the adjustment of cellular osmotic status [7], osmoprotection [8], detoxification of reactive species [9], and redox balance [10]. Some authors considered proline an antioxidant molecule that could protect against ROS, thus preventing oxidative damage [9, 11–13]. While its capacity to protect against hydroxyl radicals has been further studied and supported [14, 15], other studies have shown that proline has a poor capacity to protect

Francisco J. Corpas and José M. Palma (eds.), *ROS Signaling in Plants: Methods and Protocols*,
Methods in Molecular Biology, vol. 2798, https://doi.org/10.1007/978-1-0716-3826-2_12,

against most oxygen and nitrogen-reactive species [16, 17]. Proline accumulation often goes together with greater oxidative or nitrosative damage [18–20]. Thus, it is far from clear that proline can act as an antioxidant, and some authors have suggested that its possible role in preventing oxidative damage can be linked to its capacity to induce antioxidant enzymes [16, 21–24]. This capacity is likely because proline catabolism is known to induce respiration and potentially mitochondrial ROS generation [25, 26] which can activate the expression of antioxidant enzymes. Regardless of its cellular function, proline accumulation is a well-conserved response of plants [2–5]. Therefore, proline detection and quantification in plants can be a good marker of the stress status of plants.

1.2 Proline Metabolism and Its Possible Link to Reactive Oxygen Species

The synthesis of proline in plants occurs mainly in the cytosol, where glutamate is converted to proline by the action of two NADPH-dependent enzymes. First, the enzyme Δ1-pyrroline-5-carboxylate synthetase (P5CS) reduces glutamate to L-glutamate-γ-semialdehyde (GSAL), which in turn is spontaneously converted to Δ1-pyrroline-5-carboxylate (P5C), and then Δ1-pyrroline-5-carboxylate reductase reduces P5C to proline [27]. The conservation of proline accumulation in plants can be due to possible benefits produced by the NADPH oxidation occurring during proline biosynthesis. Many stress conditions result in a reduced availability of $NADP^+$, which is necessary as the final electron acceptor in the photosynthetic electron transport chain. Thus, by contributing to NADPH oxidation, proline biosynthesis would benefit the plant by promoting photosynthetic activity and reducing the likelihood of photosynthetic electron leakage and chloroplast ROS production [28]. In fact, *A. thaliana* mutants that are unable to accumulate proline showed increased oxidative damage under stress conditions [16, 29].

On the other hand, proline degradation occurs in the mitochondria, where proline is transformed back into glutamate by the sequential steps of proline dehydrogenase (ProDH) and Δ1-pyrroline-5-carboxylate dehydrogenase (P5CDH) enzyme activities that generate reducing power such as $FADH_2$ and NADPH respectively, delivering electrons for mitochondrial respiration [27]. The capacity of proline degradation in feeding the respiratory chain is strongly supported by the fact that *A. thaliana* leaf discs exposed to 10 mM proline led to a significant increase in respiration rates [26]. Thus, proline metabolism could act as a redox shuttle to take reduced equivalents from the chloroplast and transport them to the mitochondria, where these are degraded [28]. Moreover, enhanced ProDH expression and a concomitant increase in ROS production resulted in oxidative damage [25]. Thus, proline biosynthesis would reduce ROS production at the chloroplast level, whereas proline degradation could induce mitochondrial ROS.

Table 1
Different methods to extract and detect proline levels in higher plants. + symbols represent the complexity of the methodology from simple (+) to more complex (+++) regarding time consumption, amount and toxicity of chemicals used, and equipment required

Extraction methods			
Extracted with		**Complexity**	**References**
3% sulfosalicylic acid		+	[30]
Methanol:Chloroform:Water (spiepr A3B2 twbch ":"12:5spiepr A3B2 twbch:1)		++	[36]
Hot ethanol		+	[42]
Detection methods			
Type of method	**Reagents**	**Complexity**	**References**
Spectrophotometric	Ninhydrin:Phosphoric acid:Acetic acid	++	[48, 49]
	Ninhydrin:Phosphoric acid: Acetic acid—Extracted in toluene	++	[30]
	Ninhydrin reagent without hazardous solutions (no toluene, no phosphoric acid)	++	[25, 26, 50]
Chromatography	Isatin paper assay	+	[46]
	HPLC	+++	[46, 52]
Enzymatic activity	P5CR activity	+++	[53]
Combined methods			
Type of method	**Method**	**Complexity**	**References**
Proline sensor	Simultaneous extraction and detection in situ	+	[54]

1.3 Proline Extraction and Quantification Methods in Plants

Over the last 70 years, multiple methods to extract and quantify proline from different plant tissues have been used and reported in the literature. These methods vary in terms of complexity, sensibility, and robustness. Here, we summarize and describe the most relevant extraction and detection methods (Table 1), which we present by the categories below.

Different types of solvents have been used for proline extraction methods, an essential step for proline quantification in plants. Before the addition of the extraction solution, the plant material is generally homogenized by grinding (mechanical homogenizers, vortex, liquid N_2). Here, we briefly describe the most relevant extraction methods used.

1.3.1 Sulfosalicylic Acid Extraction Method

Proline is solubilized and extracted from plant tissues by aqueous sulfosalicylic acid solutions because it is an effective protein precipitant in aqueous solution and does not interfere with the detection by acid–ninhydrin reaction afterward [30]. Three percent aqueous sulfosalicylic acid was used to extract proline from approximately 0.5 g of plant leaves [30]; however, other concentrations have been successfully used, such as 5% [31], 8% [32], and 10% [33]. Additionally, an improvement of the sensibility of this assay was described by the incorporation of different times of incubation and heating the samples, indicating that the incubation of plant leaves on 1% sulfosalicylic acid at 100 °C for 60 min yielded the best sensitivity even to detect trace amounts of proline [34].

1.3.2 Methanol: Chloroform:Water Extraction Method

Proline extraction is also approached by a multi-phase extraction protocol relying on a two-phase separation system consisting of a mixture of chloroform, methanol, and water (MCW). After centrifugation, the lower organic phase contains chloroform, whereas the upper phase contains methanol and water, and the cellular debris remains between the two phases. Proline is a polar amino acid containing a ring structure that includes the imine group and has an anomalous hydrophilic character [35]. In this context, the upper (aqueous) phase is used for the determination of proline. MCW solution in 12:5:1 (v:v:v) proportion is commonly used to extract proline from frozen or dried plant material from *A. thaliana* [16], crops such as wheat [36] or soybean [37], also legumes [19, 20, 38, 39], and tomato [40].

1.3.3 Hot Ethanol Extraction Method

Proline is soluble in different aqueous ethanol mixtures, where the solubility decreases at 100% (v/v) ethanol concentration [41]. In addition, heating the aqueous ethanol mixtures has been reported to extract proline from plant tissue successfully [42]. For example, proline was extracted from *A. thaliana* leaves using hot ethanol 80% (v/v) at 80–95 °C, detecting proline at the order of μmol [25, 26].

1.3.4 Isatin Paper Assay Detection Method

The isatin (1H-indole-2,3-dione) paper assay is a simple and fast method to detect amino acid content in a large number of samples by a colorimetric reaction followed by a chromatography paper [43–45]. Proline and hydroxyproline are imino acids that exclusively condense with isatin under acidic conditions to yield a blue-colored product (called pyrrole blue), where the intensity of the color on the chromatography paper correlates to the proline levels. This methodology has been used to detect changes in proline content in *A. thaliana* plants under drought conditions and in mutants lacking proline catabolism [46]. Moreover, this method has allowed the screening of many samples in a relatively short time in other plant species, such as trees under abiotic stress [47].

1.3.5 Spectrophotometric Detection Methods

The colorimetric measurement using ninhydrin (2,2-dihydroxyindane-1,3-dione) is the most widely used method to quantify proline. For this method, a ninhydrin reagent is prepared, composed of 1.25 g ninhydrin in 30 mL glacial acetic acid and 20 mL of 6 M phosphoric acid [30]. The reaction of proline with ninhydrin in hot (90–100 °C) and acidic conditions forms a red product (Fig. 1) that

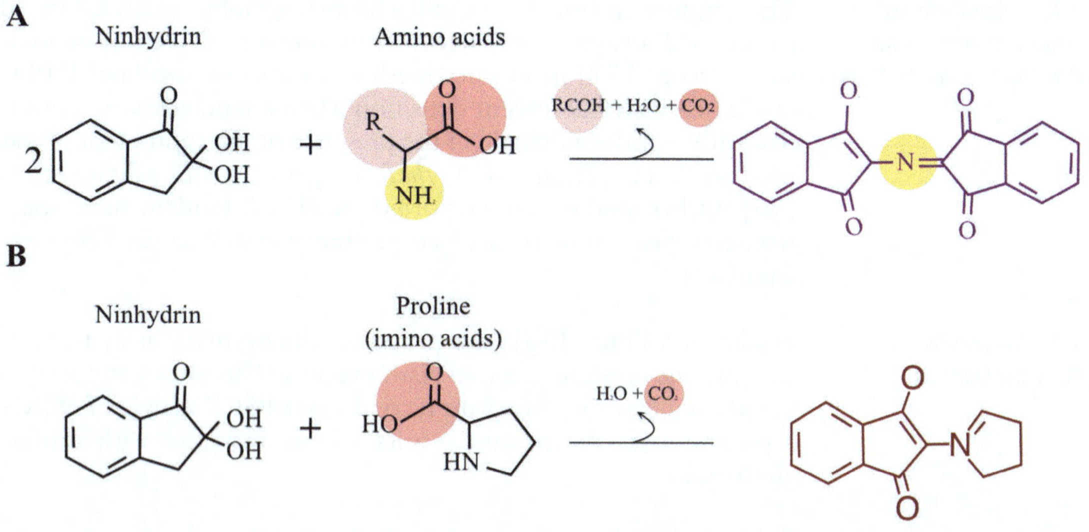

Fig. 1 Color development of amino and imino acids with ninhydrin under acidic conditions. (**a**) In the presence of a primary amino group, ninhydrin forms a blue-purple complex. The amino acid undergoes oxidative deamination when it reacts with ninhydrin, resulting in the liberation of H_2O, CO_2, RCOH, and an aldehyde along with hydrindantin that will react with a second ninhydrin molecule to form diketohydrin, also known as Ruhemann's complex (purple final product). (**b**) Ninhydrin directly oxidizes proline, producing proline decarboxylation, initially forming a yellow compound that is then turned into a red compound during color development at 90–100 °C

can be used to quantify its concentration using a spectrophotometer in the visible range [48, 49]. The colored proline–ninhydrin product has an absorbance maximum of 515 nm which can be directly measured or separated from other absorbing components using toluene, which solubilizes the proline–ninhydrin product very well but no other water-soluble molecules, and measured at 520 nm [30]. Owing to the ninhydrin's ability to react with amino acids, if proline concentrations in the sample are relatively low, proline concentration can be overestimated due to the cross-reaction of other amino acids with ninhydrin. The protocol presented by Bates et al., 1973 was described as a method to quantify proline in water-stress studies, where proline concentrations in the samples tend to be very high, making it possible to neglect the contribution in absorbance by the presence of other amino acids. Yet, it is important to bear in mind that the cross-reaction of other amino acids with ninhydrin yields a purple product with a peak of absorbance different from 515 nm. Thus, this method was widely adopted by the scientific community and not only to measure proline in drought research. Recently, this protocol has been simplified to avoid the use of hazardous organic solvents such as toluene and the use of phosphoric acid in the ninhydrin reagent, improving the sensibility of the final proline quantification on different biological systems [50] and in *A. thaliana* seedlings under drought conditions [34].

1.3.6 High-Performance Liquid Chromatography Coupled Detection Methods

The high-performance liquid chromatography (HPLC)-based amino acid analysis combined with fluorescent detection or mass spectroscopy [51] is another method to quantify proline. HPLC combined with fluorescent detection when amino acids are derivative with *o*-phthalaldehyde in the presence of mercaptoethanol was used to detect proline in *A. thaliana* plants under salinity stress [46]. Other studies have reported an HPLC-tandem mass spectrometry procedure to analyze proline metabolites in bergamot fruit [52].

1.3.7 Enzymatic Detection Methods

Forlani and Funck [53] have proposed an enzymatic assay to measure proline concentration in plant extracts. This enzymatic assay is based on measuring the plant recombinant P5CR activity. P5CR is a proline catabolic enzyme, so its activity increases with proline abundance.

1.3.8 Other Detection Methods

Recently, a multilayer paper-based microfluidic sensor based on the proline–ninhydrin reaction was developed to quantify proline in plants [54]. This method has a detection limit in the order of μM, and thus, its sensitivity is comparable to those of spectrophotometric methods, making it suitable for early drought stress detection.

1.4 Advantages and Disadvantages of Different Methods

All three extraction methods work very well in terms of extracting proline. Given its high solubility in water (162 g at 25 °C), the extraction in aqueous solution containing sulfosalicylic acid is very efficient. The use of organic solvents such as hot ethanol or the methanol:chloroform:water mix contributes to disrupting the organelle structures, therefore ensuring a better homogenate. However, the solubility of proline in these solvents is not as high as in water. Thus, considering the cost of organic solvents, the potential hazard they implicate, and, in the case of hot ethanol, the need for an additional step for heating the samples, we conclude that sulfosalicylic acid is the most recommended extraction method. Moreover, Lee et al [34] recently demonstrated that 1% of sulfosalicylic acid improves the assay sensitivity. Hence, we suggest doing 1% sulfosalicylic acid extraction instead of 3% as initially proposed by Bates et al [30].

Among the quantification methods, the P5CR activity assay, all the spectrophotometric methods, and the HPLC have the advantage of being quantitative, whereas the isatin paper assay and the proline sensor method are qualitative. Therefore, the latter are indicated for screening many samples and only to discriminate whether the samples are accumulating proline or not. However, when proline concentration needs to be determined, we recommend to use other methods. Among these quantitative methods, the HPLC and the enzymatic assay are more expensive and complex, although these can yield greater resolution. For most researchers, the spectrophotometric methods are likely to be the most accessible but also simpler to perform. Among the different variants to perform the spectrophotometric methods, we recommend the one not using phosphoric acid and not including the toluene extraction due to its simplicity, reduced toxicity, and cost.

2 Materials

Prepare all solutions using ultrapure water (18 $M\Omega^{-cm}$ at 25 °C) and analytical-grade reagents. Prepare and store all reagents at room temperature (unless indicated otherwise).

2.1 Equipment

1. Pre-weighted 1.5 μL microtubes.
2. Micro-pestles (or vortex).
3. Heating magnetic stirrer.
4. Centrifuge.
5. Spectrophotometer.
6. Plastic/glass cuvettes or 96-well plates (depending on the spectrophotometric equipment available).
7. Dry oven.
8. Precision scale.

2.2 Solutions

1. Extraction solution (1% sulfosalicylic acid in water). Weigh 1 g of sulfosalicylic acid and transfer it to a beaker glass. Add about 90 mL of water and stir on the magnetic stirrer until the sulfosalicylic acid dissolves. Transfer the solution to a graduated cylinder and add water to a volume of 100 mL. Store at 4 °C for up to 4 weeks.
2. Ninhydrin reagent (1.25% ninhydrin in 80% glacial acetic acid) (*see* **Note 1**). Weigh 1.25 g of ninhydrin (2,2-dihydroxyindane-1,3-dione) and transfer it to a beaker glass. Add 80 mL of glacial acetic acid and water to a volume of 100 mL. Dissolve the mix by heating at 60 °C and stirring the solution in a heating magnetic stirrer. Store it at room temperature in darkness or cover the bottle with aluminum foil.
3. L-proline (≥99% HPLC) standards solutions. Take 2.32 mg of proline into 10 mL of water to have a 2 mM proline standard stock. Proceed to prepare the fresh proline standard curve solutions from the 2 mM stock (Table 2).

Table 2
Preparation of proline standard solutions for the standard curve

Proline standard concentration (μM)	Volume from 2 mM proline stock (μL)	Volume of water (μL)
0	0	1000
10	5	995
100	50	950
250	125	875
500	250	750

3 Methods

3.1 Proline Detection Protocol for Spectrophotometer or Plate Reader

1. In pre-weighted 1.5 mL microtubes, add about 30 mg of pre-grinded plant sample.
2. Add 1 mL of 1% aqueous sulfosalicylic acid and homogenize (*see* **Note 2**).
3. Incubate the tubes 5 min at 65 °C under shaking at 1000 g (optional, *see* **Note 3**).
4. Centrifuge for 2 min at 6000 g at room temperature and conserved both supernatant and pellet (*see* **Note 4**).
5. For microtube size, mix 400 μL of the supernatant from samples in step 4, or proline standards, and add 800 μL of ninhydrin reagent. For multiwell plate reader size, mix 66 μL of supernatant from samples in step 4, or standards, and add 132 μL of the ninhydrin reagent (*see* **Note 5**).
6. Incubate the tubes at 95 °C in a thermoblock or hot bath (*see* **Note 6**). If using a multiwell plate, cover and incubate it in a drying oven at 95 °C for 1 h.
7. Discard the remaining supernatant from the sample tubes and dry them in a drying oven at >60 °C for at least 2 days (or until the weight remains constant) for the posterior determination of the dry weight of samples (*see* **Note 7**).
8. After completing the incubation time, place the tubes on ice to stop the reaction.
9. Read the absorbance of the standards and the samples (1 mL) at 510 nm using 1 mL in the spectrophotometer or using 200 μL in multiwell plate reader. Use sulfosalicylic acid, water, and ninhydrin reagent as the blank (*see* **Note 8**).

10. Determine the linear regression for the standard curve of proline.
11. Using the standard curve and the absorbance of your samples, calculate the concentration of proline (*see* **Notes 9** and **10**).

4 Notes

1. Ninhydrin is an irritant to the respiratory system, skin, and eyes, thus requiring adequate precautions. The use of dust mask type N95, eyeshields, and gloves is mandatory.
2. Homogenate by using either micropestles suitable for microtubes or vigorous vortexing for about 30 s.
3. If a thermoblock is available, this additional and optional step can ensure better homogenization.
4. The supernatant will be used for proline quantification, whereas the pellet containing the plant tissue will be used to determine the dry weight of samples.
5. Microtube size is suitable to read the absorbance of the reaction products analyzed by cuvette-based spectrophotometry, which requires a large sample volume (≈1 mL) and does not allow simultaneous measurement of multiple samples. In contrast, multiwell plate reader size is a useful option for low sample volume (100–200 μL) and wide-scale samples read simultaneously (up to 96 wells).
6. Make a hole in the lid of the 1.5 mL microtubes with a needle to prevent the lid bursting due to the high temperature.
7. Calculate the weight by weighing the tube containing the dry pellet and subtracting the weight of the empty tube.
8. The blank can be water and it is not relevant as the standard curve already includes a 0 mM proline solution, that will be treated as the samples with ninhydrin reagent and incubated for 1 h.
9. The calibration or standard curve is generated to determine the proline concentrations of the unknown samples. The standard curve is plotted by the reads of absorbance at 510 nm (on the *Y*-axis) vs. proline concentration (on the *X*-axis). The data fits to a linear regression, where the output is the following equation: $y = Ax + B$, where $y =$ absorbance at 510 nm and $x =$ proline concentration (μM). The coefficient of determination (R^2) quantifies the goodness of the fit (≥ 0.9).
10. For the calculation, consider the dilution 1 in 3 of **step 5**, so multiply by 3 the result obtained after calculating the concentration using the standard curve. Proline concentration is expressed as nmol proline· g^{-1} DW = [(ng proline/mL × mL extraction buffer)/115.5 ng]/g sample.

Acknowledgments

The authors S.S. and J.M. are members of the Uruguayan National System of Researchers (SNI, Uruguay) and the Basic Research Development Program (PEDECIBA, Uruguay). S.S. is supported by the CPR scheme of the International Centre of Genetic Engineering and Biotechnology (ICGEB), project number ICGEB_URY21_04_EC_2021, the CSIC I + D program (Uruguay), project number CSIC_I + D_2020_21, and CSIC I + D groups program (Uruguay), group name: "Food and Plant Biology" and group number "883431."

References

1. Chiang HH, Dandekar AM (1995) Regulation of proline accumulation in Arabidopsis thaliana (L.) Heynh during development and in response to desiccation. Plant Cell Environ 18:1280–1290
2. Verslues PE, Sharma S (2010) Proline metabolism and its implications for plant-environment interaction. Arab B 8(e0140):1–23
3. Delauney AJ, Verma DPS (1993) Proline biosynthesis and osmoregulation in plants. Plant J 4:215–223
4. Forlani G, Trovato M, Funck D, Signorelli S (2019) Regulation of proline accumulation and its molecular and physiological functions in stress defence. In: Hossain MA, Kumar V, Burritt DJ, Fujita M, Mäkelä PSA (eds) Osmoprotectant-mediated abiotic stress tolerance in plants: recent advances and future perspectives. Springer, pp 73–97. https://doi.org/10.1007/978-3-030-27423-8_3
5. Signorelli S et al (2021) GABA and Proline metabolism in response to stress. In: Hormones and plant response. Plant in challenging environments, vol 2. Springer, Cham. https://doi.org/10.1007/978-3-030-77477-6_12
6. Kavi Kishor PB et al (2005) Regulation of proline biosynthesis, degradation, uptake and transport in higher plants: its implications in plant growth and abiotic stress tolerance. Curr Sci 88:424–438
7. Szabados L, Savouré A (2010) Proline: a multifunctional amino acid. Trends Plant Sci 15: 89–97
8. Anjum F, Rishi V, Ahmad F (2000) Compatibility of osmolytes with Gibbs energy of stabilization of proteins. Biochim Biophys Acta – Protein Struct Mol Enzymol 1476:75–84
9. Smirnoff N, Cumbes QJ (1989) Hydroxyl radical scavenging activity of compatible solutes. Phytochemistry 28:1057–1060
10. Hare PD, Cress WA (1997) Metabolic implications of stress-induced proline accumulation in plants. Plant Growth Regul 21:79–102
11. Alia, Saradhi PP, Mohanty P (1991) Proline enhances primary photochemical activities in isolated thylakoid membranes of Brassica juncea by arresting photoinhibitory damage. Biochem Biophys Res Commun 181:1238–1244
12. Matysik J, Alia, Bhalu B, Mohanty P (2002) Molecular mechanisms of quenching of reactive oxygen species by proline under stress in plants. Curr Sci 82:525–532
13. Rehman AU et al (2021) Proline is a quencher of singlet oxygen and superoxide both in in vitro systems and isolated thylakoids. Physiol Plant 172:7–18
14. Signorelli S, Coitiño EL, Borsani O, Monza J (2014) Molecular mechanisms for the reaction between ˙OH radicals and proline: insights on the role as reactive oxygen species scavenger in plant stress. J Phys Chem B 118:37–47
15. Signorelli S, Dans PD, Coitiño EL, Borsani O, Monza J (2015) Connecting proline and γ-aminobutyric acid in stressed plants through non-enzymatic reactions. PLoS One 10: e0115349
16. Signorelli S et al (2016) In vivo and in vitro approaches demonstrate proline is not directly involved in the protection against superoxide, nitric oxide, nitrogen dioxide and peroxynitrite. Funct Plant Biol 43:870–879
17. Signorelli S, Arellano JB, Melø TB, Borsani O, Monza J (2013) Proline does not quench singlet oxygen: evidence to reconsider its protective role in plants. Plant Physiol Biochem 64: 80–83
18. González-Gordo S et al (2019) Nitric oxide-dependent regulation of sweet pepper fruit ripening. J Exp Bot 70:4557–4570

19. Signorelli S et al (2013) Antioxidant and photosystem II responses contribute to explain the drought-heat contrasting tolerance of two forage legumes. Plant Physiol Biochem 70:195–203
20. Signorelli S, Corpas FJ, Borsani O, Barroso JB, Monza J (2013) Water stress induces a differential and spatially distributed nitro-oxidative stress response in roots and leaves of *Lotus japonicus*. Plant Sci 201–202:137–146
21. Öztürk L, Demir Y (2002) In vivo and in vitro protective role of proline. Plant Growth Regul 38:259–264
22. Hoque MA et al (2007) Exogenous proline and glycinebetaine increase NaCl-induced ascorbate-glutathione cycle enzyme activities, and proline improves salt tolerance more than glycinebetaine in tobacco bright Yellow-2-suspension-cultured cells. J Plant Physiol 164:1457–1468
23. Hossain MA, Fujita M (2010) Evidence for a role of exogenous glycinebetaine and proline in antioxidant defense and methylglyoxal detoxification systems in mung bean seedlings under salt stress. Physiol Mol Biol Plants 16:19–29
24. Islam MM et al (2009) Exogenous proline and glycinebetaine increase antioxidant enzyme activities and confer tolerance to cadmium stress in cultured tobacco cells. J Plant Physiol 166:1587–1597
25. Oh GGK, O'Leary BM, Signorelli S, Harvey Millar A (2022) Alternative oxidase (AOX) 1a and 1d limit prolineinduced oxidative stress and aid salinity recovery in Arabidopsis. Plant Physiol 188:1521–1536
26. O'Leary BM, Khim Oh GG, Lee CP, Harvey Millar A (2020) Metabolite regulatory interactions control plant respiratory metabolism via target of rapamycin (TOR) kinase activation [open]. Plant Cell 32:666–682
27. Trovato M, Forlani G, Signorelli S, Funck DP (2019) Proline metabolism and its functions in development and stress tolerance. In: Osmoprotectant-mediated abiotic stress tolerance in plants: recent advances and future perspectives, pp 41–72. https://doi.org/10.1007/978-3-030-27423-8_2
28. Signorelli S (2016) The fermentation analogy: A point of view for understanding the intriguing role of proline accumulation in stressed plants. Front Plant Sci 7:1–6
29. Székely G et al (2008) Duplicated P5CS genes of Arabidopsis play distinct roles in stress regulation and developmental control of proline biosynthesis. Plant J 53:11–28
30. Bates LS, Waldren RP, Teare ID (1973) Rapid determination of free proline for water-stress studies. Plant Soil 39:205–207
31. Hassine AB, Ghanem ME, Bouzid S, Lutts S (2008) An inland and a coastal population of the Mediterranean xero-halophyte species Atriplex halimus L. differ in their ability to accumulate proline and glycinebetaine in response to salinity and water stress. J Exp Bot 59:1315–1326
32. Wang S et al (2004) The characteristics of Na+, K+ and free proline distribution in several drought-resistant plants of the Alxa Desert. China J Arid Environ 56:525–539
33. Sawahel WA, Hassan AH (2002) Generation of transgenic wheat plants producing high levels of the osmoprotectant proline. Biotechnol Lett 24:721–725
34. Lee MR, Kim CS, Park T, Choi YS, Lee KH (2018) Optimization of the ninhydrin reaction and development of a multiwell plate-based high-throughput proline detection assay. Anal Biochem 556:57–62
35. Gibbs PR, Radzicka A, Wolfenden R (1991) The anomalous hydrophilic character of proline. J Am Chem Soc 113:4714–4715
36. Charest C, Ton Phan C (1990) Cold acclimation of wheat (Triticum aestivum): properties of enzymes involved in proline metabolism. Physiol Plant 80:159–168
37. Casaretto E, Signorelli S, Gallino JP, Vidal S, Borsani O (2020) Endogenous •NO accumulation in soybean is associated with initial stomatal response to water deficit. Physiol Plant 172:564–576
38. Borsani O, Díaz P, Monza J (1999) Proline is involved in water stress responses of Lotus corniculatus nitrogen fixing and nitrate fed plants. J Plant Physiol 155:269–273
39. Díaz P et al (2010) Deficiency in plastidic glutamine synthetase alters proline metabolism and transcriptomic response in *Lotus japonicus* under drought stress. New Phytol 188:1001–1013
40. Borsani O, Cuartero J, Fernández JA, Valpuesta V, Botella MA (2001) Identification of two loci in tomato reveals distinct mechanisms for salt tolerance. Plant Cell 13:873–887
41. Bowden NA, Sanders JPM, Bruins ME (2018) Solubility of the Proteinogenic α-amino acids in water, ethanol, and ethanol-water mixtures. J Chem Eng Data 63:488–497
42. Gibon Y, Sulpice R, Larher F (2000) Proline accumulation in canola leaf discs subjected to osmotic stress is related to the loss of

chlorophylls and to the decrease of mitochondrial activity. Physiol Plant 110:469–476

43. Smith I (1953) Colour reactions on paper chromatograms by a dipping technique. Nature 171:43–44
44. Boctor FN (1971) An improved method for colorimetric determination of proline with isatin. Anal Biochem 70:66–70
45. Elliott RJ, Gardner DL (1976) Proline determination with isatin, in the presence of amino acids. Anal Biochem 70:268–273
46. Ábrahám E, Hourton-Cabassa C, Erdei L, Szabados L (2010) Methods for determination of proline in plants. In: Plant stress tolerance, methods in molecular biology, vol 639, pp 1–14
47. Sperling O, Secchi F, Godfrey J, Zwieniecki MA (2016) Acclimation of *Pistacia integerrima* trees to frost in semi-arid environments depends on autumn's drought. Planta 245: 671–679
48. Chinard FP (1952) Photometric estimation of proline and ornithine. J Biol Chem 199:91–95
49. Troll W, Lindsley J (1955) A photometric method for the determination of proline. J Biol Chem 215:655–660
50. Shabnam N, Tripathi I, Sharmila P, Pardha-Saradhi P (2016) A rapid, ideal, and eco-friendlier protocol for quantifying proline. Protoplasma 253:1577–1582
51. Noctor G, Foyer CH (1998) Simultaneous measurement of foliar glutathione, γ-glutamylcysteine, and amino acids by high-performance liquid chromatography: comparison with two other assay methods for glutathione. Anal Biochem 264:98–110
52. Servillo L, Giovane A, Balestrieri ML, Cautela D, Castaldo D (2011) Proline derivatives in fruits of bergamot (Citrus bergamia Risso et Poit): presence of N-methyl-L-proline and 4-hydroxy-L-prolinebetaine. J Agric Food Chem 59:274–281
53. Forlani G, Funck D (2020) A specific and sensitive enzymatic assay for the quantitation of L-Proline. Front Plant Sci 11:1–12
54. Choi YS, Im MK, Lee MR, Kim CS, Lee KH (2020) Highly sensitive enclosed multilayer paper-based microfluidic sensor for quantifying proline in plants. Anal Chim Acta 1105:169–177

Chapter 13

An Enzyme Assay Kit for GABA Quantification in Plant Tissues

Tatsuya Nishiyama, Noriko Wada, Hitoshi Kusakabe, and Kenji Ueda

Abstract

Gamma-aminobutyric acid (GABA) is an amino acid that has a role as a signaling molecule. In plants, its involvement in stress responses is widely investigated. A newly developed method of quantification of GABA is described in this chapter. The assay kit consisting of three bacterial enzymes enables easy but accurate measurement of GABA (~200 mg/mL) based on the serial enzymatic reaction leading to dye formation. The method was successfully applied to measure the GABA content in several plant tissues.

Key words Gamma-amino butyric acid, Quantification, Method, Colorimetric assay, Enzyme kit, *Streptomyces*, Aminotransferase, Glutamate oxidase, Plant tissue

1 Introduction

Gamma-aminobutyric acid (GABA) is an amino acid formed due to the decarboxylation of L-glutamic acid caused by the catalytic activity of glutamate decarboxylase (Fig. 1). GABA is not a proteinaceous amino acid, but it occurs in divergent organisms to play regulatory roles [1].

In the central nervous system of mammals, GABA acts as a major mediator of synaptic inhibition. The role of GABA as a repressive neurotransmitter is correlated with the reduction of anxiety and pain and the control of cardiovascular conditions such as blood pressure and heart rate [1]. In the synaptic cleft, GABA is sensed by two types of receptors, $GABA_A$ and $GABA_B$, which belong to pentameric ligand-gated ion channels and G protein-coupled receptors, respectively. The mechanism of signal transduction has been studied in detail [2]. The role of GABA as a neurotransmitter also occurs in divergent organisms, including invertebrates, nematodes, and even sponges [3].

Francisco J. Corpas and José M. Palma (eds.), *ROS Signaling in Plants: Methods and Protocols*, Methods in Molecular Biology, vol. 2798, https://doi.org/10.1007/978-1-0716-3826-2_13,

GABA also occurs in microorganisms including various bacteria and fungi [1]. Of these, the main producer is lactic acid bacteria. It is speculated that one of the physiological functions of GABA formation in microorganisms is the alteration of the environmental pH. The decarboxylation of glutamate that forms GABA (Fig. 1) causes an increase in intracellular and environmental pH. Thus, GABA production can be fundamental to the adaptation of producer microorganisms to the acidic environment [1]. On the other hand, there is an argument that GABA produced by some intestinal bacteria affects the physiology of the host animals and is involved in the symbiotic interaction between the host and intestinal bacteria.

GABA also serves as a signal molecule in plants [4]. The GABA content is affected by various environmental stress factors, such as temperature, availability of water, salt, heavy metals, and reactive oxygen species [5, 6]. The response of plants to low temperature is mediated by the GABA interconnection between the metabolic process and the polyamine pathway [7]. Biotic stresses that include the invasion of pathogens induce defensive mechanisms through the production of GABA [8, 9]. Furthermore, GABA controls the development of flower, root, and seed germination [4, 5]. The complex signaling network mediated by GABA underlies the regulatory mechanism governing the physiology and development.

In this chapter, we describe the newly developed method that allows for easy measurement of GABA content in biological materials. The conventional method of quantifying GABA relies on the use of high-performance liquid chromatography equipped with mass spectrometry (LC-MS), but it is costly and time consuming. On the contrary, the new method employing a series of enzymatic reactions that perform GABA-dependent dye formation is low cost and easy to handle [10]. Initially, the enzymatic kit was developed to provide a method that enables easy but accurate quantification of GABA by people who are not familiar with analytical work and/or who do not have easy access to an analytical facility. In particular, we expected that the availability of the kit would help farmers and breeders develop crops of high GABA content. However, the wide distribution of GABA and its significant roles in divergent organisms indicate that the availability of a useful measurement method will help advance basic science.

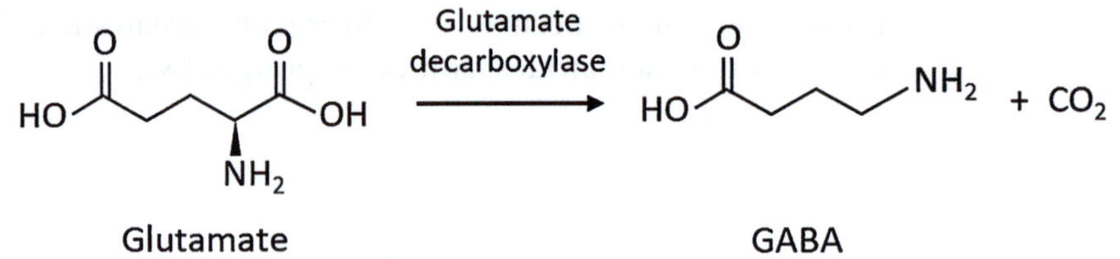

Fig. 1 Chemical structure of GABA and the enzymatic reaction of its formation from glutamate

The quantification of GABA is achieved by coupling of reactions catalyzed by the three enzymes: GABA aminotransferase, glutamate oxidase, and peroxidase (Fig. 2). GABA aminotransferase catalyzes amino transfer from GABA to 2-oxoglutarate, generating glutamate and performing GABA concentration-dependent glutamate production. The resulting glutamate is subsequently oxidized into 2-oxoglutarate, ammonium, and hydrogen peroxide, as a result of the activity of glutamate oxidase. Then, hydrogen peroxide is consumed by the reaction of peroxidase catalyzing hydrogen peroxide-dependent oxidation of Trinder's reagent in the presence of 4-aminoantipyrine. The series of enzymatic reactions allow the GABA-dependent formation of dye (i.e., Trinder oxidized reagent), which can be quantified by a photometric assay.

Originally, the coupling of glutamate oxidase and peroxidase was developed to quantify L-glutamate, the activity principle specifying Umami taste. The successful construction of the enzymatic glutamate quantification system was based on the discovery of glutamate oxidase produced by *Streptomyces*, the gram-positive filamentous bacteria [11, 12]. The high substrate specificity and productivity of hydrogen peroxide in this serial enzyme reaction enabled the construction of the easy but accurate quantification system for glutamic acid and its use as a useful analytical method to measure the tasty amino acid content in various agricultural products and foods.

Another great advantage of the glutamate quantification system is its applicability to quantifying other amino acids. The combination of the glutamate assay system with the reaction of an amino acid transferase that catalyzes amino transfer from a specific amino acid to 2-oxoglutarate allows the quantification of the corresponding amino acid through the formation of glutamate. Based on this idea, we screened for GABA aminotransferase and obtained a useful enzyme from *Streptomyces* [10]. The GABA enzyme assay kit explained below was developed by combining the GABA aminotransferase and glutamate assay system.

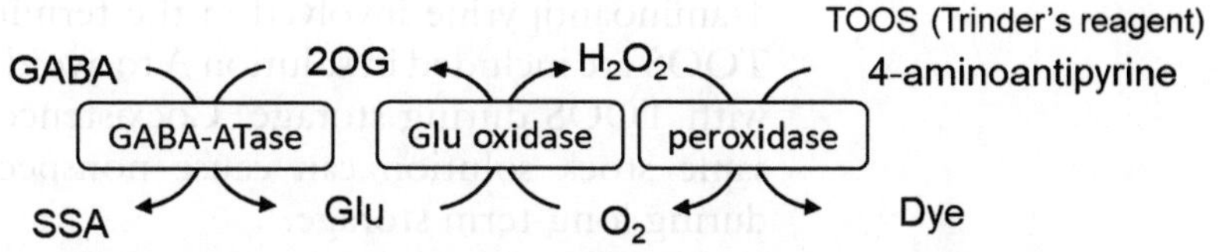

Fig. 2 Schematic representation of the serial enzymatic reaction that enables the quantification of GABA (GABA, gamma amino butyric acid; SSA, succinic semialdehyde; GABA-ATase, GABA-aminotransferase; 2OG, 2-oxoglutaric acid; Glu, glutamic acid; TOOS, *N*-ethyl-*N*-(2-hydroxy-3-sulfopropyl)-3-methylaniline)

2 Materials

2.1 Normal Assay

1. GABA enzyme assay kit (*see* **item 3.3**).
2. Homogenizer or mixer.
3. Spectrophotometer.
4. Cuvettes (capacity, 1–1.5 mL).
5. Pipets and microchips.

2.2 Microassay

1. GABA enzyme assay kit (*see* **item 3.3**).
2. Micro homogenizer.
3. Microtiter plate reader (photometric scanner for microtiter plates).
4. Microtiter plates (well volume, 100–200 mL).
5. Pipets and microchips.

2.3 GABA Enzyme Assay Kit

1. Solution A: 10 U/mL ascorbate oxidase, 0.8 U/mL glutamate oxidase, 1200 U/mL catalase, 10 U/mL peroxidase, 0.8 mM 4-aminoantipyrine, and 100 mM HEPES (pH 7.1).

2. Solution B: 2 U/mL GABA aminotransferase, 0.8 mM TOOS (Trinder's reagent), 1 mM sodium 2-oxoglutarate, 2 mM pyridoxal phosphate, 0.09% sodium azide, and 100 mM HEPES (pH 7.1).

Notes The GABA enzyme assay kit consists of two solutions: solution A (pretreatment solution for the elimination of substances that disturb the correct measurement) and solution B (solution for the main quantification reaction). Ascorbate oxidase, glutamate oxidase, and catalase are added to solution A to remove ascorbate, glutamate, and hydrogen peroxide, respectively. These substances possibly contained in the measurement sample disturb the specific detection of GABA in the main reaction. Pyridoxal phosphate is added to solution B to fulfill its requirement by GABA aminotransferase as a cofactor. Sodium azide is added to solution B to inactivate catalase included in solution A. Peroxidase and 4-aminoantipyrine involved in the terminal reaction that oxidizes TOOS are included in solution A to avoid their nonspecific reaction with TOOS during storage. Coexistence of these reagents in the same stock solution can cause nonspecific oxidation of TOOS during long-term storage.

3 Methods

3.1 Normal Assay

1. *Sample preparation*. The typical measurement of plant food materials is performed as follows. About 1–20 g of edible part is added to ten times the weight of deionized water and homogenized by electric mixer. The resulting homogenate was filtrated using filter paper and centrifuged at 3000 × g for 1 min. The resultant supernatant is subjected to the enzymatic quantification reaction.
2. *Reaction*. Mix 50 μL of a sample for measurement with 0.5 mL of solution A and incubate at 30 °C for 10 min. Then add 0.5 mL of solution B and continue incubation at 30 °C for 10 min. Then, the concentration of dye formed due to the oxidation of Trinder's reagent (Fig. 3 upper) is quantified by

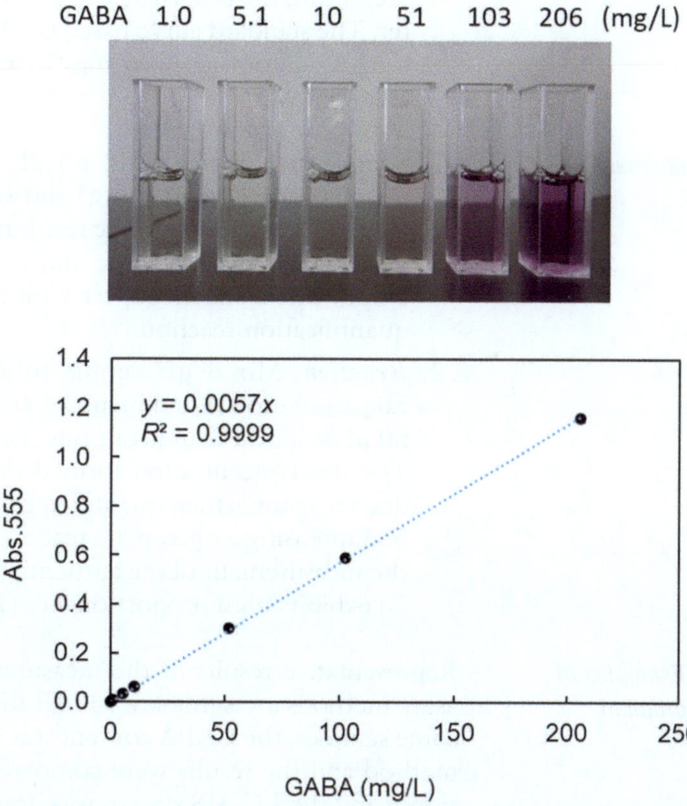

Fig. 3 The appearance of representative standard reactions by a normal assay using 1.0–206 mg/L of GABA solution (upper) and the corresponding standard curve (lower). The upper photo shows the view of the two-time scale measurement using 3 mL vol cuvettes for visibility, but the actual assay should use those with 1–1.5 mL capacity

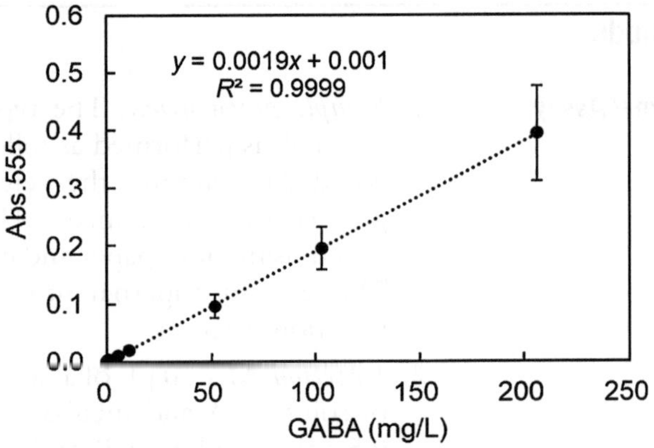

Fig. 4 A representative standard curve obtained by the microassay (data are means of measurement three times)

measuring the absorbance at 555 nm using a spectrophotometer. The standard curve based on the measurement of authentic GABA solution (0–206 mg/L) exhibits high proportionality (Fig. 3 lower).

3.2 Microassay

1. *Sample preparation.* Add 50 μL deionized water to a plant tissue sample (10–100 mg) and smash to homogeneity using a micro homogenizer. The resulting homogenate is transferred to a microcentrifuge tube and centrifuged at 10,000 × g for 1 min. The resultant supernatant is subjected to the enzymatic quantification reaction.
2. *Reaction.* Mix 5 μL sample solution for measurement with 50 μL solution A and incubate at 30 °C for 10 min. Then add 50 μL solution B and continue incubation at 30 °C for 20 min. The dye concentration formed due to the oxidation of Trinder's reagent is then quantified by measuring the absorbance at 555 nm using a microplate reader. The standard curve based on the measurement of the authentic GABA solution (0–206 mg/L) exhibits high proportionality (Fig. 4).

3.3 Examples of Measurement

Representative results of the measurement by normal and microassay methods are summarized in Tables 1 and 2, respectively. For some samples, the GABA content was also measured by the LC-MS method and the results were compared. The measurement condition using the LC-MS system was described previously [10].

The results obtained using the enzyme assay kit were comparable to those obtained by LC-MS. In particular, the measurement of tomato samples was performed accurately as previously described [10]. However, the values obtained by the enzyme system were slightly lower than those obtained by LC-MS in some samples such as grapefruit and lemon. This could be due to the effect of some

Table 1
Examples of GABA content measurement in plant materials (normal assay)

Plant sample		GABA (mg/100 g)	
Name	Tissue	Enzyme kit	LC-MS
Tomato	Fruit	26.4 ± 10.2	29.1 ± 11.1
Mikan (Japanese orange)	Fruit	52.3 ± 7.4	46.9 ± 7.4
Pink grapefruit	Fruit	41.3 ± 6.6	68.9 ± 10.6
Lemon	Fruit	29.6 ± 7.2	58.0 ± 15.0
Kiwi fruit	Fruit	12.2 ± 0.9	11.6 ± 3.4
Apple	Fruit	6.2 ± 0.3	1.5 ± 0.2
Arabidopsis thaliana	Leaf and stem Leaf and stem	7.3 4.7 ± 0.8	11.8 1.9 ± 1.0

NT not tested

Table 2
Examples of GABA content measurement in plant materials (micro assay)

Plant sample		GABA (mg/g)	
Name	Tissue	Enzyme kit	LC-MS
Tomato	Fruit	0.26 ± 0.10	0.29 ± 0.11
Mikan (Japanese orange)	Fruit	0.52 ± 0.07	0.47 ± 0.07
Radish sprout	Mixture of leaf, stem, and root Root Leaf Stem	0.07 0.05 ± 0.01 0.07 ± 0.04 0.01	0.12 0.02 ± 0.01 0.16 ± 0.06 NT
Broccoli sprout	Mixture of leaf, stem, and root Root Leaf Stem	0.03 0.03 0.04 ± 0.02 0.19	NT NT 0.12 ± 0.05 NT
A. thaliana	Leaf and stem Leaf and stem	0.04 0.09	NT NT

NT not tested

inhibitory substance(s) specifically contained in citrus fruit, but the reason for this is not yet known. The results obtained by the microassay procedure were comparable to those obtained by the normal assay.

4 Notes

Progress in scientific understanding depends on the availability of a reliable method that enables specific and accurate quantification. The enzyme kit developed in our study allows researchers to frequently measure the GABA concentration of samples of their interest. This will greatly help advance scientific understanding and its application. Furthermore, the availability of the kit will expand the possibility of measuring the GABA content even for people who are not familiar with analytical procedures in various situations. Thus, we expect that the availability of the kit will not only promote the development of new products such as high-GABA-content crops but also facilitate the study of the distribution of the nonproteinaceous amino acids of regulatory function.

Quantification based on the chromogenic enzymatic reaction is useful, but has some limitations. The method is not directly applicable to the measurement of samples of dark color or high salt content. Furthermore, some substance(s) contained in the sample may disturb the correct measurement, although the kit is designed to remove glutamate, ascorbate, and hydrogen peroxide in the pretreatment step. Thus, it is preferred that the measurement samples of those without experience are preceded by a preliminary check of the accuracy of the enzyme system by comparing the results with those obtained by the conventional LC-MS method.

The effectiveness of microassay procedure broadens the applicability of the enzyme assay system to the measurement of samples of extremely small volume. The LC-MS system can hardly measure samples whose volume is less than 10 microliters, as the sample solution is lost in prefiltration. On the contrary, as demonstrated by the quantification example (Table 2), the enzyme kit allowed the quantification of the GABA content in low volumes, such as of the small leaf of *Arabidopsis thaliana*. The applicability of the enzyme assay to such cases should broaden the range of measurement even to small samples such as specific plant tissues. We anticipate that the application of the enzyme assay kit will accelerate the study of the role and function of GABA in the complex regulatory network of plants.

Acknowledgments

The authors thank Dr. Francisco J Corpas for the opportunity to write and Dr. Tokuji Tsuchiya for providing *Arabidopsis* samples.

References

1. Sarasa SB, Mahendran R, Muthusamy G, Thankappan B, Selta DRF, Angayarkanni J (2020) A brief review on the non-protein amino acid, gamma-amino butyric acid (GABA): its production and role in microbes. Curr Microbiol 77(4):534–544. https://doi.org/10.1007/s00284-019-01839-w
2. Shaye H, Stauch B, Gati C, Cherezov V (2021) Molecular mechanisms of metabotropic GABA (B) receptor function. Sci Adv 7(22). https://doi.org/10.1126/sciadv.abg3362
3. Miller MW (2019) GABA as a neurotransmitter in gastropod molluscs. Biol Bull 236(2):144–156. https://doi.org/10.1086/701377
4. Caspi Y, Pantazopoulou CK, Prompers JJ, Pieterse CMJ, Hulshoff Pol H, Kajala K (2023) Why did glutamate, GABA, and melatonin become intercellular signalling molecules in plants? elife 12. https://doi.org/10.7554/eLife.83361
5. Li L, Dou N, Zhang H, Wu C (2021) The versatile GABA in plants. Plant Signal Behav 16(3):1862565. https://doi.org/10.1080/15592324.2020.1862565
6. Guo Z, Gong J, Luo S, Zuo Y, Shen Y (2023) Role of gamma-aminobutyric acid in plant defense response. Meta 13(6). https://doi.org/10.3390/metabo13060741
7. Xu M, Yang Q, Bai G, Li P, Yan J (2022) Polyamine pathways interconnect with GABA metabolic processes to mediate the low-temperature response in plants. Front Plant Sci 13:1035414. https://doi.org/10.3389/fpls.2022.1035414
8. Scholz SS, Malabarba J, Reichelt M, Heyer M, Ludewig F, Mithofer A (2017) Evidence for GABA-induced systemic GABA accumulation in Arabidopsis upon wounding. Front Plant Sci 8:388. https://doi.org/10.3389/fpls.2017.00388
9. Scholz SS, Reichelt M, Mekonnen DW, Ludewig F, Mithofer A (2015) Insect herbivory-elicited GABA accumulation in plants is a wound-induced, direct, systemic, and Jasmonate-independent defense response. Front Plant Sci 6:1128. https://doi.org/10.3389/fpls.2015.01128
10. Nishiyama T, Sulistyaningdyah WT, Ueda K, Kusakabe H (2020) GABA enzymatic assay kit. Biosci Biotechnol Biochem 84(1):118–125. https://doi.org/10.1080/09168451.2019.1661768
11. Arima J, Tamura T, Kusakabe H, Ashiuchi M, Yagi T, Tanaka H, Inagaki K (2003) Recombinant expression, biochemical characterization and stabilization through proteolysis of an L-glutamate oxidase from *Streptomyces* sp. X-119-6. J Biochem 134(6):805–812. https://doi.org/10.1093/jb/mvg206
12. Arima J, Sasaki C, Sakaguchi C, Mizuno H, Tamura T, Kashima A, Kusakabe H, Sugio S, Inagaki K (2009) Structural characterization of L-glutamate oxidase from *Streptomyces* sp. X-119-6. FEBS J 276(14):3894–3903. https://doi.org/10.1111/j.1742-4658.2009.07103.x

Chapter 14

Identification of Superoxide Dismutase (SOD) Isozymes in Plant Tissues

Ana Jiménez, Sandra Correa, and Francisca Sevilla

Abstract

Superoxide and hydrogen peroxide are reactive oxygen species (ROS) involved in the oxidation of multiple biological molecules and the signaling processes during plant growth and stress response. Thus, control of ROS is fundamental for cell survival and development, with superoxide dismutase (EC 1.15.1.1, SOD) being one of the main enzymes involved. Different isoforms of SOD catalyze the dismutation of superoxide ($O_2^{\cdot-}$) to hydrogen peroxide (H_2O_2) and oxygen (O_2), such as Mn-SODs, Cu,Zn-SODs, and Fe-SODs. Using non-denaturing polyacrylamide gel electrophoresis (PAGE) combined with a specific staining method for SOD activity, the protocol describes the identification of different SOD isozymes, based on their differential inhibition by KCN and H_2O_2, in different organs and plant species such as pea (*Pisum sativum* L.) leaves and pepper (*Capsicum annuum* L.) fruits.

Key words Enzyme activity assay, Isozymes, Polyacrylamide gel electrophoresis, Superoxide dismutase

1 Introduction

Superoxide dismutases (SODs, EC 1.15.1.1) are proteins belonging to a family of metalloenzymes that catalyze the dismutation of the superoxide radical ($O_2^{\cdot-}$) to O_2 and hydrogen peroxide (H_2O_2), according to the reaction:

$$2O_2^- + 2H^+ \xrightarrow{\text{SOD}} O_2 + H_2O_2$$

The discovery of the SOD enzymatic activity was made in 1969 by Joe M. McCord and Irwin Fridovich, researchers in the Department of Biochemistry at Duke University Medical Center in Durham [1]. SODs are classified depending on the metal cofactor in their reactive center: those containing Cu (II) and Zn (II) are named Cu,Zn-SODs, those containing Mn (III) are known as

Francisco J. Corpas and José M. Palma (eds.), *ROS Signaling in Plants: Methods and Protocols*,
Methods in Molecular Biology, vol. 2798, https://doi.org/10.1007/978-1-0716-3826-2_14,

Mn-SODs, and a third type containing Fe (III), known as Fe-SODs [2–4]. Subsequently, some atypical SODs containing different combinations of these metals in their molecules have been described [5]. The distribution of most SOD isozymes is multi-organellar. In plants, the Mn-SOD enzyme was initially localized in pea (*Pisum sativum* L.) leaf mitochondria and characterized as a tetrameric protein [3, 6], and it was later also identified in peroxisomes [7–9]. The enzyme Cu,Zn-SOD has been found in chloroplasts, mitochondria, peroxisomes, cytoplasm, and apoplast [10–13] and some authors have localized it in thylakoid membranes [14, 15] and the extracellular space [16, 17]. In some plant species, Fe-SOD, although mainly a chloroplastic enzyme, has been localized in the cytoplasm, mitochondria, and peroxisomes [2, 15, 18, 19].

The activity of the different SOD isozymes can be determined and localized on a polyacrylamide-bis acrylamide gel that is subjected to an electric field (PAGE, polyacrylamide gel electrophoresis). After the separation of the isoenzymes on the gel, a photochemical method can be used based on the reduction of nitroblue tetrazolium (NBT) to formazan blue, by superoxide radicals, which are promoted by the light in the presence of a solution containing N,N,N′,N′-tetramethyl ethylenediamine (TEMED) and riboflavin. Under these conditions, gels are colored by the formazan blue except in the area where SOD activity is present which remains colorless due to the enzymatic removal of $O_2^{\cdot -}$ radicals [20]. In addition, the different sensitivity of the isozymes to certain reagents is used for their identification. Thus, cyanide inhibits only Cu,Zn-SOD (and neither Mn-SOD nor Fe-SOD), while H_2O_2 inhibits both Fe-SOD and Cu,Zn-SODs, but not Mn-SOD [21].

This chapter describes an updated analytical method to identify the presence of SOD isozymes in different plant materials such as leaves of pea plants and pepper fruits, using non-denaturing PAGE and specific SOD activity staining.

2 Materials

2.1 Plant Extracts and Chemicals

1. Use MilliQ or deionized water, if possible, to dissolve chemicals.
2. Use plant material previously frozen in liquid N_2 and stored at −80 °C. We have used pea leaves (Lincoln cv, purchased from Battle company), collected from 25-day-old plants grown in vermiculite in controlled growth chambers (24/18 °C day/night, 16-h photoperiod, 80% humidity, and 250 μmol $m^{-2}s^{-2}$ light) and irrigated with Hoagland solution. We also analyzed pepper California-type fruits purchased in a local market (green, red, and yellow).

Using a mortar placed on ice and a pestle, obtain a homogenate with the sample material and the extraction buffer. To prepare 100 mL of extraction buffer, dissolve 0.1 mM ethylenedinitrilotetraacetic acid (EDTA) (0.0037 g), 5 mM cysteine (0.0606 g), 0.2% (v/v) Triton X-100 (0.2 mL), and 0.1 mM phenylmethylsulfonyl fluoride (PMSF) (0.1 mL of a 0.1 M solution, dissolve 0.017 g / 1 mL 96% ethanol) in 50 mM potassium phosphate pH 7.8 (described below). Homogenize the frozen material in a ratio of 1:2 (w/v) adding 1% soluble polyvinylpyrrolidone (PVP) (w/v) (or 5% for pepper fruits) (*see* **Note 1**).

Prepare the 50 mM potassium phosphate buffer pH 7.8 using two solutions: (1) K_2HPO_4: dissolve 0.87 g / 100 mL H_2O and (2) KH_2PO_4: dissolve 0.68 g /100 mL H_2O). Adjust the basic solution (1) with the acid one (2) till pH 7.8. Keep at 4 °C.

3. Filter the homogenate to an Eppendorf tube through a double nylon mess using a funnel.
4. Centrifuge the homogenate at 13,000 × *g* in a 1.5 mL microtube for 5 min at 4°C. Take the supernatant and separate an aliquot to measure proteins using the Bradford protein assay (BioRad, commercial, follow manufacturer instructions). Aliquot the supernatant and store it at −20/−80°C until analysis.

2.2 Non-denaturing Gel Electrophoresis Solutions

1. Non-denaturing PAGE is performed on 1.5 mm thickness 10% acrylamide gels with a stacking gel of 7.5%. You can use commercial precast gels (BioRad) or do the gel yourself as described below.
2. Gel buffer: 377 mM Tris–HCl buffer pH 8.9. To prepare 2 L, weigh 91.31 g Tris–HCl and dissolve in 1.8 L of ultrapure water. After mixing, adjust pH at 8.9 with HCl 5 M (previously prepared from the 12 M commercial one diluting it in H_2O, take **EXTRA CARE** because it burns, use a fume hood) and fill up to 2 L with water. Store at 4 °C.
3. Stacking gel buffer: 63 mM Tris–HCl, pH 6.7 (0.76 g Tris–HCl/ 100 mL H_2O MilliQ, adjust pH with HCl 5 M in H_2O). Store at 4 °C.
4. Electrode buffer: 38 mM glycine, pH 8.2 adjusted with gel buffer. Weigh 5.7 g of glycine and dissolve in 2 L water. Store at 4 °C.
5. Acrylamide:bis-acrylamide 30% solution, ratio 37.5:1 (BioRad, commercial, TOXIC). Store at 4 °C.
6. Prepare 85% glycerol solution (v/v in H_2O, Panreac).

7. Prepare 10% (w/v) ammonium persulfate in gel buffer. Dissolve 0.5 g into 5 mL gel buffer. Prepare this solution just before using it and maintain it at 4 °C.
8. N,N,N,N′-tetramethyl-ethylenediamine (TEMED) (Sigma, commercial, *see* **Note 2**). Use directly from the bottle.
9. Bromophenol blue 0.08% (w/v) containing sucrose 40% (w/v). Weigh 0.008 g of bromophenol and 4 g of sucrose and dissolve in 10 mL MilliQ H_2O. Store at 4 °C.

2.3 SOD Staining Solution

1. Prepare 2.45 mM NBT (Sigma, **Toxic).** Weigh 0.2 g / 100 mL potassium phosphate buffer 50 mM pH 7.8. Keep in darkness and at 4 °C. You can reuse it by filtering through filter paper if it turns blue.
2. Prepare 0.28 M TEMED + 2.8 × 10^{-5} M Riboflavin solution. Take 1.042 mL TEMED and 2.6 mg riboflavine and dissolve in 250 mL of 50 mM potassium phosphate buffer, pH 7.8. Prepare and keep in darkness and do not reuse.
3. Prepare 2 mM KCN (**HIGHLY TOXIC**, EXTRA CARE TO HANDLE AND WASTE!) freshly prepared. Weigh 6.5 mg / 50 mL potassium phosphate buffer 50 mM pH 7.8 containing 0.1 mM EDTA. You can use also 2 mM NaCN (0.0049 g / 50 mL). Prepare around 50 mL for each gel.
4. Prepare 5 mM H_2O_2. Use 25 μL of a 30% solution (commercial) in 50 mL potassium phosphate buffer 50 mM, pH 7.8, containing 0.1 mM EDTA. Prepare around 50 mL for each gel. Keep in darkness.

3 Methods

3.1 Polyacrylamide Gel Electrophoresis on 10% Acrylamide Gels

All the procedures should be carried out at around 4 °C (on ice).

1. Mix 15.7 mL of gel buffer, 2.86 mL of 85% glycerol, and 6.26 mL of 30% acrylamide solution in a Büchner vacuum flask. Degas with a vacuum system for 10 min. Add 0.12 mL of the ammonium persulfate solution and 60 μL of TEMED and cast the gel at room temperature within a 7.25 cm × 10 cm × 1.5 mm gel cassette (Bio-Rad Mini-Protean system) (*see* **Note 3**). Add isobutyl alcohol carefully to get the gel straight. Once polymerized (around 15 min at room temperature), wash carefully the alcohol with water and dry the top of the gel with filter paper.
2. Prepare the stacking gel on ice by mixing 3.675 mL stacking gel buffer, 1.125 mL of 30% acrylamide solution, 30 μL TEMED, 600 μL 85% glycerol, 540 μL water, and 50 mL Büchner vacuum flask. Make vacuum for 10 min.

3. Add 30 µL persulfate and mix all the components just before pouring the mixture on top of the gel. Insert a 10-well gel comb immediately avoiding air bubbles. Leave to polymerize at room temperature. Once polymerized, carefully remove the comb and clean wells with filter papers, then fill with gel buffer and leave in the cold chamber until use (maybe the next day).
4. Prepare each sample by mixing the appropriate volume of plant extract (around 100 µg protein of pepper fruit or 150 µg of pea leaf) with 0.08% (w/v) bromophenol blue in a 3:1 ratio (extract:blue), as a dye front indicator. Keep samples on ice.
5. Place the gels on the cassette (according to the instruction manual) and fill the inner and outer chambers with the electrode buffer (*see* **Note 4**).
6. Load samples carefully into each well (*see* **Note 5**).
7. Run the electrophoresis at 15 mA until the samples have entered the gel (30 min) and then continue at 40–50 mA until the dye front reaches around 1 cm above the gel edge. The procedure usually takes around 2 h. Run the electrophoresis in a cold room (about 4–6 °C) to preserve the enzyme activity (*see* **Note 6**).
8. Once finished, remove the gel cassette from the electrode buffer chamber and lift off the glass plate using a spatula or similar. Place the gel in a tray for the activity staining (*see* **Note** 7) and remove the stacking gel.

3.2 SOD Activity Staining

1. Pour the NBT staining solution into the tray until the gel is totally covered and it does not stick to the tray. Incubate for 20 min at room temperature with gentle shaking in darkness (*see* **Note 8**).
2. If using inhibitors, incubate duplicate or triplicate gels with H_2O_2 or KCN for 45 min (or 60 min) in darkness at room temperature with gentle shaking. After this, remove the solutions and incubate them in the NBT one (point 1).
3. Remove the NBT solution with a plastic pipette and keep it in darkness for reusing.
4. Incubate the gel in the TEMED + riboflavin solution in darkness for 15 min at room temperature with gentle shaking.
5. Remove the solution and add 50 mM potassium buffer pH 7.8, covering completely the gel.
6. Expose the gels to light (daylight lamps) until the bands of isozymes are visible. Time will vary depending on the activity of the different bands and the NBT solution reused. Change the buffer when it gets dark. Once developed, keep the gels in darkness at 4 °C covered by the buffer till analysis by

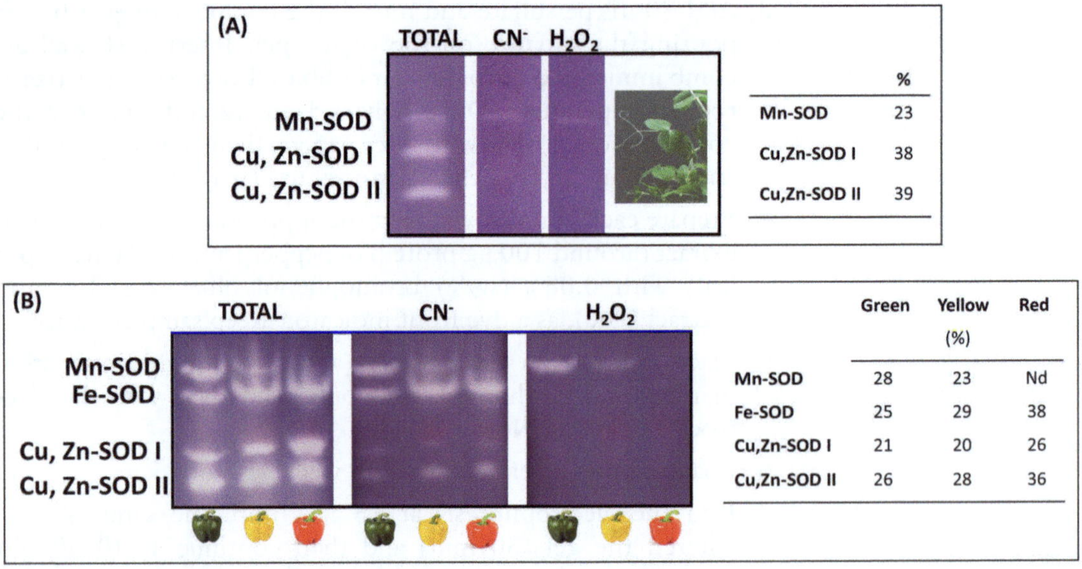

	%
Mn-SOD	23
Cu,Zn-SOD I	38
Cu,Zn-SOD II	39

	Green	Yellow (%)	Red
Mn-SOD	28	23	Nd
Fe-SOD	25	29	38
Cu,Zn-SOD I	21	20	26
Cu,Zn-SOD II	26	28	36

Fig. 1 In-gel SOD isozyme analysis in different plant organs and species. (A) SOD isozymes of pea leaves. Lane 1, total activity of the SOD isozymes of pea (*Pisum sativum* L., Lincoln cultivar) leaves (150 μg protein) after non-denaturing polyacrylamide gel electrophoresis (PAGE, 10% acrylamide) and staining with NBT (see text in 3.2 section for details). Lanes 2 and 3, SOD isozymes after incubation of the gel with KCN and H_2O_2, respectively, before the NBT staining. (B) SOD isozymes of pepper (*Capsicum annuum* L. California type) fruits (100 μg protein) in different ripening stages: unripe green and ripe red and yellow fruits. Lane 1, total activity of the SOD isozymes and lanes 2 and 3, SOD isozymes after a previous incubation with KCN and H_2O_2, respectively. Arrows on the left show the mobility of the isozymes detected in the samples. The band proportion of each SOD isozyme in the total activity gel was quantified using the Amersham ImageQuanTL 8.1 Program (Cytiva, Barcelona, Spain), and the percentage of each band is shown in the table on the right. Nd: not detected

densitometry (you can use an image analyzer and the Image J software to quantify the bands). Figure 1 shows the in-gel SOD isozyme analysis in different plant organs and species (panels A and B) with its corresponding relative quantification (table on the right) from the total activity gel, using an image analyzer (Amersham Imager 600, GE Healthcare) and the ImageQuant TL (GE Healthcare) to quantify the proportion of each isoenzyme in the different samples.

4 Notes

Be extra careful because there are some carcinogenic and toxic compounds, always wear gloves, and do not touch other instruments with contaminated gloves. Also, use the fume hood whenever possible. When using liquid N_2 **take extra care**, protect yourself with thick gloves and glasses (it burns!).

1. It is better to add PVP to the sample in the mortar directly (ratio 1 g / 100 mL buffer for leaves or 5 g / 100 mL for fruits).
2. Use a fume hood to handle it.
3. Polymerization starts with the addition of TEMED, so do it by mixing carefully and pouring the mixture fast into the gel cast.
4. Fill the inner chamber carefully and completely, to cover the sample wells.
5. Take into account the total volume of each well (around 80 μL in the 1.5 mm thickness gel). If necessary, concentrate the samples using Amicon-type filters (Merck) following the manufacturer's instructions. Always load the same amount of protein if you pretend to compare different samples. Take care of air bubbles and contamination between wells.
6. If the voltage rises above 180 V, run the electrophoresis at a constant voltage of 180 V.
7. Always handle gels with gloves and it is recommended to make a mark (cut) in a corner to identify the sequence of samples.
8. Check that the gel is not adhered to the tray and it is gently shaken. Use around 50 mL for large Petri dishes. Always work in darkness (wrap dishes with aluminum foil).

Acknowledgments

This work was supported by the Spanish grants from MICINN-FEDER (PID2021-127335NB-I00) and Fundación Séneca-Agencia de Ciencia y Tecnología de la Región de Murcia (22051/PI/22).

References

1. MCord JM, Fridovich I (1969) Superoxide dismutase: an enzymic function for erythrocuprein (hemocuprein). J Biol Chem 244:6049–6055
2. Salin ML, Bridges SM (1980) Isolation and characterization of an iron-containing superoxide dismutase from a eucaryote, *Brassica campestris*. Arch Biochem Biophys 201(2): 369–374
3. Sevilla F, Lopez-Gorgé J, Del Río LA (1982) Characterization of a manganese superoxide dismutase from the higher plant *Pisum sativum*. Plant Physiol 70:1321–1326
4. Almansa MS, Palma JM, Yarez J, del Río LA, Sevilla F (1991) Purification of an iron-containing superoxide dismutase from a citrus plant, *Citrus limonum* R. Free Radic Res Commun 12-13:319–328
5. Barkley B, Gregoryt EM (1990) Tetrameric manganese superoxide dismutases from anaerobic actinomyces. Arch Biochem Biophys 280: 192–200
6. Sandalio LM, del Río LA (1987) Localization of superoxide dismutase in glyoxysomes from *Citrullus vulgaris*. Functional implications in cellular metabolism. J Plant Physiol 127:395–409
7. Del Río LA, Sandalio LM, Palma JM, Bueno P, Corpas FJ (1992) Metabolism of oxygen radicals in peroxisomes and cellular implications. Free Rad Biol Med 13:557–580
8. Del Río LA, Sandalio LM, Altomare DA, Zilinskas BA (2003) Mitochondrial and

peroxisomal manganese superoxide dismutase: differential expression during leaf senescence. J Exp Bot 54:923–933
9. Duchêne A, Giegé P (2012) Dual localized mitochondrial and nuclear proteins as gene expression regulators in plants? Front Plant Sci 3:1–6
10. Bueno P, Varela J, Giménez-Gallego G, del Río LA (1995) Peroxisomal copper zinc superoxide dismutase. Characterization of the isoenzyme from watermelon cotyledons. Plant Physiol 108:1151–1160
11. Hernández JA, Ferrer MA, Jiménez A, Barceló AR, Sevilla F (2001) Antioxidant systems and $O_2^{\cdot-}/H_2O_2$ production in the apoplast of pea leaves. Its relation with salt-induced necrotic lesions in minor veins. Plant Physiol 127:817–831
12. Jiménez A, Romojaro F, Llanos MR, Gómez JM, León A, Sevilla F (2003) Antioxidant systems and their relationship with the response of pepper fruits to storage at 20°C. J Agric Food Chem 51:6293–6299
13. Martí MC, Camejo D, Olmos E, Sandalio LM, Fernández-García N, Jiménez A, Sevilla F (2009) Characterisation and changes in the antioxidant system of chloroplasts and chromoplasts isolated from green and mature pepper fruits. Plant Biol 11:613–624
14. Meneguzzo S, Sgherri CLM, Navari-Izzo F, Izzo R (1998) Stromal and thylakoid-bound ascorbate peroxidases in NaCl-treated wheat. Physiol Plant 104:735–740
15. Gómez LD, Noctor G, Knight MRR, Foyer CH (2004) Regulation of calcium signalling and gene expression by glutathione. J Exp Bot 55(404):1851–1859
16. Fridovich I (1995) Superoxide radical and superoxide dismutases. Annu Rev Biochem 64:97–112
17. Ogawa K, Kanematsu S, Asada K (1996) Intra- and extra-cellular localization of "cytosolic" CuZn-superoxide dismutase in spinach leaf and hypocotyl. Plant Cell Physiol 37(6): 790–799
18. Becana M, Paris FJ, Sandalio LM, del Río LA (1989) Lsoenzymes of superoxide dismutase in nodules of *Phaseolus vulgaris* (L.), *Pisum sativum* (L.), and *Vigna unguiculata* (L.) Walp. Plant Physiol 90:1286–1292
19. Droillard MJ, Bureau D, Paulin A, Daussant J (1989) Identification of different classes of superoxide dismutase in carnation petals. Electrophoresis 10:46–48
20. Weisinger RA, Fridovitch I (1973) Superoxide dismutase: organelle specificity. J Biol Chem 248:3582–3592
21. Almansa MS, del Río LA, Alcaraz CF, Sevilla F (1989) Isoenzyme pattern of superoxide dismutase in different varieties of citrus plants. Physiol Plant 76:563. https://doi.org/10.1111/j.1399-3054.1989.tb05479.x

Chapter 15

Comparative Analysis of Catalase Activity in Plants: Spectrophotometry and Native PAGE Approaches

Salvador González-Gordo, Marta Rodríguez-Ruiz, José M. Palma, and Francisco J. Corpas

Abstract

Catalase, a pivotal enzyme in plant antioxidative defense mechanisms, plays a crucial role in detoxifying hydrogen peroxide, a reactive oxygen species (ROS). In this chapter, a comparative analysis of catalase activity was conducted using two distinct methodologies: spectrophotometry and non-denaturing polyacrylamide gel electrophoresis (PAGE). The spectrophotometric approach allowed the quantification of catalase activity by measuring the breakdown rate of hydrogen peroxide, while native PAGE enabled the separation and visualization of catalase isozymes, based on their native molecular weight and charge characteristics, and specific staining assay. Both methods provide valuable insights into catalase activity, offering complementary information on the enzyme's functional diversity and distribution within different plant tissues. This study integrates different techniques, previously described, to comprehensively elucidate the role of catalase in plant metabolism. Furthermore, it provides the possibility of obtaining a holistic understanding of antioxidant defense mechanisms by considering both total activity and isoenzyme distribution of catalase enzyme.

Key words Hydrogen peroxide, Catalase, Native PAGE, Spectrophotometer, Plant antioxidants

1 Introduction

Catalase is an essential enzyme found in plants, animals, and many other living organisms. Exclusively located in peroxisomes as recognized worldwide thus far, it plays a crucial role in cellular metabolism by facilitating the breakdown of hydrogen peroxide (H_2O_2) into water (H_2O) and molecular oxygen (O_2) [1–5]. This seemingly simple reaction is of paramount importance for cellular health and survival due to the toxic nature of H_2O_2 [6, 7].

Hydrogen peroxide is a reactive oxygen species (ROS) generated as a byproduct of various metabolic processes, including

The original version of the chapter has been revised. A correction to this chapter can be found at https://doi.org/10.1007/978-1-0716-3826-2_18

Francisco J. Corpas and José M. Palma (eds.), *ROS Signaling in Plants: Methods and Protocols*,
Methods in Molecular Biology, vol. 2798, https://doi.org/10.1007/978-1-0716-3826-2_15,

respiration and photosynthesis [8, 9]. ROS are highly reactive molecules that can cause damage to cellular components, such as lipids, proteins, and DNA. When H_2O_2 accumulates in cells, it can lead to oxidative stress, resulting in cellular dysfunction and even cell death [10–14].

Through its detoxification role, catalase protects plants from oxidative stress and helps maintain cellular integrity and functionality. Moreover, catalase activity is dynamically regulated in response to environmental cues, enabling plants to adjust their antioxidant defense systems accordingly [15–18].

In addition to its primary function in H_2O_2 detoxification, catalase may also play other roles in plants. Some studies have suggested that catalase could be involved in signal transduction, cell growth, development, and fruit ripening [4, 10, 19–22]. However, the precise mechanisms and significance of these additional functions are still areas of active research.

Non-denaturing (native) PAGE (polyacrylamide gel electrophoresis) is a widely used technique in biochemistry and molecular biology to separate proteins based on their native conformations and charge-to-mass ratios. Unlike SDS-PAGE, which denatures proteins and breaks them down into their individual polypeptide chains, native PAGE preserves the native structure and interactions of proteins, providing valuable information about their native state, oligomeric forms, and activity.

In this chapter, two complementary methodologies are compiled for determining the catalase activity. This allows a comprehensive view of one of the main cellular antioxidant enzymes. This information is crucial for understanding the antioxidative defense mechanisms of plants and their responses to oxidative stress, ultimately contributing to our knowledge of plant biology and stress tolerance.

2 Materials

2.1 Extraction Buffer

1. Prepare enough volume of 50 mM Tris (2-amino-2-hydroxymethyl-propane-1,3-diol)–HCl buffer, pH 7.5 containing 0.1 mM EDTA, 0.1% (v/v) Triton X-100, and 10% (v/v) glycerol. For instance, to prepare 200 mL of this buffer, dissolve 1.21 g of Tris, 7.4 mg of EDTA, 200 μL of Triton X-100, and 20 mL of glycerol in 120–150 mL of ultrapure water. Then, carefully adjust the pH to 7.5 by adding HCl. Set the final volume to 200 mL with ultrapure water. This buffer could be stored for several months at 4 °C.
2. Add dithiothreitol (DTT) to the above solution to a final concentration of 5 mM. For instance, to a final volume of 20 mL, 15.4 mg of DTT should be added. This extraction buffer must be freshly prepared and has a shelf life of 24–48 h.

Table 1
Dilutions for the BSA standard curve

μL BSA (0.6 μg/μL)	μg Protein	μL H_2O	μL Bio-Rad protein assay
0	0	800	200
5	3	795	200
10	6	790	200
15	9	785	200
20	12	780	200
30	18	770	200

2.2 Protein Concentration in Samples

1. Bio-Rad protein assay dye reagent concentrate (Bio-Rad Laboratories, Hercules, CA, USA).
2. Prepare a 0.6 μg/μL bovine serum albumin (BSA) stock solution by dissolving 3 mg of BSA into 5 mL of ultrapure water. This solution could be stored in aliquots (500 μL) for several months at −20 °C.
3. Prepare the BSA (3–18 μg) standard curve (Table 1).
4. UV-visible spectrophotometer.
5. Plastic cuvettes (1 mL).

2.3 Spectrophotometric Assay for Catalase Activity

1. Prepare 50 mM phosphate buffer, pH 7.0. Add the required volume of 50 mM KH_2PO_4 (6.81 g/L) to a 50 mM K_2HPO_4 (8.71 g/L) solution until reaching a pH of 7.0.
2. Hydrogen peroxide solution 30% (w/w).
3. UV-visible spectrophotometer.
4. Quartz cuvettes (1 mL).

2.4 Native PAGE

1. Prepare 4X gel buffer (1.5 M Tris-HCl, pH 8.9) by dissolving 45.42 g of Tris into 800 mL of ultrapure water. Then, adjust pH to 8.9 with HCl. Fill up to 1 L with ultrapure water. This solution could be stored for several months at 4 °C.
2. Commercial 30% acrylamide/bis solution, ratio 19:1 (Bio-Rad Laboratories, Hercules, CA, USA).
3. N,N,N,N′-tetramethyl-ethylenediamine (TEMED).
4. Prepare 10% (w/v) ammonium persulfate solution (APS) by dissolving 0.10 g of ammonium persulfate in 1 mL of ultrapure water. This solution should be freshly prepared.
5. Prepare 10X Native PAGE running buffer (250 mM Tris, 1.92 M glycine, pH 8.3) by dissolving 30.30 g of Tris, and 144.10 g of glycine into 800 mL of ultrapure water. Fill up the

solution to 1 L with ultrapure water (*see* **Note 1**). This buffer can be stored for months at 4 °C. To prepare 1X Native PAGE running buffer, dissolve 100 mL of 10X into 900 mL of ultrapure water.

6. Gel handcast system: Mini-PROTEAN® Tetra Handcast Systems (Bio-Rad Laboratories, Hercules, CA, USA).
7. Electrophoresis cell: Mini-PROTEAN® Tetra Cell (Bio-Rad Laboratories, Hercules, CA, USA).

2.5 Staining Solution

2.5.1 DAB Method

1. Prepare 50 mM phosphate buffer, pH 7.0.
2. Hydrogen peroxide solution 30% (w/w).
3. Peroxidase from horseradish (Sigma-Aldrich, Burlington, Massachusetts, USA).
4. 3,3'-Diaminobenzidine (DAB) tetrahydrochloride hydrate (Sigma-Aldrich, Burlington, Massachusetts, USA).

2.5.2 Ferric Chloride/Potassium Ferricyanide Method

1. Hydrogen peroxide solution 30% (w/w).
2. Ferric chloride ($FeCl_3$).
3. Potassium ferricyanide [$K_3Fe(CN)_6$].

3 Methods

3.1 Plant Extracts Preparation

1. Grind the plant tissue in liquid nitrogen using an IKA® A11 basic analysis mill (*see* **Note 2**).
2. Mix the resulting powder with the appropriate amount of extraction buffer. The optimal plant material/buffer (w/v) ratio depends on the specific type of plant tissue (*see* **Note 3**).
3. Incubate the mixture under gentle stirring for 30 min at 4 °C.
4. Filter plant extracts through two layers of nylon.
5. Transfer crude extracts into 1.5 mL microcentrifuge tube and centrifuge at 27,000 for 30 min at 4 °C.
6. Use the collected supernatants to perform the different assays.

3.2 Protein Quantification

1. Determine the protein concentration using the appropriate sample volume (2–5 μL). Add ultrapure water until a volume of 800 μL is reached.
2. Add 200 μL of Bio-Rad protein assay dye reagent concentrate.
3. Shake gently and incubate for 10 min at room temperature.
4. Measure the absorbance at 595 nm. The measurement of the standard curve should be conducted in a similar manner.
5. Use the standard curve to calculate the protein concentration.

3.3 Spectro-photometric Catalase Assay

Spectrophotometric determination of catalase activity was performed following the method described by Aebi [23]. This method relies on the reduction in absorbance at 240 nm caused by the enzymatic breakdown of hydrogen peroxide.

1. Set the spectrophotometer to perform continuous measurements at 240 nm for 2 min. Calibrate the spectrophotometer using ultrapure water as blank.
2. Using 1 mL quartz cuvettes, add the appropriate volume (900–990 μL) of 50 mM phosphate buffer, pH 7.0 containing 10.6 mM hydrogen peroxide (add 54.1 μL of the 30% hydrogen peroxide solution to 50 mL of phosphate buffer).
3. Add plant extract (10–100 μL) and shake vigorously.
4. Calculate the enzyme activity based on the decrease in absorbance using the molar extinction coefficient of hydrogen peroxide at 240 nm (43.6 $M^{-1} \cdot cm^{-1}$). See an exemplifying plot of catalase activity in different samples obtained from different plant species and organs in Fig. 1.

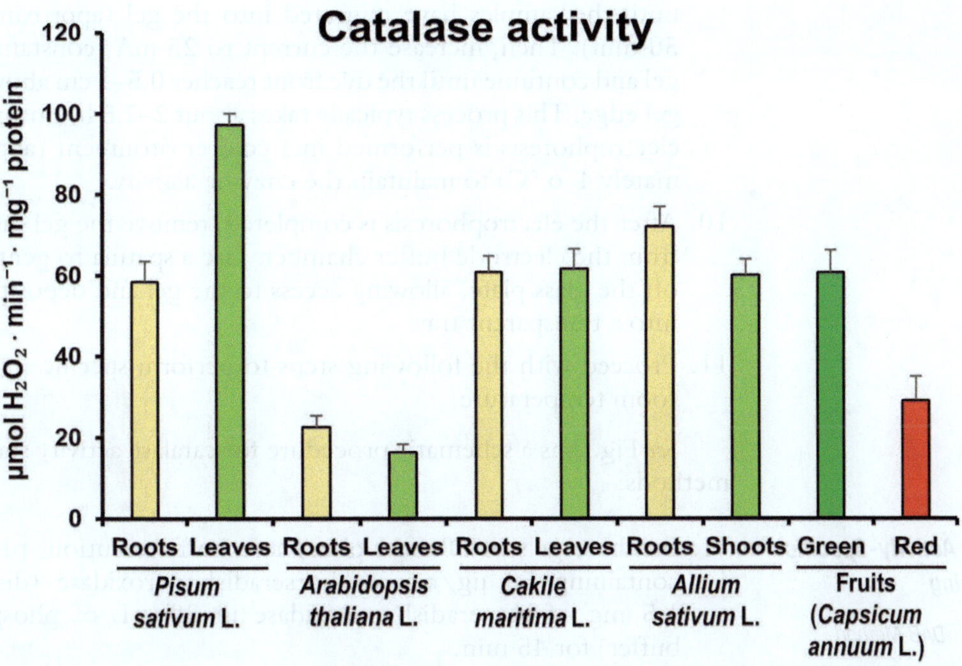

Fig. 1 Spectrophotometric analysis of total catalase activity in the organs of different plant species. For further information about the different plant sampling, *see* [24–28]

3.4 Native PAGE

1. Prepare a 5% acrylamide gel by mixing 1.66 mL of 30% commercial acrylamide solution, 2.5 mL of 4X gel buffer, and 7.45 mL of ultrapure water in a vacuum flask.
2. Degas the solution under a vacuum system for 15 min in an ice bath.
3. Meanwhile, assemble the glass cassette [10 cm (W) × 8 cm (H) × 1.5 mm (thickness)] and cast the gel.
4. After vacuuming the acrylamide solution, add 5 μL of TEMED and 50 μL of 10% APS.
5. Pour the solution into the glass cassette until it fills totally.
6. Place a 10-well gel comb and allow the gel to polymerize for 45–60 min at room temperature.
7. Prepare individual sample aliquots by combining the suitable volume of plant extracts (100 μL) with 2 μL of a 0.025% (w/v) bromophenol blue solution. This compound will be used as the dye front indicator.
8. Load the prepared samples (approximately 20–30 μg of protein) into each well.
9. Run the electrophoresis at a constant current of 15 mA per gel until the samples have migrated into the gel (approximately 30 min). Then, increase the current to 25 mA (constant) per gel and continue until the dye front reaches 0.5–1 cm above the gel edge. This process typically takes about 2–2.5 h. Ensure the electrophoresis is performed in a cold environment (approximately 4–6 °C) to maintain the enzyme activity.
10. After the electrophoresis is completed, remove the gel cassette from the electrode buffer chamber. Use a spatula to gently lift off the glass plate, allowing access to the gel and depositing it into a transparent tray.
11. Proceed with the following steps to perform specific stains at room temperature.

See Fig. 2 as a schematic procedure for catalase activity staining methods.

3.5 Activity-Specific Staining

3.5.1 DAB Method

1. Incubate gels in 50 mM phosphate buffer solution, pH 7.0 containing 50 μg/mL of horseradish peroxidase (dissolve 1.5 mg of horseradish peroxidase in 30 mL of phosphate buffer) for 45 min.
2. Add 15.3 μL of 30% hydrogen peroxide solution (final concentration of 5 mM).
3. Keep incubation for 10 min in darkness.

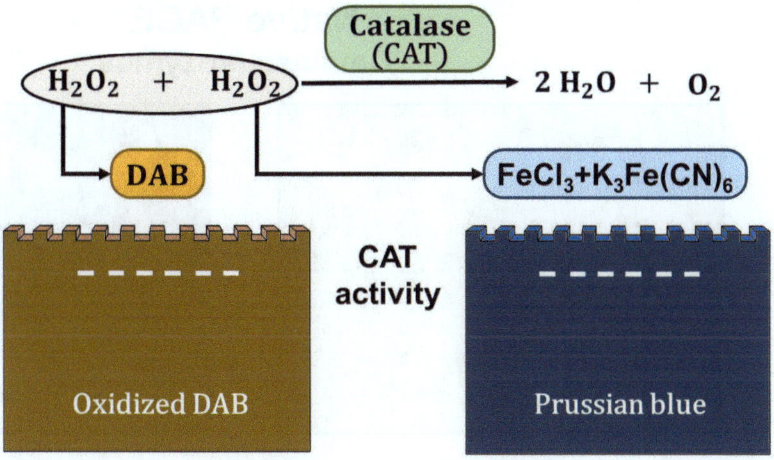

Fig. 2 Schematic representation of the catalase-catalyzed reaction. During the specific detection in gels, H_2O_2 reacts with different reagents, either DAB or $FeCl_3$ and $K_3Fe(CN)_6$ generating brown or Prussian blue background. Thus, the decomposition of H_2O_2 by catalase activity prevents the formation of these compounds, thus rendering colorless bands

4. Immediately incubate the gel in 50 mM phosphate buffer, pH 7.0, containing 0.5 mg/mL of DAB (dissolve 15 mg of DAB in 30 mL of phosphate buffer).
5. Wait until the appearance of achromatic bands on a dark brown background.
6. Once the staining process is finished, the gel can be stored in ultrapure water.

3.5.2 Ferric Chloride/ Potassium Ferricyanide Method

1. Rinse the gel in ultrapure water for 5 min.
2. Incubate the gel in a 4 mM hydrogen peroxide solution (13.5 μL of 30% hydrogen peroxide solution in 30 mL of ultrapure water).
3. Soak the gel in ultrapure water for 5 min.
4. Incubate the gel in a solution containing ferric chloride 1% (w/v) and potassium ferrocyanide 1% (w/v). To prepare this solution dissolve 0.3 g of $FeCl_3$ and 0.3 g of $K_3Fe(CN)_6$ in 30 mL of ultrapure water.
5. Wait till the appearance of achromatic bands on a Prussian blue background. Avoid excessive staining.
6. Once the staining process is finished, the gel can be stored in ultrapure water.

Representative images of in-gel catalase activity staining are depicted in Fig. 3.

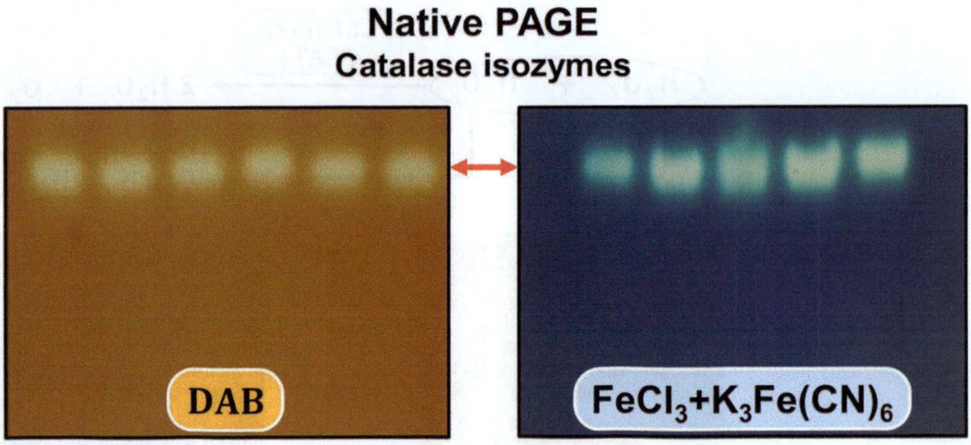

Fig. 3 Isoenzymatic pattern of catalase in sweet pepper fruits at different ripening stages. After non-denaturing PAGE, the gels were stained with DAB (left) and Ferric chloride/Potassium ferricyanide (right). Protein of 20 μg was loaded per sample. The red arrow indicates the specific bands of catalase activity. (The DAB gel image was reproduced with permission from [4])

4 Notes

1. No pH adjustment of this buffer is necessary.
2. Any instrument that allows obtaining a fine powder, such as a mortar and pestle, could be used.
3. The ratio of plant material to extraction buffer utilized can differ depending on the plant species and type of plant tissue. For instance, we used a ratio of 1:1 for pepper fruits, 1:3 for Arabidopsis leaves, and 1:4 for pea organs.

Acknowledgments

The research of the authors Francisco J. Corpas and José M. Palma is supported by the European Regional Development Fund with cofinanced grants from the Spanish Ministry of Science and Innovation (PID2019-103924GB-I00) and Junta de Andalucía (P18-FR-1359), Spain. Salvador González-Gordo also acknowledges a contract associated with the grant PID2019-103924GB-I00. Marta Rodríguez-Ruiz acknowledges a postdoctoral contract (MCIN/AEI/10.13039/501100011033) associated with the grant CPP2021-008703.

References

1. Baker A, Lin C-C, Lett C et al (2023) Catalase: a critical node in the regulation of cell fate. Free Radic Biol Med 199:56–66
2. Mhamdi A, Noctor G, Baker A (2012) Plant catalases: peroxisomal redox guardians. Arch Biochem Biophys 525:181–194

3. Palma JM, Mateos RM, López-Jaramillo J et al (2020) Plant catalases as NO and H_2S targets. Redox Biol 34:101525
4. Rodríguez-Ruiz M, González-Gordo S, Cañas A et al (2019) Sweet pepper (*Capsicum annuum* L.) fruits contain an atypical Peroxisomal catalase that is modulated by reactive oxygen and nitrogen species. Antioxidants (Basel) 8:374
5. Scibior D, Czeczot H (2006) Catalase: structure, properties, functions. Postepy Hig Med Dosw (Online) 60:170–180
6. Marzo ND, Chisci E, Giovannoni R (2018) The role of hydrogen peroxide in redox-dependent signaling: homeostatic and pathological responses in mammalian cells 7
7. Mukherjee S and Corpas FJ (2023) H_2O_2 , NO, and H_2S networks during root development and signalling under physiological and challenging environments: Beneficial or toxic? Plant Cell Environ 46:688–717
8. Slesak I, Libik M, Karpinska B et al (2007) The role of hydrogen peroxide in regulation of plant metabolism and cellular signalling in response to environmental stresses. Acta Biochim Pol 54:39–50
9. Quan L-J, Zhang B, Shi W-W et al (2008) Hydrogen peroxide in plants: a versatile molecule of the reactive oxygen species network. J Integr Plant Biol 50:2–18
10. Schieber M, Chandel NS (2014) ROS Function in Redox Signaling and Oxidative Stress 24:R453
11. Hasanuzzaman M, Bhuyan MHMB, Parvin K et al (2020) Regulation of ROS metabolism in plants under environmental stress: a review of recent experimental evidence. Int J Mol Sci 21
12. Nadarajah KK (2020) ROS homeostasis in abiotic stress tolerance in plants. Int J Mol Sci 21
13. Petrov V, Hille J, Mueller-Roeber B et al (2015) ROS-mediated abiotic stress-induced programmed cell death in plants. Front Plant Sci 6:69
14. Hasanuzzaman M, Bhuyan MHMB, Zulfiqar F et al (2020) Reactive oxygen species and antioxidant Defense in plants under abiotic stress: revisiting the crucial role of a universal defense regulator. Antioxidants (Basel) 9:681
15. Huang H, Ullah F, Zhou D-X et al (2019) Mechanisms of ROS regulation of plant development and stress responses. Front Plant Sci 10:800
16. Considine MJ, Foyer CH (2021) Stress effects on the reactive oxygen species-dependent regulation of plant growth and development. J Exp Bot 72:5795–5806
17. You J, Chan Z (2015) ROS regulation during abiotic stress responses in crop plants. Front Plant Sci 6:1092
18. Choudhury S, Panda P, Sahoo L et al (2013) Reactive oxygen species signaling in plants under abiotic stress. Plant Signal Behav 8: e23681
19. Su T, Wang P, Li H et al (2018) The Arabidopsis catalase triple mutant reveals important roles of catalases and peroxisome-derived signaling in plant development. J Integr Plant Biol 60: 591–607
20. Fichman Y, Rowland L, Oliver MJ et al (2023) ROS are evolutionary conserved cell-to-cell stress signals. Proc Natl Acad Sci USA 120: e2305496120
21. Van Breusegem F, Vranová E, Dat JF et al (2001) The role of active oxygen species in plant signal transduction. Plant Sci 161:405–414
22. González-Gordo S, Palma JM, Corpas FJ (2022) Peroxisomal proteome mining of sweet pepper (*Capsicum annuum* L.) fruit ripening through whole isobaric tags for relative and absolute quantitation analysis. Front Plant Sci 13:893376
23. Aebi H (1984) Catalase in vitro. Methods Enzymol 105:121–126
24. Rodríguez-Ruiz M, Aparicio-Chacón MV, Palma JM et al (2019) Arsenate disrupts ion balance, sulfur and nitric oxide metabolisms in roots and leaves of pea (*Pisum sativum* L.) plants. Environ Exp Bot 161:143–156
25. Corpas FJ, Aguayo-Trinidad S, Ogawa T et al (2016) Activation of NADPH-recycling systems in leaves and roots of *Arabidopsis thaliana* under arsenic-induced stress conditions is accelerated by knock-out of Nudix hydrolase 19 (*AtNUDX19*) gene. J Plant Physiol 192: 81–89
26. Houmani H, Debez A, Freitas-Silva L de, et al (2022) Potassium (K^+) starvation-induced oxidative stress triggers a general boost of antioxidant and NADPH-generating systems in the halophyte *Cakile maritima*. Antioxidants (Basel) 11:401
27. Ruíz-Torres C, Feriche-Linares R, Rodríguez-Ruíz M et al (2017) Arsenic-induced stress activates sulfur metabolism in different organs of garlic (*Allium sativum* L.) plants accompanied by a general decline of the NADPH-generating systems in roots. J Plant Physiol 211:27–35
28. Palma JM, Terán F, Contreras-Ruiz A et al (2020) Antioxidant profile of pepper (*Capsicum annuum* L.) fruits containing diverse levels of Capsaicinoids. Antioxidants (Basel) 9: 878

Chapter 16

Detection of Ascorbate Peroxidase (APX) Activity in Plant Tissues: Using Non-denaturing PAGE and Spectrophotometric Assay

Marta Rodríguez-Ruiz, Hayet Houmani, María A. Muñoz-Vargas, José M. Palma, and Francisco J. Corpas

Abstract

At the cellular level, the generation of reactive oxygen species (ROS), such as hydrogen peroxide (H_2O_2), due to different abiotic or biotic stress, causes oxidative stress that induces an imbalance in the metabolism. Among the different H_2O_2-scavenging enzymatic antioxidants, ascorbate peroxidase (APX) is a heme-peroxidase that plays an important role in the ascorbate-glutathione pathway using ascorbate to reduce H_2O_2 to water. Using non-denaturing polyacrylamide gel electrophoresis (PAGE) in combination with a spectrophotometric assay for APX activity, the protocol allows identifying diverse APX isozymes present in different organs and plant species.

Key words Ascorbate Peroxidase, Enzyme activity assay, Non-denaturing polyacrylamide gel electrophoresis, Hydrogen peroxide, Isozymes, Ascorbate

1 Introduction

H_2O_2 is an essential nonradical reactive oxygen species (ROS) produced in different cellular compartments. It is a key component of the signaling pathway's network, involved in plant biological/physiological processes, and cellular responses to environmental factors, at low concentrations. However, at high concentrations, this ROS could cause damage to plants, and in these conditions should be controlled and scavenged by an antioxidative defense system [1, 2]. However, there are cases in which a high production of H_2O_2 is used as a local defense mechanism against pathogens. The amount of cell H_2O_2 could be a good indicator of oxidative

The original version of the chapter has been revised. A correction to this chapter can be found at https://doi.org/10.1007/978-1-0716-3826-2_18

Francisco J. Corpas and José M. Palma (eds.), *ROS Signaling in Plants: Methods and Protocols*,
Methods in Molecular Biology, vol. 2798, https://doi.org/10.1007/978-1-0716-3826-2_16,

Corrected Publication 2024

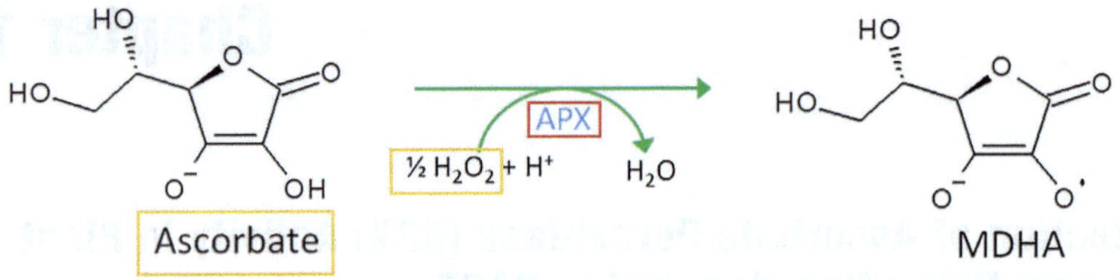

Fig. 1 *Ascorbate peroxidase reaction.* Hydrogen peroxide (H_2O_2) is removed by ascorbate peroxidase (APX) by converting to water in the presence of ascorbate. (Figure modified from [3])

stress whenever it is associated with oxidative damage to cellular components such as oxidation of lipids or proteins. Consequently, the main enzymatic H_2O_2 scavenging mechanism in plants, such as ascorbate peroxidase (APX), catalase (CAT), peroxiredoxins (PRX), or glutathione peroxidase (GPX), is crucial. In green plants, H_2O_2 removal is catalyzed by a specialized family of heme-containing ascorbate peroxidases (APXs) [3].

Ascorbate peroxidase (APX; EC 1.11.1.11) is a member of Class I of heme-peroxidases enzyme that is part of the plant ascorbate-glutathione (AsA-GSH) pathway and requires an ascorbate and glutathione (GSH) regeneration system. It is the first reaction of this cycle (Fig. 1) [1], and it appears that a fundamental difference between mammals and plants is an additional high-capacity H_2O_2 ascorbate-based scavenging system to complement catalase [3]. This enzyme reacts with H_2O_2, and this reaction is slow under physiological conditions; APX uses two molecules of ascorbate (AsA) as the specific electron donor to reduce H_2O_2 to water with a concomitant generation of two molecules of monodehydroascorbate (MDHA) (Fig. 1). Different from other hemeproteins, the heme group of the APX is non-covalently bound to the enzyme but is strictly necessary for their activity [4].

APX is encoded by a multigene family generating diverse isozymes whose number and expression differ depending on plant species, as well as according to their tissue/organ distribution, subcellular compartmentalization, stage of growth, and development, and environmental condition [5–7]. In plants, APX isozymes are found in distinct subcellular compartments, including cytosol, mitochondria, peroxisomes, and chloroplasts [7, 8]. Around eight *APX* genes occur in *Arabidopsis thaliana*, and the proteins encoded by these genes are located in different subcellular compartments (two in the chloroplast, three in the cytosolic, three microsomal isoforms, one thylakoidal membrane, other at the stroma of chloroplast, and the latter has also been found in the mitochondrion) [9–11]. In tomato (*Solanum lycopersicum* L.), there are seven *APX* genes encoded by three in the cytosol, two in peroxisome, and two in chloroplast [12]. In pepper fruit (*Capsicum annuum* L.), six APX isozymes were identified (CaAPX1-CaAPX6) (two peroxisomal, two plastidial/mitochondrial, and two cytosolic) [2].

Even though the relevance of these hemeproteins has been associated with photo-oxidative stresses or high-temperature stress, some scientists reported enhanced activity of APX in response to abiotic stresses such as drought, salinity, chilling, metal toxicity, and UV irradiation [4, 8]. Furthermore, this enzyme is one of the potential targets of post-translational modifications (PTMs) such as tyrosine nitration, *S*-nitrosation, and persulfidation. Diverse works showed that APX is activated by *S*-nitrosation and its activity is also affected by nitration and carbonylation [3, 13].

Considering the relevance of APX as a component in the metabolism of ROS, this chapter describes an update of two analytical methods to identify the presence of APX isozymes and spectrophotometric detection of total APX activity in different plant tissues from diverse plant species such as pepper (*Capsicum annuum* L.) fruit, or the halophyte *Cakile maritima*, using non-denaturing polyacrylamide gels and measurement the change in absorbance of specific wavelength of light.

2 Materials

All solutions should be prepared with ultrapure water and analytical-grade reagents.

2.1 Plant Material and Extraction Buffer

1. Powder the plant sample under liquid N_2 using a mortar with a pestle.
2. Prepare a solution of 0.1 M Tris–HCl, pH 7.8, 0.1 mM EDTA, Triton-X 100 0.1% (v/v), and 10% glycerol (v/v). This solution must be stored at 4 °C.
3. Add just before the extraction, **2 mM ascorbate** to avoid the inactivation of the enzyme (*see* **Note 1**).

2.2 Non-denaturing Polyacrylamide Solution

1. Resolving gel buffer: 1.5 M Tris–HCl, pH 8.9. Store at 4 °C. Weigh 18.7 g of Tris-Base and transfer to a 100 mL glass beaker. Dissolve it in about 80 mL of ultrapure water and adjust pH with HCl to 8.9. Finally, add the solution into a graduated cylinder and adjust it to 100 mL.
2. Ammonium persulfate (APS) 7% (w/v) in resolving gel buffer; this solution could be stored at −20 °C (*see* **Note 2**). Dissolve 0.07 g into 1 mL of resolving gel buffer.
3. Non-denaturing PAGE performed at 10% acrylamide with 1.5 mm of thickness: 3.37 mL of resolving gel buffer, 4.5 mL of acrylamide native 30% (acrylamide bis solution 19:1) from Bio-Rad, 1.35 mL of glycerol 10% (v/v), 725 μL of 7% APS (w/v), 100 μL of N,N,N,N′-tetramethyl-ethylenediamine (TEMED), and 3.57 mL ultrapure water. Since TEMED and

APS catalyze the polymerization of acrylamide, add them after 10 min of vacuum and only before use in the flow chamber.

4. Bromophenol blue (BPB) solution: Dissolve 0.1 g BPB in 100 mL water.
5. Cathode (running buffer): 38 mM glycine, 5 mM Tris–HCl, pH 8.2, and **2 mM ascorbate**. Weigh 0.18 g of Tris-Base, 0.86 g of glycine, and 0.35 g of ascorbate and transfer to a 500 mL graduated cylinder or glass beaker. Dissolve into about 200 mL ultrapure water and adjust pH with HCl. Fill up to 300 mL with water. Store at 4 °C.
6. Anode (electrode buffer): 38 mM glycine, and 5 mM Tris–HCl, pH 8.2. Weight 1.21 g of Tris-Base and 5.7 g of glycine and transfer to 1 L glass beaker. Dissolve it in about 1.5 L of ultrapure water and adjust pH with HCl. Fill up to 2 L and store at 4 °C.

2.3 APX Activity Staining Solution

1. 50 mM potassium phosphate (T-K) buffer:

   Basic solution: Weigh 8.7 g of dipotassium hydrogen phosphate (K_2HPO_4) and dissolve in 1 L of ultrapure water.

   Acid solution: Weigh 3.4 g of potassium dihydrogen phosphate (KH_2PO_4) and dissolve in 0.5 mL of ultrapure water.

   Prepare one solution to pH 7.0 by mixing both the basic and the acid solutions, and another to pH 7.8.
2. Equilibration solution: 50 mM potassium phosphate pH 7.0 and **2 mM ascorbate**. Dissolve 0.021 g of ascorbate in 60 mL of potassium buffer.
3. Solution of H_20_2: Dilute 12 μL of H_2O_2 30% (commercial) into 10 mL of ultrapure water (**SL1**). From this solution (**SL1**) take 10 μL and dilute in 10 mL of ultrapure water (**SL2**), and finally pipette 940 μL of this solution **(SL2)** and add it into the incubation solution.
4. Incubation solution: 50 mM potassium phosphate pH 70. 4 mM ascorbate, and 0.5 μM of H_2O_2. Dissolve 0.014 g of ascorbate in 20 mL of T-K (pH 7.0) and add 940 μL of **SL2** from the solution of H_2O_2.
5. Staining solution: 50 mM potassium phosphate pH 7.8, 12 mM TEMED, and 2 mM nitroblue tetrazolium chloride (NBT). Dissolve 0.033 g of NBT in 20 mL T-K (pH 7.8) and add 35.82 μL of TEMED.

2.4 Spectrophotometric Solution

1. Reaction buffer: 100 mM HEPES-NaOH, pH 7.6. Weigh 0.238 g of HEPES, transfer to a glass beaker, dissolve and adjust the pH with NaOH to 7.6.
2. **2 mM Ascorbate**: Dissolve 1 mg of ascorbate in 3 mL of ultrapure water (*see* **Note 1**).

3. Hydrogen peroxide (H_2O_2) 3 mM: add 3.07 μL of H_2O_2 30% in 10 mL of ultrapure water in a falcon tube and protect it from the light.
4. Quartz cuvette.

2.5 Determination of Protein Concentration by Bradford Assay

1. Protein assay dye reagent concentrate from Bio-Rad.
2. Bovine serum albumin (BSA) protein standard.
3. Spectrophotometer.
4. Cuvettes for spectrophotometer.

3 Methods

3.1 Protein Extraction

1. Plant fresh material is frozen in liquid nitrogen and ground to a fine powder (*see* **Note 3**).
2. Then 500 mg of ground tissue is suspended in 0.5 mL of extraction buffer in a 1.5 mL microtube, with **2 mM ascorbate**, added just before use (*see* **Note 4**), vortexed, and keep the sample on ice.
3. Centrifuge at 27,000 *g* for 27 min at 4 °C. The supernatant is recovered carefully to avoid contact with the pellet generated after centrifugation.
4. Samples must be kept below 4 °C.

3.2 Determination of Protein Concentration by Bradford Assay

Bradford assay is based on an observed shift in the absorbance maximum when Coomassie Brilliant Blue G-250 dye reacts with proteins. This is achieved by measuring at 595 nm.

1. Prepare five concentrations of bovine serum albumin (BSA) as a standard curve with a range from 3 to 18 μg.
2. Transfer the correct volume of each standard into a 1.5 mL microtube and add (deionized) ddH_2O to a final volume of 800 μL.
3. Add 200 μL of Bradford reagent and mix well.
4. Incubate the sample and standards for 5–10 min at RT and transfer it into cuvettes for spectrophotometer.
5. Measure the absorbance at 595 nm. Collect the data for at least three independent replicates.
6. Estimate the concentration of protein using the equation of the standard curve on the data.

3.3 Native Polyacrylamide Gel Electrophoresis

1. Clean the short plate and spacer plate with ddH_2O, and 70% (v/v) ethanol. Transfer the glass plates into the casting cassettes and these into the casting stand.

2. Pipette 4.5 mL of acrylamide native 30% (acrylamide Bis solution 19:1) solution, 3.37 mL of resolving gel buffer, 1.35 mL of 10% (v/v) glycerol, and 3.56 mL ddwater in a 50 mL Büchner vacuum flask. Make vacuum with a water pump for 10 min. Add 0.750 mL of 7% (w/v) ammonium persulfate and 100 μL of TEMED (*see* **Note 5**), shake softly and pour into the gap between the plates, and insert 10-well gel combs of 1.5 mm.
3. Once the gel is set, take the glass plate from the casting cassette and place it into the running module (without combs) inside the Mini-PROTEAN 3 multi-casting chamber.
4. Pour the cathode buffer (with **2 mM ascorbate**) into the running module until it covers the wells, and pour the anode buffer outside until the outer chamber reaches the required levels.
5. Gels (10% acrylamide) are pre-run for 30 min at 120 V to allow the ascorbate, present in the cathode buffer, to enter the gel before loading the samples.
6. Prepare each sample with the appropriate volume of extract (about 30 μg) and add 1 μL 0.025% (w/v) bromophenol blue (BPB).
7. Load the samples into the wells (*see* **Note 6**).
8. Cover the lid and set it at 15 mA until the samples have entered the gel and then continue at 25 mA per gel to 4 °C, till the dye front (from the BPB dye in the samples) reaches the bottom of the gel.

3.4 Detection of APX Activity in Gel

The steps described below should be performed at room temperature (RT, 20–25 °C) [14].

1. Following electrophoresis, take off the gel from the glass plates using a spatula and transfer it to a Petri dish.
2. Equilibrate the gel with 20 mL of 50 mM potassium phosphate buffer pH 7.0, and **2 mM ascorbate** for 30 min. Change the equilibration buffer every 10 min (*see* **Note 7**).
3. Between each change, wash the gel with ultrapure water.
4. Incubate the gel with 20 mL of 50 mM potassium phosphate pH 7.0, **4 mM ascorbate**, and 940 μL of 0.5 μM H_2O_2 (*see* **Note 8**).
5. Wash the gel for 1–2 min with 50 mM potassium phosphate pH 7.8.
6. Incubate the gel in 20 mL of 50 mM potassium phosphate pH 7.8, 2 mM NBT (nitroblue tetrazolium), and 12 mM of TEMED with gentle shaking (*see* **Note 9**).
7. Expose to light and see the achromatic bands on a purple-blue background (*see* **Note 10**).

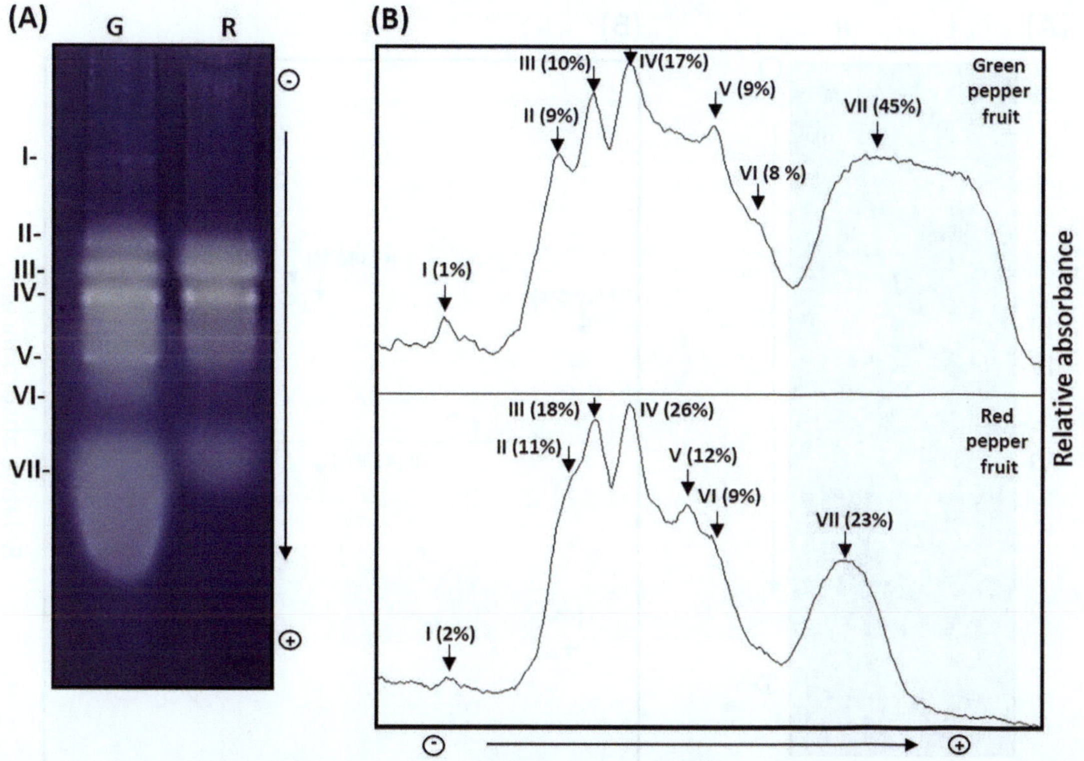

Fig. 2 *Detection and quantification of ascorbate peroxidase (APX) in pepper fruit at different stages of ripening.* (**a**) APX isozymes. G, Green pepper fruit and R, red pepper fruit (*Capsicum annuum* L.). Protein samples (30 μg per lane) were separated by non-denaturing polyacrylamide gel electrophoresis and activity was detected by the nitro blue tetrazolium (NBT)-reducing method. APX isozymes were labeled I–VII (on the left) according to their increasing electrophoretic mobility. (**b**) The band intensity of each APX isozyme in the different fruit samples was quantified using Image J 1.45 software and the percentage of each band is shown in parentheses

8. Stop the reaction by rinsing briefly with deionized water (Figs. 2 and 3).

3.5 Detection of APX Spectrophotometrically

The steps described below should be performed with a UV spectrophotometer at 25 °C, measuring at 290 nm the oxidation of ascorbic acid [15].

1. Use a quartz cuvette to measure the activity.
2. This assay requires different controls before measuring the samples (*see* **Note 11**):
   I. **Blank**: Add 500 μL of 100 mM HEPES-NaOH, pH 7.6, and 500 μL of ultrapure water.
   II. **Sample control**: Add the appropriate volume of sample, 500 μL of 100 mM HEPES-NaOH, pH 7.6, and complete to a final volume of 1 mL with ultrapure water.

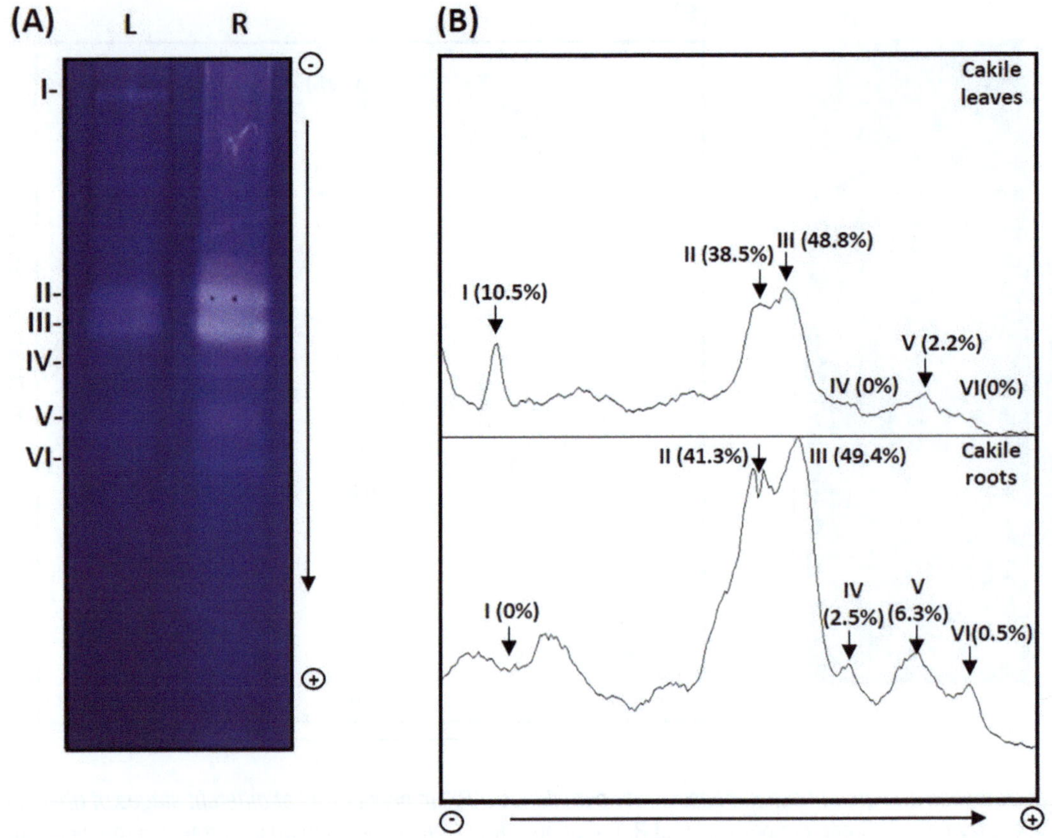

Fig. 3 *Detection and quantification of ascorbate peroxidase (APX) in Cakile maritima plant organs.* (**a**) APX isozymes. L, leaves and R, roots (*C. maritima*). Protein samples (40 μg per lane) were separated by non-denaturing polyacrylamide gel electrophoresis and activity was detected by the nitro blue tetrazolium (NBT)-reducing method. APX isozymes were labeled I-VI (on the left) according to their increasing electrophoretic mobility. (**b**) The band intensity of each APX isozyme in the different fruit samples was quantified using Image J 1.45 software and the percentage of each band is shown in parentheses

III. **AsA control**: Add the appropriate volume of sample, 500 μL of 100 mM HEPES-NaOH, 100 μL of **2 mM** AsA, and complete to 1 mL.

IV. **H_2O_2 control**: Add the appropriate volume of sample, 500 μL of 100 mM HEPES-NaOH, pH 7.6, 100 μL of 0.3 mM H_2O_2, and complete to 1 mL.

3. Samples: Add to the cuvette the reaction mixture in a final volume of 1 mL, in the following order (*see* **Note 12**):

   I. Add 500 μL of 100 mM HEPES-NaOH, pH 7.6 (*see* **Note 13**).

   II. Add the appropriate volume of **sample** (*see* **Note 14**).

   III. Add 100 μL of **2 mM ascorbate** (*see* **Note 15**).

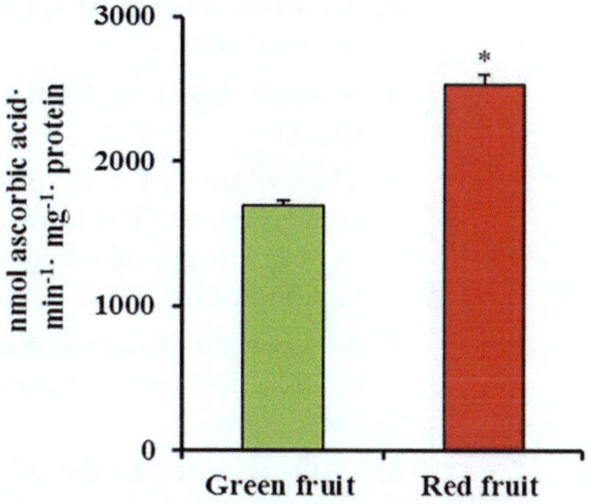

Fig. 4 *Ascorbate peroxidase (APX) activity in sweet pepper fruits.* Spectrophotometric assay of APX activity from green and red fruits. Values are means of at least three independent experiments. *Differences were significant (Student's *t*-test)

IV. Finally, add 100 μL of 0.3 mM H_2O_2 to initiate the reaction.

V. Complete to 1 mL with ultrapure water if necessary.

4. Measure for 1 min at 25 °C.
5. Estimate the oxidation of AsA using an absorbance extinction coefficient (ε) of 2.8 m $M^{-1} \cdot cm^{-1}$. Ascorbate peroxidase activity is expressed as nmol of oxidized AsA $min^{-1} \cdot mg^{-1}$ protein (Fig. 4). ΔDO/min: sample absorbance per min; ε_{290}: absorbance extinction coefficient of ascorbate at 290 nm; V: final volume of the reaction; v: sample volume; P: quantification of sample protein.

$$\text{nmol AsA} : \frac{\Delta\text{DO}/\min}{\varepsilon} \frac{V(\text{mL})}{v(\text{mL})P(\text{mg protein/mL})}.$$

4 Notes

1. Ascorbate is a physiological antioxidant and light-sensitive, used in fresh and just before protein extraction, electrophoresis procedure, and spectrophotometric assay.
2. It is recommended that this solution be prepared fresh each time.
3. The power should be homogenous and fine to ensure an efficient protein extraction.

4. Ascorbate should be applied just before use in all the steps of the method.
5. A fume hood is recommended to add the TEMED and the APS.
6. The volume of each plant extract depends on the protein concentration. The same amount of protein should be loaded to perform a good comparison among samples and experimental conditions.
7. As ascorbate is light-sensitive, all the steps during the method of detection must be performed by covering the Petri dish to avoid light.
8. H_2O_2 is added to the solution immediately before incubation with the gel.
9. Observe the gel frequently during agitation to see the achromatic bands.
10. Longer incubation with TEMED/NBT solution could result in a darker purple-blue background that, in turn, leads to a clouding of the achromatic activity bands.
11. Scheme of controls (Table 1).
12. Scheme of sample measure (Table 2).
13. The final concentration of HEPES buffer in the reaction mixture should be 50 mM.
14. The selected volume for each sample depends on the type of sample and should be adjusted for each condition.
15. The final concentration of ascorbate in the reaction mixture should be 0.3 mM.

Table 1
Scheme of control measurement for spectrophotometric APX assay

Reagents	Blank	Sample control	AsA control	H_2O_2 control
50 mM HEPES-NaOH, pH 7.6	500 µL	500 µL	500 µL	500 µL
Sample	–	15–150 µL	–	–
0.2 mM ascorbate	–	–	100 µL	–
0.3 mM H_2O_2	–	–	–	100 µL
H_2O ultrapure water	500 µL	X	400 µL	400 µL
Final volume	1 mL	1 mL	1 mL	1 mL

Table 2
Scheme of sample measure for spectrophotometric APX assay

Reagents	Volume
50 mM HEPES-NaOH, pH 7.6	500 μL
Sample	15–150 μL
0.2 mM ascorbate	100 μL
0.3 mM H_2O_2	100 μL
Ultrapure water	500 μL
Final volume	1 mL

Acknowledgments

The research is supported by the European Regional Development Fund which cofinanced grants from the Ministry of Science and Innovation (PID2019-103924GB-I00 and CPP2021-008703), Spain. Marta Rodríguez-Ruiz acknowledges a postdoctoral contract (MCIN/AEI/10.13039/501100011033) associated with the grant CPP2021-008703.

References

1. Sofo A, Scopa A, Nuzzaci M, Vitti A (2015) Ascorbate peroxidase and catalase activities and their genetic regulation in plants subjected to drought and salinity stresses. Int J Mol Sci 16: 13561–13578. https://doi.org/10.3390/ijms160613561
2. González-Gordo S, Rodríguez-Ruiz M, López-Jaramillo J et al (2022) Nitric oxide (NO) differentially modulates the ascorbate peroxidase (APX) isozymes of sweet pepper (*Capsicum annuum* L.) fruits. Antioxidants 11:765. https://doi.org/10.3390/antiox11040765
3. Smirnoff N (2018) Ascorbic acid metabolism and functions: a comparison of plants and mammals. Free Radic Biol Med 122:116–129. https://doi.org/10.1016/j.freeradbiomed.2018.03.033
4. Palma JM, Gupta DK, Corpas FJ (2013) Metalloenzymes involved in the metabolism of reactive oxygen species and heavy metal stress. In: Gupta DK, Corpas FJ, Palma JM (eds) Heavy metal stress in plants. Springer, Berlin/Heidelberg, pp 1–17. https://doi.org/10.1007/978-3-642-38469-1_1
5. Asada K (1992) Ascorbate peroxidase–a hydrogen peroxide-scavenging enzyme in plants. Physiol Plant 85:235–241. https://doi.org/10.1111/j.1399-3054.1992.tb04728.x
6. Zhang J, Jia X, Wang G-F et al (2022) Ascorbate peroxidase 1 confers resistance to southern corn leaf blight in maize. J Integr Plant Biol 64:1196–1211. https://doi.org/10.1111/jipb.13254
7. Shigeoka S, Ishikawa T, Tamoi M et al (2002) Regulation and function of ascorbate peroxidase isoenzymes. J Exp Bot 53:1305–1319. https://doi.org/10.1093/jexbot/53.372.1305
8. Anjum NA, Sharma P, Gill SS et al (2016) Catalase and ascorbate peroxidase-representative H_2O_2-detoxifying heme enzymes in plants. Environ Sci Pollut Res Int 23:19002–19029. https://doi.org/10.1007/s11356-016-7309-6
9. Li ZQ, Li JT, Bing J, Zhang GF (2019) The role analysis of APX gene family in the growth and developmental processes and in response to abiotic stresses in *Arabidopsis thaliana*. Yi Chuan 41:534–547. https://doi.org/10.16288/j.yczz.19-026

10. Panchuk II, Zentgraf U, Volkov RA (2005) Expression of the *APX* gene family during leaf senescence of *Arabidopsis thaliana*. Planta 222:926–932. https://doi.org/10.1007/s00425-005-0028-8
11. Santos M, Gousseau H, Lister C et al (1996) Cytosolic ascorbate peroxidase from *Arabidopsis thaliana* L. is encoded by a small multigene family. Planta 198:64–69. https://doi.org/10.1007/BF00197587
12. Najami N, Janda T, Barriah W et al (2008) Ascorbate peroxidase gene family in tomato: its identification and characterization. Mol Gen Genomics 279:171–182. https://doi.org/10.1007/s00438-007-0305-2
13. Begara-Morales JC, Sánchez-Calvo B, Chaki M et al (2014) Dual regulation of cytosolic ascorbate peroxidase (APX) by tyrosine nitration and S-nitrosylation. J Exp Bot 65:527–538. https://doi.org/10.1093/jxb/ert396
14. Bieker S, Riester L, Stahl M et al (2012) Senescence-specific alteration of hydrogen peroxide levels in *Arabidopsis thaliana* and oilseed rape spring variety *Brassica napus* L. cv. Mozart. J Integr Plant Biol 54:540–554. https://doi.org/10.1111/j.1744-7909.2012.01147.x
15. Hossain MA, Asada K (1984) Inactivation of ascorbate peroxidase in spinach chloroplasts on dark addition of hydrogen peroxide: its protection by ascorbate. Plant Cell Physiol 25:1285–1295. https://doi.org/10.1093/oxfordjournals.pcp.a076837

Chapter 17

Inactivation of Three Stacked Genes of Cytosolic Peroxiredoxins by Genome Editing

Lara Vogelsang and Karl-Josef Dietz

Abstract

A set of peroxidases detoxifies H_2O_2 and mediates H_2O_2-dependent signal propagation. The peroxidases include peroxiredoxins, glutathione peroxidases, ascorbate peroxidases, and catalases. This at least partial redundancy impedes addressing individual proteins in living plant cells so that the protein functions are often studied by biochemical assays in vitro. In vivo analysis frequently relies on transgenic insertion lines resulting in the knockdown or knockout of the protein of interest. However, many proteins have multiple isoforms in close genomic arrangement so that even crossing of transgenic lines does not allow for a knockdown of all isoforms. The genes encoding for the three cytosolic peroxiredoxins PRXIIB, C, and D in *Arabidopsis thaliana* are located in close vicinity on chromosome 1 so that crossing over between the genes most rarely occurs and successful crossing of the plants appears impossible. Genome editing instead allows targeting of multiple isoforms and knocks out several genes at once. This chapter describes how to inactivate the three cytosolic peroxiredoxins by CRISPR/Cas9 in *A. thaliana*.

Key words Peroxiredoxins, CRISPR/Cas9, Genome editing, Isogenes, *Arabidopsis thaliana*

1 Introduction

Peroxidases decompose and sense reactive molecular species (RMS) such as reactive oxygen species (ROS). In particular, H_2O_2 serves as signaling molecule under normal growth conditions, but with increased significance in response to abiotic and biotic stress. Peroxiredoxins (PRX) and glutathione peroxidases (GPXs) belong to the group of thiol peroxidases and contribute to H_2O_2 signaling as redox sensors in the cytosol [24]. Both families are present with multiple cytosolic isoforms in the cytosol like GPX2 and GPX8 or four PRX isoforms, i.e., PRXIIA-D, where PRXIIA presumably is a pseudogene. Their biochemical properties can be addressed by enzyme assays with recombinant or purified proteins, but the physiological relevance for the plant cannot be fully assessed in such assays since they lack the multiple interactions and competing side reactions present in vivo.

Francisco J. Corpas and José M. Palma (eds.), *ROS Signaling in Plants: Methods and Protocols*, Methods in Molecular Biology, vol. 2798, https://doi.org/10.1007/978-1-0716-3826-2_17,

On the other hand, one can use gene-encoded life cell imaging probes that allow for assessing, e.g., NADPH/NADP$^+$ ratios and H_2O_2 amounts in specific compartments of living cells [15, 19]. However, changes observed in these central parameters cannot be assigned to a single class of thiol peroxidases, since they form a complicated redox network together with thioredoxins (TRXs), glutaredoxins (GRXs), and other proteins.

Transgenic lines allow for the analysis of protein functions in living plants, but collections of transgenic lines are available for only a few plants. But even if they exist as in the case of *A. thaliana*, they certainly do not cover the entire genome [16]. In addition, the copy number of genes encoding the cytosolic PRXIIA-D and their position in proximity on chromosome 1 in *A. thaliana* (AT1G65990, AT1G65980; AT1G65970, AT1G60740) devalue the typical benefit of the insertion line collections for *A. thaliana* genes. In this case, the selection of a complete knockout plant based on insertion collections appears challenging or even impossible.

Genome editing by CRISPR/Cas9 (clustered regularly interspaced short palindromic repeats/CRISPR-associated protein 9) offers a way out of this dilemma by simultaneously editing more than one gene. Cas9 is the CRISPR-associated protein 9 from *Streptococcus pyogenes*, which belongs to class II of the Cas-proteins [7, 12]. In case of high identity of the genomic sequences, a single sgRNA can already affect all genes of closely related proteins, and a second one can result in large deletions within all genetic copies of those proteins. The expression of two or more sgRNAs, so-called multiplexing, also allows for the knockout of unrelated proteins such as GPX and PRX in one transgenic line.

The protocol, presented in this chapter, is based on the plasmids and procedures published by Hahn et al. [8] and Schumacher et al. [21]. In both plasmids, Cas9 is under control of the constitutive *A. thaliana* ubiquitin promoter. The central difference between the plasmids of these publications is the codon usage of Cas9, which is optimized for *Chlamydomonas reinhardtii* [8] or *A. thaliana* [21], the latter carries the additional intron IV2 from potato to avoid expression in *Escherichia coli* and *Agrobacterium tumefaciens*. Both Cas9 vectors finally carry the sgRNA scaffold and the sgRNA under control of the promoter of the U6 small nucleolar RNA 26 gene (U6-26 snRNA), which are provided by preparing the vectors pFH6_new and AtU6-26-V4, respectively. The fluorescent protein DsRed, under the control of a seed-specific promoter, was introduced into the Cas9-constructs as a plant selection marker.

2 Materials

2.1 Identification of Targets and sgRNA Design

1. The Arabidopsis Information Resource (TAIR, www.arabidopsis.org).
2. Nucleotide BLAST (https://blast.ncbi.nlm.nih.gov/Blast.cgi).
3. Clustal Omega (https://www.ebi.ac.uk/Tools/msa/clustalo/).
4. sgRNAs can either be designed manually or by online tools (*see* Table 1).

2.2 Cloning of sgRNA Cassettes and Multiplexing

1. pfH6_new [8].
2. AtU6-26-V4 [21].
3. Restriction enzymes Bbs1-HF, Kpn1-HF, Spe1-HF, and Xba1.
4. T4 DNA ligase.
5. *E. coli* strain TOP10 (*see* **Note 1**).
6. CCMB80: 10 mM potassium acetate, 80 mM $CaCl_2$, 20 mM $MnCl_2$, 10 mM $MgCl_2$, and 10% (v/v) glycerol; adjust to pH 7.0.
7. LB media: 0.5% yeast extract, 1% tryptone, 1% NaCl.
8. LB-Agar plates: 1.5% agar in LB with 100 μg/mL ampicillin (stock: 100 mg/mL).
9. Phusion polymerase (*see* **Note 2**).
10. dNTP.
11. Oligonucleotides (*see* Table 2).
12. Agarose.
13. 50-fold TAE: 50 mM EDTA disodium salt, 2 M Tris, 1 M glacial/acetic acid.
14. Midori Green Xtra.
15. 6× loading buffer: 30% (v/v) glycerol, 0.25% (w/v) bromophenol blue, 0.25% (w/v) xylene cyanol FF.
16. 1 kb ladder (*see* **Note 3**).
17. Gel extraction kit.
18. Miniprep kit.

2.3 Optimizing the Cas9 Vectors

1. pHUGE-RedSeed [23].
2. pUB-Cas9 [8].
3. UBQ10::pcoCas9p1300 [21].
4. pET28a (Novagen, *see* **Note 4**).
5. Phusion polymerase (*see* **Note 2**).

Table 1
Tools for sgRNA design and off-target search

Tool	URL and reference
CHOPCHOP	https://chopchop.cbu.uib.no/ [10]
E-CRISP	http://www.e-crisp.org/E-CRISP/designcrispr.html [9]
CRISPick	https://portals.broadinstitute.org/gppx/crispick/public [20]
CRISPR-ERA	http://crispr-era.stanford.edu/index.jsp [11]
CRISPRdirect	http://crispr.dbcls.jp/ [14]
CRISPRscan	https://www.crisprscan.org/ [13]
SSC—Sequence scan for CRISPR	http://cistrome.org/SSC/ [26]
sgRNA scorer 2.0	https://frederick.cancer.gov/resources/repositories/sgrnascorer [2]
CRISPOR	http://crispor.tefor.net/ [3]
CRISPR MultiTargeter	http://www.multicrispr.net/index.html [17]
Cas-OFFinder	http://www.rgenome.net/cas-offinder/ [1]
CRISPR-offinder	http://www.biootools.com/software.html [25]

6. dNTP.
7. Restriction enzymes NdeI and HindIII-HF.
8. Antarctic phosphatase.
9. T4 DNA ligase.
10. Isothermal reaction buffer: 500 mM Tris–HCl, pH 7.5, 50 mM DTT, 50 mM $MgCl_2$, 1 mM of each dNTP, 5 mM nicotinamide adenine dinucleotide (NAD), 25% PEG-8000.
11. Gibson Mastermix: 320 μL isothermal reaction buffer, 0.64 μL T5 exonuclease, 20 μL phusion polymerase (*see* **Note 5**), 160 μL Taq DNA ligase, 699.36 μL H_2O. Store aliquots of 100 μL at −20 °C.
12. *E. coli* strain TOP10 (*see* **Note 1**).

Table 2
Primer sequences

Primer	Sequence
H_M13F[1]	GTAAAACGACGGCCAG
H_M13R[1]	CAGGAAACAGCTATGAC
H_FH41[1]	AAACGACGGCCAGTGCCAGAATTGGGCCCGACGTCG
H_FH42[1]	TACTGACTCGTCGGGTACCAAGCTATGCATCCAACGCG
H_FH254[1]	GCCCAATTCCAAGCTATGCATCCAACGCG
H_FH255[1]	CATAGCTTGGAATTGGGCCCGACGTCG
H_FH179[1]	TATTACTGACTCGTCGGGTA
H_FH61[1]	CTGCCCAACGAGAAGGTGC
H_FH201[1]	AGTCGGACAGGCGGTTGATG
H_sg1-F	ATTGCAGAGGAGCTGAAGTCAAA
H_sg1-R	AAACTTTGACTTCAGCTCCTCTG
H_sg2-F	ATTGAGAATGACTTTTTTACCGG
H_sg2-R	AAACCCGGTAAAAAAGTCATTCT
S_M13F[2]	TGTAAAACGACGGCCAGT
S_M13R[2]	CAGGAAACAGCTATGACC
S_K197[2]	CTGTTAATCAGAAAAACTCAG
S_FH61	GAACCTTCCAAACGAGAAGGTTCTTCC
S_FH201	CAGAAAGTCTGTTGATGTCAAGCTCCTG
S_sg1-F	GATTGCAGAGGAGCTGAAGTCAAA
S_sg1-R	AAACTTTGACTTCAGCTCCTCTGC
S_sg2-F	GATTGAGAATGACTTTTTTACCGG
S_sg2-R	AAACCCGGTAAAAAAGTCATTCTC
GA1_F	GTATTGGCTAGAGCAGCTTGCCATTCGGAGCTCTCTTCATCGGTG
GA2_R	GAATTAACGCCGAATTAATTCGGGGGAGGCCCGATCTAGTAACATAGATG
GA3_F	CACCGATGAAGAGAGCTCCGAATGGCAAGCTGCTCTAGCCAATAC
GA4_R	GTGTTAGTTTCTAGTTTGTGCGATCGAATTTGTAGATTAATCTGAG
GA5_F	CTCAGATTAATCTACAAATTCGATCGCACAAACTAGAAACTAACAC
GA6_R	CATCTATGTTACTAGATCGGGCCTCCCCCGAATTAATTCGGCGTTAATTC
GAseq1	GCTCAATAGCGTCAATCCTTAC
GAseq2	CCACCATCTGTTCCTTATGTAC
GAseq3	CTTCGATTTTGTTTTCGTTCG
rS_mut_F	CACAATACCGTAAAGCTGAAGGTAACCAAG
rS_mut_R	CTTGGTTACCTTCAGCTTTACGGTATTGTG
rS_NdeI_F	AAAACATATGCTCGAATTCGATTCGGAGCTCTCTTC
rS_HindIII_R	TTTTAAGCTTCAGGAGGCCCGATCTAGTAACATAG
rS_seq_F	CAAGTATTTTCAAACCGTTCG
rS_seq_R	GCTTCAGAGCCTTATGAATC
T7F	TAATACGACTCACTATAGGG
T7R	TAGTTATTGCTCAGCGGTGG
PRXIIB_F	ACACAGACCTTACCTACCACCT
PRXIIB_R	TCGATCCAAATGACGGCGTA
PRXIIC_F	TCATATGGGCGGGCTAGGAT
PRXIIC_R	GCGTGTATTCTCCAGACCCA
PRXIID_F	TCATGTCTCTTGCCAACTTCCA
PRXIID_R	ACCAAGACCCTTGTCCTTGAG

Primer sequences starting with H are specific for pfH6_new/pUB-Cas9 and those starting with S for AtU6–26-V4/UBQ10::pcoCas9p1300. [1]Hann et al. [8] and [2]Schumacher et al. [21]

13. CCMB80: 10 mM potassium acetate, 80 mM $CaCl_2$, 20 mM $MnCl_2$, 10 mM $MgCl_2$, 10% (v/v) glycerol; adjust to pH 7.0.
14. LB media: 0.5% yeast extract, 1% tryptone, 1% NaCl.
15. LB agar plates: 1.5% agar in LB with 50 μg/mL kanamycin (stock: 50 mg/mL).
16. Oligonucleotides (*see* Table 2).
17. Agarose.
18. 50-fold TAE: 50 mM EDTA disodium salt, 2 M Tris, 1 M glacial/acetic acid.
19. Midori Green Xtra.
20. 6× loading buffer: 30% (v/v) glycerol, 0.25% (w/v) bromophenol blue, 0.25% (w/v) xylene cyanol FF.
21. 1 kb ladder (*see* **Note 3**).
22. Gel extraction kit.
23. Miniprep kit.
24. Primers for sequencing (*see* Table 2).

2.4 Preparing the Cas9 Vectors

1. pUB-Cas9 (redSeed).
2. pfH6_new-sgRNA.
3. UBQ10::pcoCas9p1300 (redSeed).
4. AtU6-26-V4-sgRNA.
5. Restriction enzymes KpnI-HF, HindIII-HF, and SbfI-HF.
6. Antarctic phosphatase.
7. T4 DNA ligase.
8. Isothermal reaction buffer: 500 mM Tris–HCl, pH 7.5, 50 mM DTT, 50 mM $MgCl_2$, 1 mM of each dNTP, 5 mM nicotinamide adenine dinucleotide (NAD), 25% PEG-8000.
9. Gibson master mix: 320 μL isothermal reaction buffer, 0.64 μL T5 exonuclease, 20 μL phusion polymerase (*see* **Note 5**), 160 μL Taq DNA ligase, 699.36 H_2O.
10. *E. coli* strain TOP10 (*see* **Note 1**).
11. CCMB80: 10 mM potassium acetate, 80 mM $CaCl_2$, 20 mM $MnCl_2$, 10 mM $MgCl_2$, 10% (v/v) glycerol; adjust to pH 7.0.
12. LB media: 0.5% yeast extract, 1% tryptone, 1% NaCl.
13. LB agar plates: 1.5% agar in LB with 50 μg/mL kanamycin (stock: 50 mg/mL).
14. Phusion polymerase (*see* **Note 2**).
15. dNTP.
16. Oligonucleotides (*see* Table 2).
17. Miniprep kit.

18. Primers for colony PCR (*see* Table 2).
19. Agarose.
20. 50-fold TAE: 50 mM EDTA disodium salt, 2 M Tris, 1 M glacial/acetic acid.
21. Midori Green Xtra.
22. 6× loading buffer: 30% (v/v) glycerol, 0.25% (w/v) bromophenol blue, 0.25% (w/v) xylene cyanol FF.
23. 1 kb ladder (*see* **Note 3**).
24. Gel extraction kit.

2.5 Transformation of *A. thaliana*

1. pUB-Cas9-sgRNA (redSeed).
2. UBQ10::pcoCas9p1300-sgRNA (redSeed).
3. *Arabidopsis thaliana* seeds (Col-0), *see* **Note 6**.
4. *Agrobacterium tumefaciens* strain GV3101::pMP90.
5. YEP media: 1% yeast extract, 1% tryptone, 0.5% NaCl.
6. YEP agar plates with 100 μg/mL rifampicin (stock: 100 mg/mL), 25 μg/mL gentamycin (stock: 25 mg/mL), and 50 μg/mL kanamycin (stock: 50 mg/mL): 1% yeast extract, 1% peptone, 0.5% NaCl, 2% agar.
7. 20 mM $CaCl_2$, ice cold.
8. Phusion polymerase (*see* **Note 2**).
9. dNTP.
10. Oligonucleotides (*see* Table 2).
11. Agarose.
12. 50-fold TAE: 50 mM EDTA disodium salt, 2 M Tris, 1 M glacial/acetic acid.
13. Midori Green Xtra.
14. 6× loading buffer: 30% (v/v) glycerol, 0.25% (w/v) bromophenol blue, 0.25% (w/v) xylene cyanol FF.
15. 1 kb marker (*see* **Note 3**).
16. Transformation media: 5% (w/v) sucrose, 0.02% Silwet L-77 (*see* **Note 7**).

2.6 Screening for Positive Plants

1. Fluorescence microscope or fluorescence stereo microscope with mCherry filter set. In this work, a Zeiss Axioskop 2 was applied with an EC-Plan-NEOFLUAR 5× objective and the Zeiss filter set 31 (BP 565/30 Exc., FT 585 beamsplitter and BP 620/60 Em.). Images were obtained with a Zeiss ICc1 camera and Zeiss AxioVision 4.7.
2. Edwards Buffer: 200 mM Tris-HCl, pH 7.5, 250 mM NaCl, 25 mM EDTA, 0.5% (w/v) SDS.

3. Isopropanol.
4. Phusion polymerase (*see* **Note 2**).
5. dNTP.
6. Oligonucleotides (*see* Table 2).
7. Agarose.
8. 50-fold TAE: 50 mM EDTA disodium salt, 2 M Tris, 1 M glacial/acetic acid.
9. Midori Green Xtra.
10. 6× loading buffer: 30% (v/v) glycerol, 0.25% (w/v) bromophenol blue, 0.25% (w/v) xylene cyanol FF.
11. 1 kb ladder (*see* **Note 3**).

2.7 Confirmation of Positive Plants

1. Kit for extracting high molecular weight DNA (HMW DNA).
2. Oxford Nanopore Sequencing Technologies; for more information, *see* Pugh [18].

3 Methods

3.1 Identification of Targets and sgRNA Design

sgRNA can be designed manually or with online tools, many of the latter are specific for defined species or kingdoms. It should be kept in mind that the sgRNA depends on the nuclease CRISPR-associated protein (Cas); in this protocol, we used Cas9. For the identification of targets, a nucleotide BLAST should be performed against the genome of the plant species to test for similar genes which either have to be silenced, too, or represent unwanted off-targets. In case of targeting multiple genes with a single sgRNA, the sgRNA design goes along with a nucleic acid alignment. *See* **Notes 8–11** and Fig. 1.

1. Retrieve the genomic sequences from the Arabidopsis Information Resource (TAIR, www.arabidopsis.org).
2. Perform a nucleotide BLAST against the genome of your plant species to identify related genes of high similarity and putative off-targets. (https://blast.ncbi.nlm.nih.gov/Blast.cgi).
3. Perform an alignment of related genes by Clustal Omega in case of manual sgRNA design (https://www.ebi.ac.uk/Tools/msa/clustalo/).
4. Automated design of sgRNA can be done using multiple online tools (*see* Table 1). The procedure for CHOPCHOP [10] is described as follows.
5. Submit the AGI of your gene and choose *Arabidopsis thaliana* as the organism as well as Cas9 as the enzyme. State "for knock-out." Default parameters are a PAM-3′ of NGG, off-targets

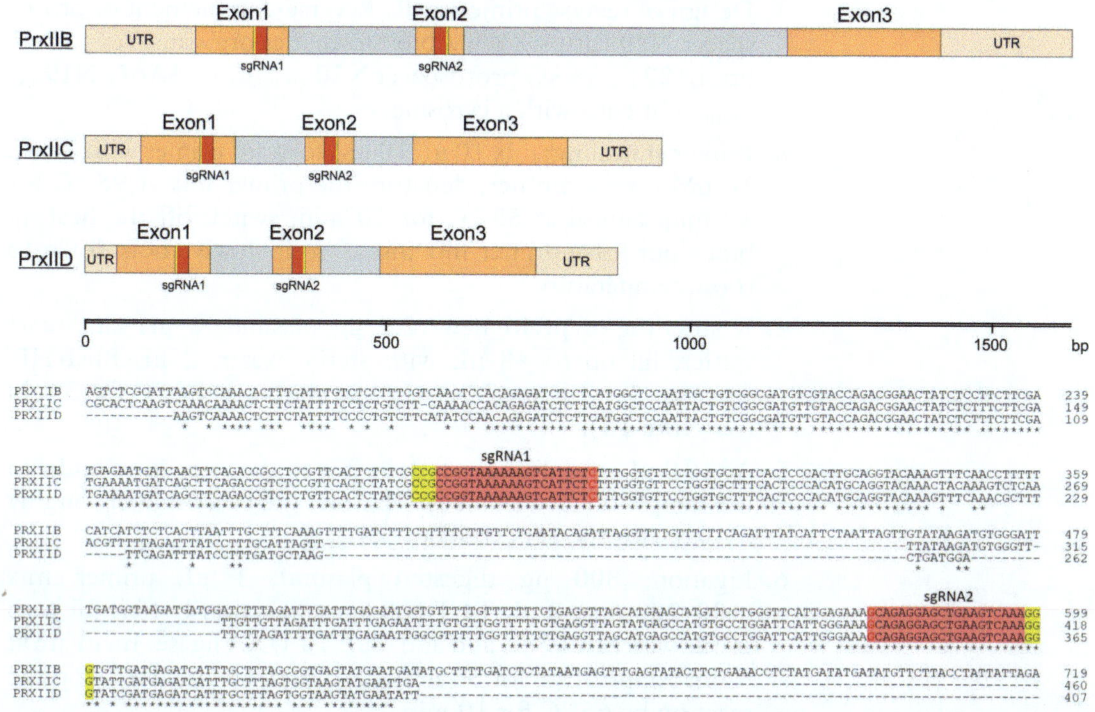

Fig. 1 Alignment of PRXIIB, C, and D genomic DNA and localization of sgRNAs (red) and PAM (yellow). The sgRNA1 resides in Exon 1 and sgRNA2 in Exon 2. Both sgRNAs affect all three PRXIIB, C, and D

with up to 3 mismatches in the protospacer, efficiency scores according to Doench et al. [5], an mESC repair profile production, no 5′-requirements for sgRNA, and check for self-complementarity and complementarity against the standard backbone.

6. For manual sgRNA design, screen the exons for “NGG” motifs on both strands. The sgRNA consists of a protospacer-adjacent motif (PAM) (usually NGG) and a gene-specific protospacer of 20 base pairs (N_{20}NGG) as target site, optimally starting with a G for transcription by the U6–26 promoter (*see* **Note 12**).
7. Identify off-targets to avoid knocking out genes that share the target site using online tools like Cas-OFFinder (*see* Table 1 and **Note 13**).

3.2 Cloning of sgRNA Cassettes

Two vector sets were applied and compared, one relying on conventional cloning [21], and the other on Gibson assembly in the final cloning step [8]. However, the cloning of sgRNA is similar for both vector sets (*see* **Note 14**).

3.2.1 Cloning of sgRNA into pfH6_new According to Hahn et al. [8]

1. Design of forward-primer sg-F: ATTG-protospacer N20 or ATTG-N19, if starting with a guanine.

2. Design of reverse primer sg-R: Reverse-complement of protospacer N20 (https://www.bioinformatics.org/sms/rev_comp.html, [22]); AAAC-protospacer $N20_{rev\text{-}compl}$ or $AAAC\text{-}N19_{rev\text{-}compl}$, if it ends with a cytosine.
3. Primer annealing: mix 10 μL 10 μM forward primer with 10 μL 10 μM reverse primer, denature the primer mix at 98 °C for 10 min, anneal at 55 °C for 10 min, switch off the heating block but leave primer mix inside, so it slowly cools down to room temperature.
4. Restriction of pfH6_new: 2.5 μg plasmid, 5 μL CutSmart buffer, fill up to 48 μL with sterile water, 2 μL BbsI-HF. Incubate for 2 h at 37 °C, then inactivate at 80 °C for 5 min (*see* **Note 15**).
5. Add 10 μL of sixfold loading buffer to the restriction and run the sample on an agarose gel. Extract the backbone of roughly 3.6 kb with the gel extraction kit.
6. Ligation: 300 ng digested plasmid, 1 μL primer mix (100–200 ng), 2 μL T4 DNA ligase reaction buffer; fill with sterile water to 19 μL and add 1 μL T4 DNA ligase. Incubate at room temperature for 30 min (*see* **Notes 15** and **16**), stop the reaction by 65 °C for 10 min.
7. Transformation into *E. coli* Top10: Add 3 μL of ligation mix to competent cells, incubate for 30 min on ice. Perform a 30 s heat shock at 42 °C, keep the cells on ice for 2 min. Incubate the cells for 1 h at 37 °C upon shaking. Spread 75 μL of transformed cells on agar plates with 100 μg/mL ampicillin (*see* **Note 17**). Keep the plates at 37 °C overnight.
8. Positive *E. coli* colonies are identified by colony PCR. Transfer single colonies to 50 μL sterile water (*see* **Note 18**).
9. The PCR master mix contains 9.1 μL sterile water, 4 μL phusion HF buffer, 1 μL forward primer (sg-F), 1 μL reverse primer (M13R), 0.4 μL dNTPs, and 0.5 μL phusion polymerase per reaction. Transfer 16 μL of the master mix to a 0.2 mL PCR tube and add 4 μL of resuspended colony. For primer sequences, *see* Table 2.
10. Run a PCR with 36 cycles: Initial denaturation at 98 °C for 10 min, the cycles comprise three steps: 10 s at 98 °C, 25 s at 55 °C (*see* **Note 19**), and 20 s at 72 °C. Final elongation was performed at 72 °C for 5 min.
11. Add 4 μL of loading buffer to the PCR products and run them on an agarose gel.
12. Identify positive clones with strong PCR products of 281 bp (false positive PCR products with empty vector as template show low intensity) and transfer the corresponding cells in

sterile water (*see* **step 8**) to 20 mL LB supplemented with 20 μL ampicillin stock (final concentration 100 μg/mL).

13. Incubate the culture in a shaking incubator at 120 rpm and 37 °C overnight.
14. Perform a miniprep using the kit.
15. Send plasmids for sequencing with the M13R primer.

3.2.2 Cloning of sgRNA into AtU6-26-V4 According to Schumacher et al. [21]

1. Design of forward-primer sg-F: GATTG-protospacer N20. If the first nucleotide of the protospacer is different from a guanine, replace it with a guanine.
2. Design of reverse primer sg-R: reverse-complement of protospacer N20 (https://www.bioinformatics.org/sms/rev_comp.html, [22]); AAAC-protospacer $N20_{rev\text{-}compl}$. If the last nucleotide of the protospacer is different from a cytosine, replace it according to **step 1**.
3. Primer annealing: Mix 10 μL 10 μM forward primer with 10 μL 10 μM reverse primer, denature the primer mix at 98° for 3 min, anneal by decreasing temperature from 98 °C to 22 °C, ramp is 30 s/1 °C.
4. Restriction of AtU6-26-V4: 2.5 μg plasmid, 5 μL CutSmart buffer, fill up to 48 μL with sterile water, 2 μL BbsI-HF. Incubate at 37 °C for 2 h, then inactivate at 80 °C for 5 min (*see* **Note 15**).
5. Add 10 μL of sixfold loading buffer to the restriction and run the sample on an agarose gel. Extract the backbone of roughly 3.5 kb with the gel extraction kit.
6. Ligation: 50 ng digested plasmid, 1 μL T4 DNA ligase reaction buffer, primer mix in a molar ratio of 7:1 for primer mix and plasmid (2 μL 1:100 diluted primer mix, 2–4 ng), fill with sterile water to 9.5 μL, add 0.5 μL T4 DNA ligase. Incubate at room temperature for 30 min (*see* **Notes 15** and **16**), stop the reaction by 65 °C for 10 min.
7. Transformation into *E. coli* Top10: Add 3 μL of ligation to competent cells, incubate for 30 min on ice. Perform a 30 s heat shock at 42 °C, keep the cells on ice for 2 min. Incubate the cells at 37 °C for 1 h with shaking. Spread 75 μL of transformed cells on agar plates with 100 μg/mL ampicillin (*see* **Note 17**). Keep the plates at 37 °C overnight.
8. Positive *E. coli* colonies are identified by colony PCR. Transfer single colonies to 50 μL sterile water (*see* **Note 18**).
9. The PCR master mix contains 9.1 μL sterile water, 4 μL phusion HF buffer, 1 μL forward primer (sg-R), 1 μL reverse primer (M13F), 0.4 μL dNTPs, and 0.5 μL phusion polymerase per reaction. Transfer 16 μL of the master mix to a 0.2 mL

PCR tube and add 4 µL of resuspended colony. For primer sequences, *see* Table 2.

10. Run a PCR with 36 cycles: Initial denaturation at 98 °C for 10 min, the cycles comprise three steps: for 10 s at 98 °C, 25 s at 55 °C (*see* **Note 19**), and 20 s at 72 °C. Final elongation occurs at 72 °C for 5 min.
11. Add 4 µL of loading buffer to the PCR products and run them on an agarose gel.
12. Identify positive clones with strong PCR products of 581 bp (false positive PCR products with empty vector as template show low intensity) and transfer the corresponding cells in sterile water (*see* **step 8**) to 20 mL LB supplemented with 20 µL ampicillin stock (final concentration 100 µg/mL).
13. Incubate the culture in a shaking incubator at 120 rpm and 37 °C overnight.
14. Perform a miniprep using the kit.
15. Send plasmids for sequencing with the M13R primer.

3.3 Multiplexing

More than one sgRNA can be used to induce large deletions in the target genome or to destroy two or more different genes at once (*see* **Note 20**).

3.3.1 Multiplexing According to Hahn et al. [8]

1. Both sgRNAs must be cloned separately into pFH6_new as described above.
2. The sgRNA cassettes are amplified to be integrated into pUB-Cas9 in a single step.

3.3.2 Multiplexing According to Schumacher et al. [21]

Two sgRNAs can be integrated into one vector starting from two independent AtU6-26-V4 vectors carrying either sgRNA1 or sgRNA2. These steps can be repeated to add more sgRNAs.

1. Digest AtU6-26-V4:sgRNA1 with KpnI and SpeI to open the vector: 2 µg plasmid, 5 µL CutSmart buffer, fill with sterile water to 47 µL, and add 1.5 µL KpnI-HF and 1.5 µL SpeI. Incubate at 37 °C for 2 h. Dephosphorylate the open plasmid: add 6 µL Antarctic phosphatase buffer and 3 µL Antarctic phosphatase, incubate at 37 °C for 40 min, stop the reaction at 80 °C for 2 min (*see* **Note 15**). Add 12 µL of sixfold loading buffer to the sample.
2. Cut the sgRNA2 out of AtU6–26-V4:sgRNA2 with KpnI and XbaI: 2 µg plasmid, 5 µL CutSmart buffer, fill with sterile water to 47 µL, and add 1.5 µL KpnI-HF and 1.5 µL XbaI. Incubate at 37 °C for 2 h, and stop the reaction at 80 °C for 2 min (*see* **Note 15**). Add 10 µL of sixfold loading buffer to the sample.

3. Run the reactions on an agarose gel.
4. Extract the backbone of roughly 3.5 kb (from **step 1**) and the insert of 643 bp (from **step 2**) with the gel extraction kit.
5. Ligation, *see* **Note 21**: 50 ng digested plasmid, 1 μL T4 DNA ligase reaction buffer, insert in a molar ratio of 7:1 (64.6 ng) and 3:1 (27.7 ng) for insert and plasmid, fill with sterile water to 9.5 μL, 0.5 μL T4 DNA ligase. Incubate at room temperature for 40 min (*see* **Notes 15** and **16**), stop the reaction by 65 ° C for 10 min.
6. Transformation into *E. coli* Top10: add 3 μL of ligation to competent cells, incubate for 30 min on ice. Perform a 30 s heat shock at 42 °C, keep the cells on ice for 2 min. Incubate the cells at 37 °C for 1 h with shaking. Spread 75 μL of transformed cells on agar plates with 100 μg/mL ampicillin (*see* **Note 17**). Keep the plates at 37 °C overnight.
7. Positive *E. coli* colonies are identified by colony PCR. Transfer single colonies to 50 μL sterile water (*see* **Note 18**).
8. The PCR master mix contains 9.1 μL sterile water, 4 μL phusion HF buffer, 1 μL forward primer (sg-R of last added sgRNA), 1 μL reverse primer (M13F), 0.4 μL dNTPs, and 0.5 μL phusion polymerase per reaction. Transfer 16 μL of the master mix to a 0.2 mL PCR tube and add 4 μL of resuspended colony. For primer sequences, *see* Table 2.
9. Run a PCR with 36 cycles: Initial denaturation at 98 °C for 10 min, the cycles comprise three steps: 10 s at 98 °C, 25 s at 56 °C (*see* **Note 19**), and 40 s at 72 °C. Final elongation occurs at 72 °C for 5 min.
10. Add 4 μL of loading buffer to the PCR products and separate them on an agarose gel.
11. Identify positive clones with PCR products of 581 bp (for two added sgRNAs) and transfer the corresponding cells in sterile water (*see* **step** 7) to 20 mL LB supplemented with 20 μL ampicillin stock (final concentration 100 μg/mL).
12. Incubate the culture in a shaking incubator at 120 rpm and 37 °C overnight.
13. Perform a miniprep using the kit.
14. Send plasmids for sequencing with the M13R primer.

3.4 Optimizing and Modifying Cas9 Vectors

An efficient knockout of genes might result in lethality so that screening by antibiotics is less promising to identify positive seeds. In addition, screening by antibiotics is time consuming. To optimize the screening, the hygromycin resistance was replaced by a cassette for expression of DsRed under control of the seed-specific NAP-promoter, controlling the expression of seed storage protein

Napin (NAP) in *Brassica napus*, and tNOS (Origin: pHUGE-RedSeed). Furthermore, DsRed as reporter allows for checking, if the T-DNA has been successfully removed from the genome of the finally generated mutant. This is one example for an improvement of the plasmids and the procedure is shown exemplarily.

Modify the NAP-DsRed cassette to remove an unwanted HindIII restriction site that interferes with cloning into the final vector pUB-Cas9 (rS):

1. Perform PCRs A and B with overlapping primers: (A) 32.5 μL sterile water, 10 μL phusion HF buffer, 2.5 μL forward primer (Rs_NdeI_F), 2.5 μL reverse primer (Rs_mut_R), 1 μL dNTPs, 0.5 μL 100 ng/μL pHUGE-RedSeed, 1 μL phusion DNA polymerase. (B) 32.5 μL sterile water, 10 μL phusion HF buffer, 2.5 μL forward primer (rS_mut_F), 2.5 μL reverse primer (rS_HindIII_R), 1 μL dNTPs, 0.5 μL 100 ng/μL pHUGE-RedSeed, 1 μL phusion DNA polymerase. Run the PCR with 36 cycles: Initial denaturation at 98 °C for 30 s, the cycles comprise the 10 s at 98 °C, 20 s at 65 °C, and 60 s at 72 °C. Final elongation was performed at 72 °C for 5 min. For primer sequences, *see* Table 2.
2. Add 10 μL of loading buffer to the sample.
3. Run the reactions on agarose gel.
4. Extract the PCR products of 1319 bp (A) and 851 bp (B) with the gel extraction kit.
5. Adjust the concentration of the extracted PCR products to 100 ng/μL.
6. Run an overlap-extension PCR: 32 μL sterile water, 10 μL phusion HF buffer, 2.5 μL forward primer (rS_NdeI_F), 2.5 μL reverse primer (rS_HindIII_R), 1 μL dNTPs, 0.5 μL PCR product A, 0.5 μL PCR product B, 1 μL phusion DNA polymerase. For 36 cycles: Initial denaturation at 98 °C for 30 s, the cycles comprise the 10 s at 98 °C, 20 s at 65 °C, and 90 s at 72 °C. Final elongation was performed for 5 min at 72 °C. For primer sequences, *see* Table 2.
7. Add 10 μL loading buffer to the sample.
8. Run the reaction on an agarose gel.
9. Extract the PCR product of 2120 bp with the gel extraction kit.
10. Digest the PCR product with NdeI and HindIII: 1.5 μg PCR product, 5 μL CutSmart buffer, fill up with sterile water to 48 μL, add 1 μL NdeI and 1 μL HindIII-HF. Incubate at 37 °C for 2 h, stop the reaction at 80 °C for 20 min (*see* **Note 15**).
11. Digest the plasmid pET28a (*see* **Note 4**) with NdeI and HindIII: 2 μg plasmid, 5 μL CutSmart buffer, adjust with sterile water to 48 μL, add 1 μL NdeI and 1 μL HindIII-HF. Incubate

for 2 h at 37 °C. Dephosphorylate the plasmid: add 6 μL Antarctic phosphatase buffer and 3 μL Antarctic phosphatase, incubate at 37 °C for 40 min, stop the reaction at 80 °C for 5 min (*see* **Note 15**). Add 12 μL of sixfold loading buffer to the sample.

12. Run the digested plasmid on an agarose gel.
13. Extract the backbone of 5.3 kb with the gel extraction kit.
14. Ligation, *see* **Note 21**: 50 ng digested plasmid, 1 μL T4 DNA ligase reaction buffer, insert in a molar ratio of 7:1 (140 ng) and 3:1 (60 ng) for insert and plasmid, fill up with sterile water to 9.5 μL, 0.5 μL T4 DNA ligase. Incubate at room temperature for 40 min (*see* **Notes 15** and **16**), stop the reaction by 65 °C for 10 min.
15. Transformation into *E. coli* Top10: Add 3 μL of ligation to competent cells, incubate for 30 min on ice. Perform a 30 s heat shock at 42 °C, keep the cells on ice for 2 min. Incubate the cells for 1 h at 37 °C upon shaking. Spread 75 μL of transformed cells on agar plates with 50 μg/mL kanamycin (*see* **Note 17**). Keep the plates at 37 °C overnight.
16. Positive *E. coli* colonies were identified by colony PCR. Transfer single colonies to 50 μL sterile water (*see* **Note 18**).
17. The PCR master mix contained 9.1 μL sterile water, 4 μL phusion HF buffer, 1 μL forward primer (T7F), 1 μL reverse primer (T7R), 0.4 μL dNTPs, and 0.5 μL phusion polymerase per reaction. Transfer 16 μL of the master mix to a 0.2 mL PCR tube and add 4 μL of resuspended colony. For T7F and T7R primer sequences, *see* Table 2.
18. Run a PCR with 36 cycles: Initial denaturation at 98 °C for 10 min; the cycles comprised three steps: 10 s at 98 °C, 25 s at 55 °C, and 2 min at 72 °C. Final elongation was performed at 72 °C for 5 min.
19. Add 4 μL of loading buffer to the PCR products and run them on an agarose gel.
20. Identify positive clones with a 2378-bp PCR product size and transfer the corresponding cells in sterile water (*see* **step 16**) to 20 mL LB supplemented with 20 μL kanamycin (final concentration of 50 μg/mL).
21. Incubate the culture in a shaking incubator at 120 rpm and 37 °C overnight.
22. Perform a miniprep using the kit.
23. Send plasmids for sequencing with the rS_seq_F, rS_seq_R, T7F, and T7R primers to cover the entire modification.

Replace the hygromycin resistance by pNAP-DsRed-tNOS applying Gibson Assembly, the target plasmids pUB-Cas9 and UBQ::pcoCas9p1300 were amplified by PCRs A and B (*see* **Note 22**), the new screening cassette by PCR C.

24. (A) 33 μL sterile water, 10 μL phusion HF buffer, 2.5 μL forward primer (GA3_F), 2.5 μL reverse primer (GA4_R), 1 μL dNTPs, 0.5 μL 100 ng/μL Cas9-plasmid, 1 μL phusion DNA polymerase. (B) 31.5 μL sterile water, 10 μL phusion HF buffer, 2.5 μL forward primer (GA5_F), 2.5 μL reverse primer (GA6_R), 1 μL dNTPs, 0.5 μL 100 ng/μL Cas9-plasmid, 1.5 μL DMSO, 1 μL phusion DNA polymerase. C) 33 μL sterile water, 10 μL phusion HF buffer, 2.5 μL forward primer (GA1_F), 2.5 μL reverse primer (GA2_R), 1 μL dNTPs, 0.5 μL 100 ng/μL mutagenized redSeed in pET28a, 1 μL phusion DNA polymerase. For primer sequences, *see* Table 2. Run the PCR with 36 cycles: Initial denaturation at 98 °C for 30 s; the cycles comprised 10 s at 98 °C, 30 s at 70 °C, and 3 min at 72 ° C for (A) and (B) but 90 s at 72 °C for C. Final elongation was performed at 72 °C for 5 min.
25. Run the reactions on an agarose gel.
26. Extract the PCR products of 4.9 kb (A), 7.3 kb (B) and 2.1 kb (C) with the gel extraction kit.
27. Mix the three PCR products at equimolar ratios to a final amount of 0.1 pmol. Add 5 μL of the PCR products to 15 μL Gibson master mix (*see* **Note 5**). Incubate for 1 h at 50 °C, then keep on ice.
28. Transformation into *E. coli* Top10: add 3 μL of Gibson assembly mix to competent cells, incubate for 30 min on ice. Perform a 30 s heat shock at 42 °C, keep the cells on ice for 2 min. Incubate the cells at 37 °C for 1 h upon shaking. Spread 75 μL of transformed cells on agar plates with 50 μg/mL kanamycin (*see* **Note 17**). Keep the plates at 37 °C overnight.
29. Positive *E. coli* colonies were identified by colony PCR. Transfer single colonies to 50 μL sterile water (*see* **Note 18**).
30. The PCR master mix contains 9.1 μL sterile water, 4 μL phusion HF buffer, 1 μL forward primer (FH61), 1 μL reverse primer (Gaseq1), 0.4 μL dNTPs, and 0.5 μL phusion polymerase per reaction. Transfer 16 μL of the master mix to a 0.2 mL PCR tube and add 4 μL of resuspended colony. For primer sequences, *see* Table 2.
31. Run a PCR with 36 cycles: Initial denaturation at 98 °C for 10 min; the cycles comprised three steps: 10 s at 98 °C, 25 s at 56 °C, and 2 min at 72 °C. Final elongation was performed at 72 °C for 5 min.
32. Add 4 μL of loading buffer to the PCR products and run them on an agarose gel.

33. Identify positive clones with a 3.5 kb PCR product size and transfer the corresponding cells in sterile water (*see* **step 29**) to 20 mL LB supplemented with 20 μL kanamycin (final concentration of 50 μg/mL).
34. Incubate the culture in a shaking incubator at 120 rpm and 37 °C overnight.
35. Perform a miniprep using the kit.
36. Send plasmids for sequencing with Gaseq1, Gaseq2, and Gaseq3 primers.

3.5 Preparing the Cas9 Vectors

The two vector systems differ in the insertion of the sgRNA cassette (*see* Fig. 2). Gibson assembly is applied to insert the cassette into pUB-Cas9 [8] while classical cloning is recommended for UBQ10::pcoCas9p1300 [21].

3.5.1 Cloning into pUB-Cas9

1. Digest 2 μg pUB-Cas9 (RedSeed) with KpnI and HindIII: 2 μg plasmid, 5 μL CutSmart buffer, fill with sterile water to 48 μL, add 1 μL KpnI-HF and 1 μL HindIII-HF. Incubate for 2 h at 37 °C (*see* **Note 15**). Add 10 μL of sixfold loading buffer to the sample.
2. Run the digested plasmid on an agarose gel.
3. Extract the backbone of 14 kb with the gel extraction kit.
4. Run a PCR on the sgRNA cassette: 32.5 μL sterile water, 10 μL phusion HF buffer, 2.5 μL forward primer (FH41), 2.5 μL reverse primer (FH42), 1 μL dNTPs, 0.5 μL 100 ng/μL sgRNA in pFH6_new, 1 μL phusion DNA polymerase. For primer sequences, *see* Table 2.

   For 36 cycles: Initial denaturation at 98 °C for 30 s; the cycles comprised 10 s at 98 °C, 25 s at 68 °C, and 30 s at 72 °C. Final elongation was achieved at 72 °C for 5 min.
5. For cloning two different sgRNAs into pUB-Cas9 (multiplexing), run the PCR on the previously generated sgRNA1-pFH6_new with primers FH41/FH254 and on sgRNA2-pFH6_new with primers FH42/FH255. For primer sequences, *see* Table 2.
6. Run the reaction(s) on an agarose gel.
7. Extract the PCR product(s) of 771 bp with the gel extraction kit.
8. Gibson assembly: Mix digested vector and purified PCR product(s) equimolar at a final amount of 0.1 pmol. Add 5 μL of the DNA mix to 15 mL Gibson assembly master mix (*see* **Note 5**). Incubate the sample at 50 °C for 1 h, then keep on ice.
9. Transformation into *E. coli* Top10: add 3 μL of Gibson assembly mix to competent cells, incubate for 30 min on ice. Perform

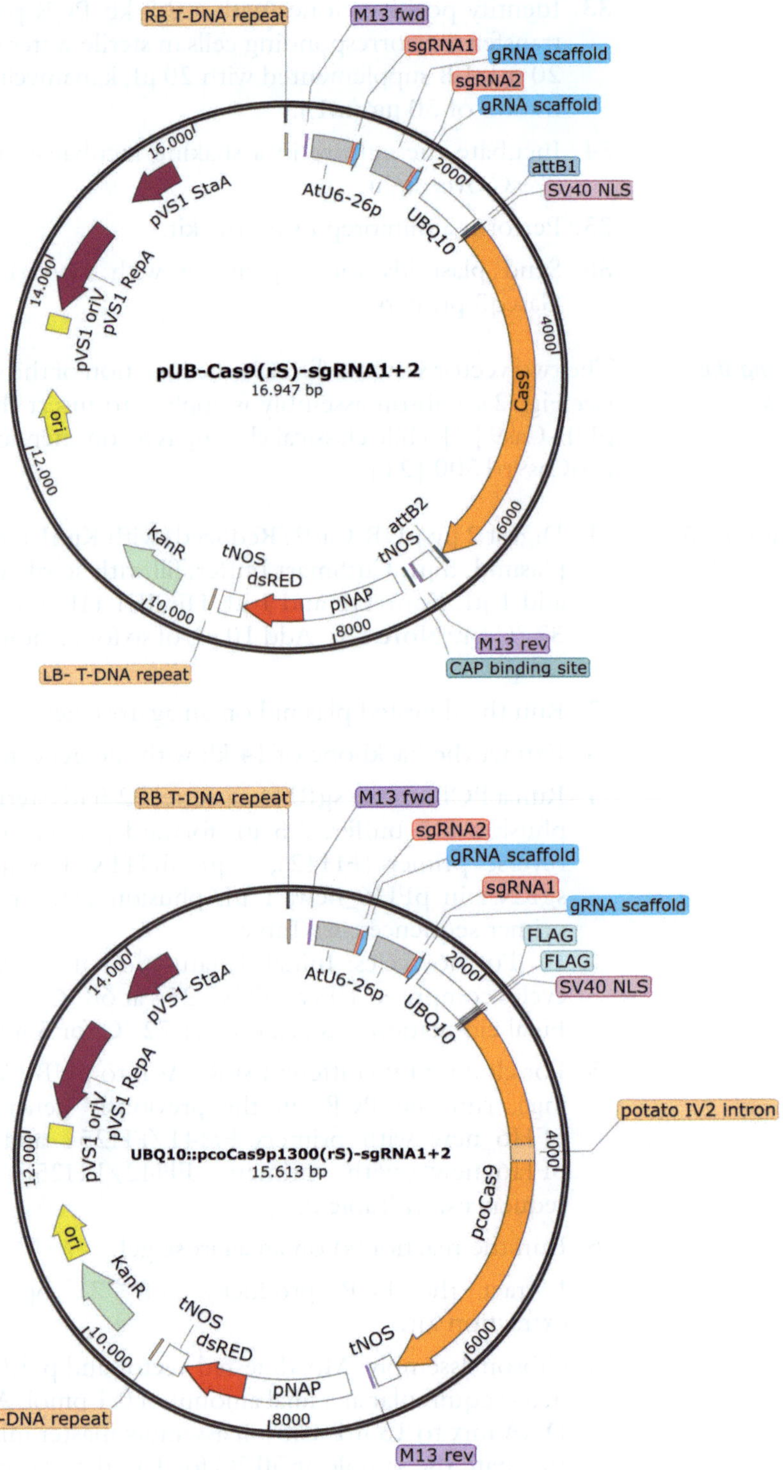

Fig. 2 Vector map of red seed plasmids. The map shows the plasmids with two inserted sgRNAs. pcoCas9 has a potato IV2 intron to suppress its expression in bacteria. Both vectors share similar sgRNA cassettes, the novel red seed feature (plant selection marker), and a kanamycin resistance (bacterial selection)

a 30 s heat shock at 42 °C, keep the cells on ice for 2 min. Incubate the cells at 37 °C for 1 h with shaking. Spread 75 μL of transformed cells on agar plates with 50 μg/mL kanamycin (*see* **Note 17**). Keep the plates at 37 °C overnight.

10. Transfer single colonies to 50 μL sterile water (*see* **Note 18**).
11. The colony PCR master mix contained 9.1 μL sterile water, 4 μL phusion HF buffer, 1 μL forward primer (M13F), 1 μL reverse primer (FH179), 0.4 μL dNTP and 0.5 μL phusion polymerase per reaction. Transfer 16 μL of the master mix to a 0.2 mL PCR tube and add 4 μL of resuspended colony. For primer sequences, *see* Table 2.
12. Run a PCR with 36 cycles: Initial denaturation at 98 °C for 10 min, the cycles comprise three steps: 10 s at 98 °C, 25 s at 55 °C, and 60 s (90 s for two sgRNAs) at 72 °C. Final elongation was performed at 72 °C for 5 min. For primer sequences, *see* Table 2.
13. Add 4 μL of loading buffer to the PCR products and run them on an agarose gel.
14. Identify positive clones with a 779 bp (1512 bp for two sgRNAs) PCR product size and transfer the corresponding cells in sterile water (*see* **step 10**) to 20 mL LB supplemented with 20 μL kanamycin stock (final concentration 50 μg/mL).
15. Incubate the culture in a shaking incubator at 120 rpm and 37 °C overnight.
16. Perform a miniprep using the kit.
17. Send plasmids for sequencing with M13F and FH179 primers.

3.5.2 Cloning into UBQ10::pcoCas9p1300 (RedSeed)

1. Digest 5 μg AtU6–26-V4-sgRNA(s) with KpnI and HindIII to cut out the sgRNA including U6-promoter: 5 μg plasmid, 5 μL CutSmart buffer, fill with sterile water to 46 μL, add 2 μL KpnI-HF and 2 μL SbfI-HF. Incubate for 3 h at 37 °C (*see* **Note 15**).
2. Digest 1.5 μg UBQ10::pcoCas9p1300 (RedSeed) with KpnI and HindIII: 1.5 μg plasmid, 5 μL CutSmart buffer, fill up with sterile water to 47 μL, add 1.5 μL KpnI-HF and 1.5 μL SbfI-HF. Incubate for 2 h at 37 °C. Dephosphorylate the plasmid: Add 6 μL Antarctic phosphatase buffer and 3 μL Antarctic phosphatase, incubate at 37 °C for 30 min, and stop the reaction at 80 °C for 5 min (*see* **Note 15**). Add 12 μL of sixfold loading buffer to the sample.
3. Run the restriction reactions on an agarose gel.
4. Extract the backbone of UBQ10::pcoCas9p1300 (RedSeed) of 14.2 kb and the sgRNA cassette (660 bp for one sgRNA; 1281 bp for two sgRNAs) with the gel extraction kit.

5. Ligation, *see* **Note 21**: 50 ng digested plasmid, 1 μL T4 DNA ligase reaction buffer, add at a molar ratio of 7:1 (16.3 ng for one sgRNA; 31.5 ng for two sgRNAs) and 3:1 (7 ng for one sgRNA; 13.5 ng for two sgRNAs) for insert and plasmid, fill with sterile water to 9.5 μL, add 0.5 μL T4 DNA ligase. Incubate at room temperature for 40 min (*see* **Notes 15** and **16**), stop the reaction by 65 °C for 10 min.
6. Transformation into *E. coli* Top10: add 3 μL of ligation to competent cells, incubate on ice for 30 min. Perform a 30 s heat shock at 42 °C, keep the cells on ice for 2 min. Incubate the cells at 37 °C with shaking for 1 h. Spread 75 μL of transformed cells on agar plates with 50 μg/mL kanamycin (*see* **Note 17**). Keep the plates at 37 °C overnight.
7. Positive *E. coli* colonies were identified by colony PCR. Transfer single colonies to 50 μL sterile water (*see* **Note 18**).
8. The PCR master mix contained 9.1 μL sterile water, 4 μL phusion HF buffer, 1 μL forward primer (sg-F, specific forward primer for first-added sgRNA), 1 μL reverse primer (K197), 0.4 μL dNTPs, and 0.4 μL phusion polymerase per reaction. Transfer 16 μL of the master mix to a 0.2 mL PCR tube and add 4 μL of resuspended colony. For primer sequences, *see* Table 2.
9. Run a PCR with 36 cycles: Initial denaturation at 98 °C for 10 min; the cycles comprised three steps: 10 s at 98 °C, 22 s at 58 °C (*see* **Note 19**), and 30 s at 72 °C. Final elongation was performed at 72 °C for 5 min.
10. Add 4 μL of loading buffer to the PCR products and separate them on an agarose gel.
11. Identify positive clones with PCR products of 801 bp and transfer the corresponding cells in sterile water (*see* **step 7**) to 20 mL LB supplemented with 20 μL kanamycin stock (final concentration 50 μg/mL).
12. Incubate the culture in a shaking incubator at 120 rpm and 37 °C overnight.
13. Perform a miniprep using the kit.
14. Send plasmids for sequencing with M13F primer.

3.6 Transformation of *A. thaliana*

The Agrobacteria strain GV3101 pMP90 was used for stable transformation of *A. thaliana*. This strain carries rifampicin and gentamycin resistance. For transformation, the *A. thaliana* plants had to be in a late developmental state with flowers and buds. Siliques should be removed (*see* **Note 6**).

3.6.1 Preparation of Competent Agrobacteria

1. Inoculate 10 mL of YEP-media supplemented with 10 μL rifampicin stock (final concentration of 100 μg/mL) and

10 μL gentamycin stock (final concentration of 25 μg/mL). Cultivate the bacteria at 30 °C and 120 rpm in a shaking incubator overnight, e.g., Infors HT Multitron.

2. Transfer 1 mL of overnight culture to 50 mL YEP-media with 50 μL rifampicin stock (final concentration of 100 μg/mL) and 50 μL gentamycin stock (final concentration of 25 μg/mL) and cultivate until the OD_{600nm} reaches the range of 0.8–1.
3. Transfer the cells to a 50 mL reaction tube, incubate them on ice for 10 min, and sediment them at 3000 rpm and 4 °C for 5 min.
4. Resuspend the pellet with 1 mL ice-cold 20 mM $CaCl_2$.
5. Split the culture into 100 μL aliquots in 2 mL reaction tubes, keep the aliquots on ice for immediate transformation; otherwise, freeze them in liquid nitrogen and store the competent cells at −80 °C.

3.6.2 Transformation of Agrobacteria

1. For the transformation, keep the cells on ice for 30 min after the resuspension in $CaCl_2$.
2. Add 2 μL of the final Cas9 vector with sgRNA cassette(s).
3. Freeze the cells in liquid nitrogen for 5 min.
4. Perform a 5 min heat shock at 37 °C.
5. Incubate the cells on ice for 5–30 min.
6. Add 1 mL of YEP media.
7. Shake the cells for 2 h at 28 °C.
8. Streak out 100 μL of cells on agar plates with 100 μg/mL rifampicin, 25 μg/mL gentamycin and 50 μg/mL kanamycin.
9. Incubate the agar plates at 28 °C for 2 d.
10. Check the transformation by colony PCR.
11. Transfer single colonies to 75 μL sterile water (*see* **Note 18**).
12. The PCR master mix contains 9.1 μL sterile water, 4 μL phusion HF buffer, 1 μL forward primer, 1 μL reverse primer, 0.4 μL dNTPs, and 0.5 μL phusion polymerase per reaction. Transfer 16 μL of the master mix to a 0.2 mL PCR tube and add 4 μL of resuspended colony. Two different primer combinations are possible (A: Gaseq2 + GA2_R, B: FH61 + FH201). For primer sequences, *see* Table 2.
13. Run a PCR with 36 cycles: Initial denaturation at 98 °C for 10 min; the cycles comprised three steps: 10 s at 98 °C, 25 s at 58 °C, and 15 s at 72 °C for A or 1 min at 72 °C for B. Final elongation was performed at 72 °C for 5 min.
14. Add 4 μL of loading buffer to the PCR products and run them on an agarose gel.

3.6.3 Preparation of Agrobacteria for Transformation of A. thaliana

1. Continue with positive Agrobacteria (PCR product of A: 326 bp and B: 1 kb) resuspended in sterile water (see Subheading 3.6.2, **step 11**).
2. Inoculate 20 mL YEP-media supplemented with 100 μg/mL rifampicin, 25 μg/mL gentamycin, and 50 μg/mL kanamycin with a positive colony resuspended in sterile water.
3. Incubate the culture at 28 °C and 120 rpm for 2 days.
4. Transfer 2 mL of the pre-culture to 500 mL YEP supplemented with 100 μg/mL rifampicin, 25 μg/mL gentamycin and 50 μg/mL kanamycin.
5. Incubate the culture at 28 °C and 120 rpm for 2 days.

3.6.4 Transformation of A. thaliana

1. Transfer the plants in a late developmental state with flowers and buds (siliques should be removed) to 400 μmol photons/m^2·s for 2 h to induce flower opening.
2. Add 5% sucrose and 0.02% Silwet L-77 (*see* **Note 7**) to the 500 mL culture.
3. Shake the culture for 10 min.
4. Transfer the culture to a 600 mL beaker with a magnetic stirring bar, place it on a magnetic stirrer.
5. Switch on the stirrer for a few seconds, switch it off, and place one plant head first into the suspension for 3 min, move it up and down.
6. Place the pot with the plant horizontally in a tray, and repeat the procedure with additional plants as desired (*see* **Note 23**). In between, repeat the stirring (*see* **Note 24**).
7. Repeat **steps 2–6** with all constructs designed for transformation of *A. thaliana* (*see* **Note 25**).
8. Cover the plants with plastic foil and another tray—it should not be airtight.
9. Incubate the plants in the dark overnight.
10. Unwrap the plants and position them upright, keep the plants in the dark until they are dry (approximately for 2 days).
11. Transfer the plants to a growth chamber with regular day–night cycle of 16 and 8 h, respectively.
12. Grow the plants until siliques are formed, cover the plants with bags, harvest the seeds when the siliques are dry.

3.7 Screening for Positive Plants

The transgenic plants express the fluorescent protein DsRed in the seeds, so that a rapid selection is enabled by fluorescence microscopy.

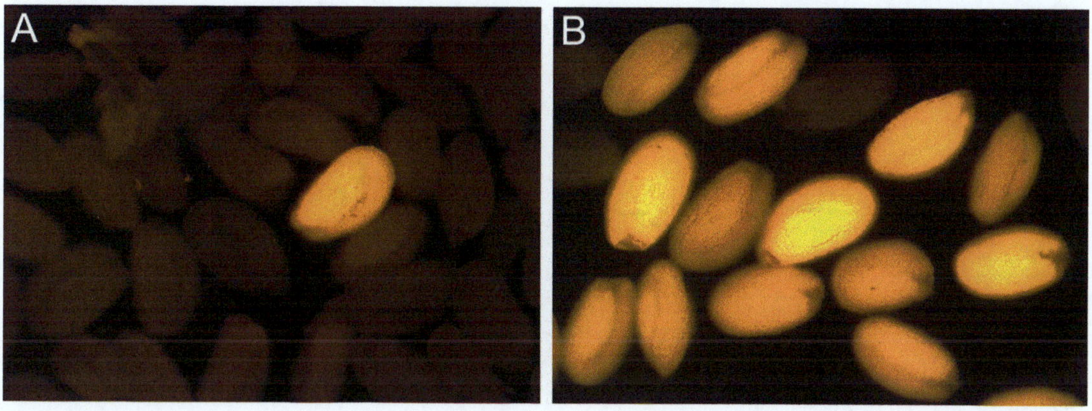

Fig. 3 Seeds expressing DsRed. Screening was performed by fluorescence microscopy with mCherry filter set. (Zeiss Axioskop 2 was applied with an EC-Plan-NEOFLUAR 5× objective and the Zeiss filter set 31: BP 565/30 Exc., FT 585 Beamsplitter, and BP 620/60 Em. Images were obtained with a Zeiss ICc1 camera and Zeiss AxioVision 4.7.) The red-fluorescent protein DsRed is expressed in the seeds of successfully transformed plants. (**a**) Seeds of generation F_1 and (**b**) seeds of generation F_2

1. Spread the seeds on a slide.
2. Use an upright fluorescence microscope with an air-objective of low magnification (five- to tenfold) and an appropriate filter set for DsRed (for instance, excitation in the range of 540–550 nM, beam splitter at 560 nM, emission in the range of 570–620 nM) to screen for fluorescent seeds (*see* Fig. 3 and **Note 26**).
3. Sow glowing seeds and grow the plants on soil under long day conditions (*see* **Note 27**).

3.8 Confirmation of Positive Plants

The number of generations that must be screened can vary and must be continued until the mutation(s) turn(s) homozygous. Initially, chimeras might occur, editing is not equally distributed in the organs and demands continued screening of further generations (*see* **Note 28**). Depending on the importance of the edited gene, screening can be facilitated by obvious phenotypes (*see* **Note 29**). Positive plants must be confirmed by PCR (*see* **Note 30**) and by sequencing. For PCR, the DNA has to be isolated according to Edwards et al. [6]. For sequencing, a commercial kit can be applied.

1. Transfer one leaf of a 4–6-week-old plants to a 1.5 mL reaction tube and grind it with a micro-pestle.
2. Add 500 μL of Edwards extraction buffer, and continue grinding.
3. Vortex the sample.
4. Centrifuge the sample at 13,000 rpm for 2 min.
5. Transfer 300 μL of the supernatant to a fresh 1.5 mL reaction tube.

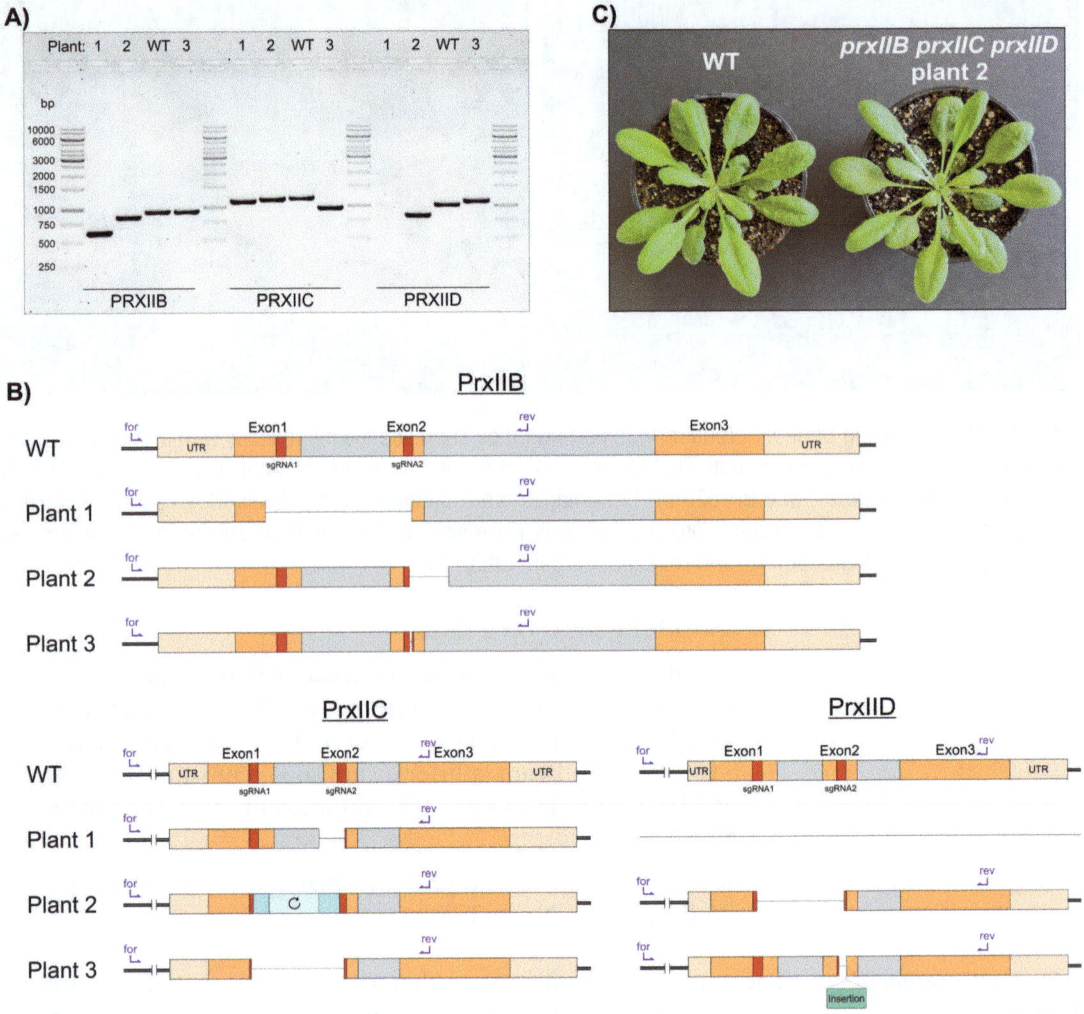

Fig. 4 Genotype and phenotype of *prxIIB*, *prxIIC*, and *prxIID*. (**a**) PCR on genomic DNA of the new lines. PCR results are shown for three transgenic plants in comparison to the wildtype. All PCR products differ from the WT in length, though with differently pronounced shift. (**b**) The schemes show the genotypes of the three plants for the genes encoding for *PrxIIB*, *C*, and *D*. Plant 1, 2, and 3 show deletions in *PrxIIB*, *PrxIID*, and *PrxIIC*, respectively. The deletion in plant 1 covers the entire gene *PRXIID*. In *PrxIIC* of plant 2, one part of the gene was removed by Cas9 and inserted again in opposite orientation. An insertion occurred in *PrxIID* of plant 3. Plant 1 has been generated *with* pUB-Cas9-sgRNA1 + 2 (redSeed) and plants 2 and 3 with UBQ10:: pcoCas9p1300-sgRNA1 + 2 (redSeed). (**c**) Apparently, the mutant does not show a phenotype distinct from the wildtype

6. Add 310 μL isopropanol.
7. Incubate the sample at room temperature for 10 min.
8. Centrifuge the sample at 13,000 rpm for 7 min.
9. Remove the supernatant, and air-dry the pellet.

10. Resuspend the pellet with 50–100 μL sterile water, centrifuge it again and adjust the supernatant to a concentration of 50–100 ng/μL.
11. Perform a PCR on genomic DNA to check for mutations in the target genes and possible off-targets.
12. The PCR master mix contains 12.1 μL sterile water, 4 μL phusion HF buffer, 1 μL gene-specific forward primer and 1 μL gene-specific reverse primer which cover the target site and its flanking region of ~300 bp up- and downstream, 0.4 μL dNTPs, and 0.5 μL phusion polymerase per reaction. Transfer 19 μL of the master mix to a 0.2 mL PCR tube and add 1 μL of genomic DNA (50–100 ng). For primer sequences (PRXIIB/C/D_F and PRXIIB/C/D_R), *see* Table 2.
13. Run a PCR with 36 cycles: Initial denaturation at 98 °C for 30 s; the cycles comprised three steps 5 s at 98 °C, 10 s at 64.5 °C, and 1 min at 72 °C. Final elongation was achieved at 72 °C for 5 min.
14. Add 4 μL of loading buffer to the PCR products and separate them on an agarose gel (*see* Fig. 4).
15. Keep the plants that showed a size of the PCR product that is different from the wildtype.
16. For sequencing of PCR products, add 1 μL Exo-CIP A and 1 μL Exo-CIP B to 5 μL PCR product (Exo-CIP™ Rapid PCR Cleanup, New England Biolabs), *see* **Note 31**.
17. Incubate the sample at 37 °C for 4 min and at 80 °C for 1 min.
18. Send the PCR product for sequencing.
19. For sequencing the entire genome of plants, employ the high molecular weight DNA kit for genomic DNA isolation.
20. Sequencing according to Oxford Nanopore Technologies is recommended.
21. Locate the T-DNA in your plant and check if it has an intergenic localization or affects the function of other genes. With increasing number of generations, the probability of homozygous T-DNA increases. Then, removal of the T-DNA demands crossing with wildtype plants (*see* **Note 32**).
22. Confirm the knockout by checking the transcript level of the mutated gene with a qPCR and, if available, use a specific antibody to check for protein abundance in the generated plant.

4 Notes

1. Competent cells can be easily produced: Transfer 0.5 mL of an overnight culture to 50 mL LB, incubate until Abs. 600 nM reaches 0.25. Then harvest the cells (5 min 4 °C, 5 min. 2500 rpm) and resuspend the pellet in 15 mL CCMB80, keeping the cells on ice. Centrifuge them again (5 min. 4 °C, 5 min. 2500 rpm) and resuspend the cells in 2 mL CCMB80, keep the cells on ice and prepare 50 μL aliquots. Store the competent cells at −80 °C.
2. Some labs use self-made phusion polymerases (*see* https://barricklab.org/twiki/bin/view/Lab/ProtocolsReagentsPfuSso7d). It is recommended to use more of the self-made polymerase for large PCR products.
3. For PCR products and linearized plasmids of more than 10 kb, the 1 kb plus ladder is recommended.
4. One example of a vector is pET28a. Any plasmid can be used to ensure the mutagenized red seed cassette. Alternatively, one can run a PCR with primers for Gibson assembly directly on the mutagenized PCR product.
5. Do not use self-made phusion polymerase for Gibson assembly.
6. Instead of wildtype Col-0, you can use other ecotypes or even transgenic lines for gene editing.
7. The 100 μL Silwet-77 can be replaced by 200–300 μL Silwet Gold (per 500 mL culture).
8. The sgRNAs display differences in their efficiency of genome editing. It is recommended to use at least two sgRNAs to gain deletions in the genome and thus, an efficient inactivation of the gene.
9. Multiple sgRNAs can be used to knockout more than one gene. For multiplexing with more than one target gene, it is recommended to use target-specific primers for screening.
10. If a restriction site is present in the mutagenic region 3 bp upstream of the PAM site, mutations lead to cleavage resistant PCR amplicons. This effect can be applied for screening by restriction analysis.
11. Manual design of sgRNA is recommended instead of comparing gene-specific results of online tools for knockout of more than one gene with a single sgRNA. To this end, perform an alignment and search for N20NGG motifs in conserved regions of exons, if you want to knockout all isoforms, or for N20NGG motifs in gene-specific exon regions to knockout only one isoform. Check both strands for the motifs.

12. Be aware that the U6-26 promoter demands a guanine at the 5′-end of the sgRNA. This is introduced automatically by the vector pfH6_new but has to be included in the sgRNA when using AtU6-26-V4.
13. Check the chosen sgRNA for compatibility with the off-targets. If there are no mismatches between off-target and the sgRNA and it fits perfectly to the off-target, a different sgRNA should be used. A few mismatches in the range of 1–3 might also result in unwanted mutagenesis of the off-targets, especially if these are at the 5′-end of the sgRNA.
14. Both protocols for primer annealing and ligation work for both pfH6_new and AtU6-26-V4.
15. Enzymes from most suppliers allow for shorter incubation times, for instances 15 min for a restriction digest and 10 min for ligation of sticky ends.
16. Perform the ligation at 16 °C overnight, if ligation was not successful with short incubations at RT.
17. Streak out more than 75 μL, if no or a low number of colonies were visible.
18. Adjust the bacteria suspension to an OD_{600} of 0.2, if the colony PCR does not work.
19. Annealing temperature of PCRs with sgF/R-Primers depends on target sequence.
20. Multiplexing with UBQ10::pcoCas9p1300 theoretically allows for an unlimited number of sgRNA, whereas using pUB-Cas9likely is limited to two sgRNAs (three might be possible by designing a new pair of primers, but more than three might lead to non-specific Gibson assembly).
21. Calculate the molar ratio of plasmid and insert DNA before ligation, to this end consider the length in bp (Eq. 1: required mass insert (ng) = desired insert/vector molar ratio · mass of vector (ng) · ratio of insert to vector lengths).
22. For Gibson assembly with vectors of more than 10 kb it is recommended to amplify the backbones in more than one PCR. To this end use more than one primer pair to split and amplify the backbone in more than one run instead of amplifying the entire backbone in one run.
23. We have used five plants per construct.
24. Stirring is important to avoid sedimentation of the Agrobacteria.
25. Carefully clean the workspace and materials between transformations with different constructs to avoid contaminations.
26. Screening for red fluorescent seeds can be facilitated using a fluorescence stereo microscope.

27. Use successors of at least three independently transformed plants for screening to compensate for variable efficiency of gene editing. Keep in mind that the efficiency can differ between cells and tissues. Pooling samples from one plant balance such effects.
28. Driving the Cas9-expression by a female gamete-specific promoter enhances generation and screening of transgenic plants.
29. Some proteins are of central importance for the plant, so that a loss of function causes lethality. Such proteins are very often encoded by single-copy genes. Once a protein is less important or other, unrelated but functionally similar proteins exist, the number of genes encoding for the protein might increase [4]. In this case, the phenotype of knockout plants might be less pronounced.
30. Deletions can be easily identified by PCR.
31. In case of multiple bands, a gel extraction or subcloning of individual bands is recommended for sequencing of the PCR product.
32. During the screening process, check for T-DNA in each generation to ensure the presence of Cas9. Check for red fluorescent seeds and confirm the genotype by PCR. If possible, avoid plants that are homozygous for T-DNA bearing Cas9. If they are homozygous, a crossing with wildtype plants is required to remove Cas9.

Acknowledgments

This work was supported by the Deutsche Forschungsgemeinschaft (DI 346/23).

References

1. Bae S, Park J, Kim JS (2014) Cas-OFFinder: a fast and versatile algorithm that searches for potential off-target sites of Cas9 RNA-guided endonucleases. Bioinformatics 30(10): 1473–1475
2. Chari R, Yeo NC, Chavez A, Church GM (2017) sgRNA Scorer 2.0: a species-independent model to predict CRISPR/Cas9 activity. ACS Synth Biol 6(5):902–904
3. Concordet JP, Haeussler M (2018) CRISPOR: intuitive guide selection for CRISPR/Cas9 genome editing experiments and screens. Nucleic Acids Res 46(W1):W242–W245
4. De Smet R, Adams KL, Vandepoele K, Van Montagu MC, Maere S, Van de Peer Y (2013) Convergent gene loss following gene and genome duplications creates single-copy families in flowering plants. Proc Natl Acad Sci 110(8):2898–2903
5. Doench JG, Fusi N, Sullender M, Hegde M, Vaimberg EW, Donovan KF et al (2016) Optimized sgRNA design to maximize activity and minimize off-target effects of CRISPR-Cas9. Nat Biotechnol 34(2):184–191
6. Edwards K, Johnstone C, Thompson C (1991) A simple and rapid method for the preparation of plant genomic DNA for PCR analysis. Nucleic Acids Res 19(6):1349
7. Gupta D, Bhattacharjee O, Mandal D, Sen MK, Dey D, Dasgupta A et al (2019) CRISPR-Cas9 system: a new-fangled dawn in gene editing. Life Sci 232:116636

8. Hahn F, Eisenhut M, Mantegazza O, Weber AP (2017) Generation of targeted knockout mutants in Arabidopsis thaliana using CRISPR/Cas9. Bio-protocol 7(13):e2384–e2384
9. Heigwer F, Kerr G, Boutros M (2014) E-CRISP: fast CRISPR target site identification. Nat Methods 11(2):122–123
10. Labun K, Montague TG, Krause M, Torres Cleuren YN, Tjeldnes H, Valen E (2019) CHOPCHOP v3: expanding the CRISPR web toolbox beyond genome editing. Nucleic Acids Res 47(W1):W171–W174
11. Liu H, Wei Z, Dominguez A, Li Y, Wang X, Qi LS (2015) CRISPR-ERA: a comprehensive design tool for CRISPR-mediated gene editing, repression and activation. Bioinformatics 31(22):3676–3678
12. Ma X, Zhu Q, Chen Y, Liu YG (2016) CRISPR/Cas9 platforms for genome editing in plants: developments and applications. Mol Plant 9(7):961–974
13. Moreno-Mateos MA, Vejnar CE, Beaudoin JD, Fernandez JP, Mis EK, Khokha MK, Giraldez AJ (2015) CRISPRscan: designing highly efficient sgRNAs for CRISPR-Cas9 targeting in vivo. Nat Methods 12(10):982–988
14. Naito Y, Hino K, Bono H, Ui-Tei K (2015) CRISPRdirect: software for designing CRISPR/Cas guide RNA with reduced off-target sites. Bioinformatics 31(7): 1120–1123
15. Nietzel T, Elsässer M, Ruberti C, Steinbeck J, Ugalde JM, Fuchs P et al (2019) The fluorescent protein sensor ro GFP 2-Orp1 monitors in vivo H2O2 and thiol redox integration and elucidates intracellular H2O2 dynamics during elicitor-induced oxidative burst in Arabidopsis. New Phytol 221(3):1649–1664
16. O'Malley RC, Barragan CC, Ecker JR (2015) A user's guide to the Arabidopsis T-DNA insertion mutant collections. In: Plant functional genomics: methods and protocols, pp 323–342
17. Prykhozhij SV, Rajan V, Gaston D, Berman JN (2015) CRISPR multitargeter: a web tool to find common and unique CRISPR single guide RNA targets in a set of similar sequences. PLoS One 10(3):e0119372
18. Pugh J (2023) The current state of nanopore sequencing. In: Nanopore sequencing: methods and protocols, pp 3–14
19. Sallin O, Reymond L, Gondrand C, Raith F, Koch B, Johnsson K (2018) Semisynthetic biosensors for mapping cellular concentrations of nicotinamide adenine dinucleotides. elife 7: e32638
20. Sanson KR, Hanna RE, Hegde M, Donovan KF, Strand C, Sullender ME et al (2018) Optimized libraries for CRISPR-Cas9 genetic screens with multiple modalities. Nat Commun 9(1):5416
21. Schumacher J, Kaufmann K, Yan W (2017) Multiplexed GuideRNA-expression to efficiently mutagenize multiple loci in arabidopsis by CRISPR-Cas9. Bio-protocol 7(5):e2166–e2166
22. Stothard P (2000) The sequence manipulation suite: JavaScript programs for analyzing and formatting protein and DNA sequences. BioTechniques 28(6):1102–1104
23. Untergasser A, Bijl GJ, Liu W, Bisseling T, Schaart JG, Geurts R (2012) One-step agrobacterium mediated transformation of eight genes essential for rhizobium symbiotic signaling using the novel binary vector system pHUGE. PLoS One 7(10):e47885
24. Vogelsang L, Dietz KJ (2022) Plant thiol peroxidases as redox sensors and signal transducers in abiotic stress acclimation. Free Radic Biol Med
25. Xie S, Shen B, Zhang C, Huang X, Zhang Y (2014) sgRNAcas9: a software package for designing CRISPR sgRNA and evaluating potential off-target cleavage sites. PLoS One 9(6):e100448
26. Xu H, Xiao T, Chen CH, Li W, Meyer CA, Wu Q et al (2015) Sequence determinants of improved CRISPR sgRNA design. Genome Res 25(8):1147–1157

Correction to: ROS Signaling in Plants

Francisco J. Corpas and José M. Palma

Correction to:
Francisco J. Corpas and José M. Palma (eds.), *ROS Signaling in Plants: Methods and Protocols*, Methods in Molecular Biology, vol. 2798,
https://doi.org/10.1007/978-1-0716-3826-2

The original version of the book was inadvertently published with a wrong affiliation of Professor José M. Palma in the front matter and chapters 1, 15 and 16.

The correct affiliation of Professor José M. Palma is

Spanish National Research Council, Estación Experimental del Zaidín, Granada, Spain

The correction chapters and the book have been updated with these changes.

The updated versions of these chapters can be found at
https://doi.org/10.1007/978-1-0716-3826-2_1
https://doi.org/10.1007/978-1-0716-3826-2_15
https://doi.org/10.1007/978-1-0716-3826-2_16

Francisco J. Corpas and José M. Palma (eds.), *ROS Signaling in Plants: Methods and Protocols*,
Methods in Molecular Biology, vol. 2798, https://doi.org/10.1007/978-1-0716-3826-2_18,

Correction to: Noninvasive *In Planta* Live Measurements of H_2O_2 and Glutathione Redox Potential with Fluorescent roGFPs-Based Sensors

Stefano Buratti, Matteo Grenzi, Giorgia Tortora, Sara Paola Nastasi, Elisa Dell'Aglio, Andrea Bassi, and Alex Costa

Correction to:
Chapter 4 in: Francisco J. Corpas and José M. Palma (eds.),
ROS Signaling in Plants: Methods and Protocols,
Methods in Molecular Biology, vol. 2798,
https://doi.org/10.1007/978-1-0716-3826-2_4

The chapter "Noninvasive *In Planta* Live Measurements of H_2O_2 and Glutathione Redox Potential with Fluorescent roGFPs-Based Sensors", written by Stefano Buratti, Matteo Grenzi, Giorgia Tortora, Sara Paola Nastasi, Elisa Dell'Aglio, Andrea Bassi, and Alex Costa, was originally published without open access. The copyright of the chapter changed on 10-Mar-2025 to © The Author(s) 2025 and the chapter is now distributed under the terms of the Creative Commons Attribution 4.0 International License (https://creativecommons.org/licenses/by/4.0/), which permits use, sharing, adaptation, distribution and reproduction in any medium or format, as long as you give appropriate credit to the original author(s) and the source, provide a link to the Creative Commons licence and indicate if changes were made.

The images or other third party material in this chapter are included in the chapter's Creative Commons licence, unless indicated otherwise in a credit line to the material. If material is not included in the chapter's Creative Commons licence and your intended use is not permitted by statutory regulation or exceeds the permitted use, you will need to obtain permission directly from the copyright holder.

The updated version of this chapter can be found at
https://doi.org/10.1007/978-1-0716-3826-2_4

Francisco J. Corpas and José M. Palma (eds.), *ROS Signaling in Plants: Methods and Protocols*,
Methods in Molecular Biology, vol. 2798, https://doi.org/10.1007/978-1-0716-3826-2_19,

Index

Francisco J. Corpas and José M. Palma (eds.), *ROS Signaling in Plants: Methods and Protocols*,
Methods in Molecular Biology, vol. 2798, https://doi.org/10.1007/978-1-0716-3826-2,

MIX
Papier aus verantwortungsvollen Quellen
Paper from responsible sources
FSC® C105338
FSC
www.fsc.org

If you have any concerns about our products,
you can contact us on
ProductSafety@springernature.com

In case Publisher is established outside the EU,
the EU authorized representative is:
Springer Nature Customer Service Center GmbH
Europaplatz 3, 69115 Heidelberg, Germany

Printed by Libri Plureos GmbH
in Hamburg, Germany